Advances in Fluid Dynamics

Maurice Holt

W.F. Ballhaus, Jr. M.Y. Hussaini

Advances in Fluid Dynamics

Proceedings of the Symposium in Honor of Maurice Holt on His 70th Birthday

With 139 Illustrations in 201 Parts

Springer-Verlag New York Berlin Heidelberg
London Paris Tokyo Hong Kong

W.F. Ballhaus, Jr.
NASA Ames Research Center
Moffet Field, CA 94035 USA

M.Y. Hussaini
Institute for Computer
Applications in Science
and Engineering
NASA Langley Research Center
Hampton, VA 23665-5335 USA

Library of Congress Cataloging-in-Publication Data

Advances in fluid dynamics : proceedings of the symposium in honor of Maurice Holt on his 70th birthday / [edited by] W.F. Ballhaus, Jr., M.Y. Hussaini.
p. cm.
"Symposium held on June 26, 1988 a Williamsburg, Virginia"—Pref.
1. Fluid dynamics—Congresses. 2. Holt, Maurice. I. Ballhaus, W.F. (William F.) II. Hussaini, M. Yousuff.
QA911.A38 1989
532'.05—dc20 89-21849

Printed on acid-free paper

Softcover reprint of the hardcover 1st edition 1989

ISBN-13: 978-1-4612-8201-3 e-ISBN-13: 978-1-4612-3684-9
DOI: 10.1007/978-1-4612-3684-9
Camera-ready copy supplied by authors.

9 8 7 6 5 4 3 2 1

Dedication

Yousuff Hussaini and I are grateful for the opportunity to organize this symposium in honor of Professor Maurice Holt. This symposium is being held in conjunction with the 11th International Conference of Numerical Methods in Fluid Dynamics. All of us who have gathered here today dedicate this symposium to Professor Maurice Holt on the occasion of his seventieth birthday and in recognition of his lifetime contributions to applied mathematics, fluid dynamics, and the production of two generations of computational fluid dynamicists.

The famous American, Mark Twain, once said, "Get your facts straight, and then you can distort them as much as you please." Let me first relate a few facts about Professor Holt.

Professor Maurice Holt was born on May 16, 1918 in Wildboarclough, Cheshire, England. He was educated at the Manchester Grammar School and the University of Manchester. He did his Masters thesis with T. G. Cowling in 1944, and his doctorate with Sydney Goldstein in 1948. Immediately after graduation, he joined the University of Liverpool as a lecturer in mathematics, and a year later moved to the University of Sheffield in England. In 1952, he joined the Ministry of Supply where he served as the Principal Scientific Officer in Charge of the Theoretical Aerodynamics Section of the Applied Mathematics Division, Armament Research and Development Establishment, Fort Halstead, Kent. In 1955, he was a visiting lecturer in the Mathematics Department at Harvard University at the invitation of Garrett Birkhoff. In 1956, he joined the faculty of the Division of Applied Mathematics at Brown University. Since 1960, he has been Professor of Aeronautical Sciences at the University of California at Berkeley.

His technical interests include supersonic and transonic aerodynamics, blast waves, underwater explosions, and supersonic separated flows. He is a pioneer in the field of computational fluid dynamics. He has published over a hundred technical papers with contributions to nine books. He is a fellow of the American Society of Mechnical Engineers, the American Physical Society and the American Institute of Aeronautics and Astronautics. He has trained over three dozen students, including Yousuff and me.

Professor Holt is one of the cofounders of the International Conference on Numerical Methods in Fluid Dynamics. The meeting was originally conceived in 1968 when Academician Belotserkovskii proposed a conference between the Soviet Union and the US in computational fluid dynamics. The first meeting was held in Novosibirsk in 1969, chaired by the late Professor Yanenko. Later, it was decided to hold the meeting during even numbered years and to include other countries. Professor Holt hosted the second meeting at the University of California, Berkeley in 1970. I was his principal graduate student slave labor assistant helping in organizing the meeting and editing the papers—my first real exposure to computational fluid dynamics. I have many fond memories of that conference. For example, students were recruited as bartenders for the reception. Our pay was that we could take home the bottles that had been opened; all the rest had to be returned. I can assure you that every bottle was opened, and the UC graduate students enjoyed many shots of good liquor that none of them could have afforded to buy in those days. I also remember that Maurice and I had a large banner printed to cover the front of the building where the conference was held. We had to inspect it on a day when neither of us had a car. So Maurice rode on the back of my motorcycle racing through the streets of Berkeley to go see our sign. The only mistake we made was telling Mrs. Holt about the ride afterwards.

In the years since, this meeting has attracted the top contributors to computational fluid dynamics, and has included many milestone papers. I have had the pleasure of participating in eight of them, and I believe Maurice has been to all of them but one in 1980 when he had a broken leg.

A few personal comments about Maurice. He is a colleague, mentor, and a friend to all of us. He is a kind man who has more friends around the world than anyone else I know. He is a man who knows how to enjoy his work; he is a man who knows how to enjoy his family and he is a man who knows how to enjoy his friends. In short, he is a man who knows how to enjoy life. Maurice is a man who has a house in America and a house in France; yet he is a man who speaks both American and French with an English accent. He is a man who enjoys good whiskey, good wine, and a good story. He is a man who, because he ran out of wine, nearly missed the most exciting conclusion to a college football game ever. It was the 1985 Big Game between Stanford and Cal; Cal scored with one minute left to take the lead; Stanford, with no timeouts left, marched the length of the field under the leadership of John Elway and kicked a field goal to go ahead. There were four seconds left, and Maurice wanted to leave because of a shortage of wine. The rest is history: Cal took the ball on the ensuing kickoff and, using the Stanford Band to run interference, scored the winning touchdown as the gun sounded to end the game.

It is fitting indeed that we should gather from all corners of the world today to honor this great man. He has touched all of our lives professionally and personally. We wish Maurice and Eileen many years of health and happiness, and many grandchildren.

W.F. Ballhaus, Jr.

Preface

This special volume contains the proceedings of the Symposium held on June 26, 1988 at Williamsburg, Virginia, in honor of Professor Maurice Holt on the occasion of his seventieth birthday. There were more than two dozen participants from eleven countries. They were either his past students or his colleagues whose careers crossed his at some point.

The twenty-one papers in this volume are the written version of the presentations at this Symposium; they are mostly in the area of computational fluid dynamics (CFD), a field in which Professor Holt is a pioneer. These papers cover almost all aspects of CFD including numerical analysis, symbolic analysis, and grid generation. They cover diverse topics such as complex plume flows, shock waves and shock focussing, coronary circulation, free surface flows, direct containment heating in nuclear reactors, and uranium enrichment. There is also an article on the progress and future directions in CFD by one of the true experts in this area. In addition to CFD papers, there is an experimental paper on the flow of spherical glass beads in airflow in a 90° vertical-to-horizontal bend, as well as a historical paper on seventy years of fluid dynamic research at Aerodynamisches Institut at Aachen. It is worth pointing out that there is also an article on the simple fluid concept by a world-renowned authority on continuum mechanics.

The editors would like to take this opportunity to thank all the participants for making the Symposium a success. Thanks are also due to Jeffrey Robbins and his colleagues at Springer-Verlag for their cooperation in producing this volume.

W.F. Ballhaus, Jr
M.Y. Hussaini

Seated from left to right: J.S. Meng, C.A.J. Fletcher, Soubbaramayer, R. Temam, M.Y. Hussaini, M. Holt, W. Ballhaus, Jr., J.-J. Chattot, H. Cabannes, A. Jameson, K. Morton.

Standing from left to right: R. Collins, M. Tulin, S. Berger, G. Klopfer, F.G. Zhuang, K. Oshima, E. Krause, B. Yim, P. Zandbergen, P. Kutler, T. Taylor, J.-P. Boujot, R.-J. Yang, M. Hafez, H. Dwyer, W.G. Habashi, T.A. LU, L. Turchak.

Contents

Non-unique Solutions of the Euler Equations

G. Baruzzi, W.G. Habashi, and M.M. Hafez

1. Introduction

In this paper Finite Element (FE) solutions of the Euler equations are obtained using an artificial viscosity method. The FE discretization is based on Galerkin's weighted residual approximation with the same shape functions for the unknowns, ρ, u and v; the pressure being eliminated by assuming constant total enthalpy. The algebraic system is solved using a Newton method and a direct solver to avoid convergence problems. A quasi-1D solution is used as the initial guess.

It is well known that the velocity potential may produce non-unique solutions under certain circumstances [1,2]. To demonstrate the non-uniqueness of the steady Euler equations, a compressible flow through a nozzle is considered. The quasi-1-D formulation admits two solutions for supersonic inlet conditions in a certain range of prescribed back pressures. Finite Difference (FD) solutions of this problem have been recently reported by Hafez et al. [3]. While there are other well-known examples of non-unique solutions of the Euler equations (e.g. weak and strong attached shock waves over wedges and cones), they will not be examined here.

We consider a 2-D symmetric convergent-divergent nozzle with a smooth area variation to avoid the generation of oblique shock waves for supersonic inlet conditions. The numerical FE solution demonstrates the non-uniqueness problem under the same conditions predicted by the quasi-1-D FD analysis. The 2-D effects are clearly demonstrated in the shock structure due to the curvature of the walls. Both stable and unstable solutions are obtainable with the present fully implicit technique unlike time-marching procedures, which produce only the stable solution.

The stability of the 2-D shocks is investigated by solving the truly unsteady Euler equations (continuity, momentum and energy) with a backward Euler finite difference discretization of the time dependent terms. At each time step, the four equations are solved fully coupled, in a similar manner to the steady-state approach. Shocks are then perturbed around the steady-state position and the shock motion examined for each case.

The artificial viscosity needed for the stability of the Galerkin approximation in this numerical experiment is intentionally chosen to be very simple and is the Laplacian of ρ, u and v added to the continuity and momentum equations, respectively. The coefficients of these terms are kept uniform throughout the domain. The reason for the choice of this primitive form is to avoid the possible spurious non-unique solutions of the discrete equations associated with more complicated artificial viscosity forms. The current trend in solving the Euler equations using nonlinear artificial viscosity terms, with limiters, may lead to illegitimate solutions which are not related to the solution of the differential equation [4].

The present method is only first order accurate and shocks may be spread over a few elements but neither the quality of the solution (shock sharpness) nor its efficiency (order of the truncation error) is the issue of this study.

In the following, the Finite Element aspects of the 2-D steady Euler representation are presented together with the numerical procedures employed to obtain the non-unique solution. Numerical results are presented for the steady and unsteady formulations.

2. Two-Dimensional Steady Euler Equations

Consider the steady 2-D Euler equations:

$$\frac{\partial}{\partial x}(\rho u)+\frac{\partial}{\partial y}(\rho v)=0 \tag{2.1a}$$

$$\frac{\partial}{\partial x}(\rho u^2+p)+\frac{\partial}{\partial y}(\rho uv)=0 \tag{2.1b}$$

$$\frac{\partial}{\partial x}(\rho uv)+\frac{\partial}{\partial y}(\rho v^2+p)=0 \tag{2.1c}$$

$$p=\frac{\gamma-1}{\gamma}\rho\left(H_\infty-\frac{u^2+v^2}{2}\right) \tag{2.1d}$$

Eqs. (2.1) are formulated under the assumption of constant total enthalpy H_∞ and the specific heat ratio is constant and set as $\gamma=1.4$. After the explicit addition of linear artificial viscosity terms the system of equations in non-dimensional form becomes:

$$\frac{\partial}{\partial x}(\rho u)+\frac{\partial}{\partial y}(\rho v)=\upsilon_1\left(\frac{\partial^2\rho}{\partial x^2}+\frac{\partial^2\rho}{\partial y^2}\right) \tag{2.2a}$$

$$\frac{\partial}{\partial x}(\rho u^2+p)+\frac{\partial}{\partial y}(\rho uv)=\upsilon_2\left(\frac{\partial^2 u}{\partial x^2}+\frac{\partial^2 u}{\partial y^2}\right) \tag{2.2b}$$

$$\frac{\partial}{\partial x}(\rho uv)+\frac{\partial}{\partial y}(\rho v^2+p)=\upsilon_2\left(\frac{\partial^2 v}{\partial x^2}+\frac{\partial^2 v}{\partial y^2}\right) \tag{2.2c}$$

$$p=\frac{\gamma-1}{\gamma}\rho\left(H_\infty-\frac{u^2+v^2}{2}\right) \tag{2.2d}$$

where:

$$\upsilon_1=\frac{\mu}{U_\infty L} \qquad \upsilon_2=\frac{\mu}{\rho_\infty U_\infty L}$$

are the artificial viscosity coefficients. The Galerkin form of the residuals of Eqs. (2.2) is given by Eqs. (2.3):

$$\iint_A W_i\left[\frac{\partial}{\partial x}(\rho u)+\frac{\partial}{\partial y}(\rho v)-\upsilon_1\left(\frac{\partial^2\rho}{\partial x^2}+\frac{\partial^2\rho}{\partial y^2}\right)\right]dA=0 \tag{2.3a}$$

$$\iint_A W_i\left[\frac{\partial}{\partial x}(\rho u^2+p)+\frac{\partial}{\partial y}(\rho uv)-\upsilon_2\left(\frac{\partial^2 u}{\partial x^2}+\frac{\partial^2 u}{\partial y^2}\right)\right]dA=0 \tag{2.3b}$$

$$\iint_A W_i\left[\frac{\partial}{\partial x}(\rho uv)+\frac{\partial}{\partial y}(\rho v^2+p)-\upsilon_2\left(\frac{\partial^2 v}{\partial x^2}+\frac{\partial^2 v}{\partial y^2}\right)\right]dA=0 \tag{2.3c}$$

where W_i are the weight functions. The Galerkin principle is the most compact of the residual minimization formulations because the weight functions are also the finite element shape functions. The weak formulation of Eqs. (2.3) requires a less stringent criterion of flux conservation at the element boundaries without penalties and allows the recovery of the natural boundary conditions in the form of contour integrals. After the integration by parts and the elimination of pressure from the momentum equations using the energy equation, Eq. (2.2d), the system of equations can be written in the form:

$$\iint_A\left[(\rho u)\frac{\partial W_i}{\partial x}+(\rho v)\frac{\partial W_i}{\partial y}\right]dA-\upsilon_1\iint_A\left(\frac{\partial\rho}{\partial x}\frac{\partial W_i}{\partial x}+\frac{\partial\rho}{\partial y}\frac{\partial W_i}{\partial y}\right)dA$$

$$=\int_S W_i\,\rho(\bar{q}\cdot\bar{n})dS \tag{2.4a}$$

$$\int_A\int\left[\left(\frac{\gamma+1}{2\gamma}\rho u^2-\frac{\gamma-1}{2\gamma}\rho v^2+\frac{\gamma-1}{\gamma}\rho H_\infty\right)\frac{\partial W_i}{\partial x}+(\rho uv)\frac{\partial W_i}{\partial y}\right]dA-\upsilon_2\int_A\int\left(\frac{\partial u}{\partial x}\frac{\partial W_i}{\partial x}+\frac{\partial u}{\partial y}\frac{\partial W_i}{\partial y}\right)dA$$

$$=\int_S W_i\,\rho u(\bar{q}\cdot\bar{n})\,dS+\int_S W_i\,p\frac{dy}{dS}\,dS \qquad (2.4b)$$

$$\int_A\int\left[(\rho uv)\frac{\partial W_i}{\partial x}+\left(\frac{\gamma+1}{2\gamma}\rho v^2-\frac{\gamma-1}{2\gamma}\rho u^2+\frac{\gamma-1}{\gamma}\rho H_\infty\right)\frac{\partial W_i}{\partial y}\right]dA-\upsilon_2\int_A\int\left(\frac{\partial v}{\partial x}\frac{\partial W_i}{\partial x}+\frac{\partial v}{\partial y}\frac{\partial W_i}{\partial y}\right)dA$$

$$=\int_S W_i\,\rho v(\bar{q}\cdot\bar{n})\,dS-\int_S W_i\,p\frac{dx}{dS}\,dS \qquad (2.4c)$$

where the contour integrals are now divided into flux and pressure components, while the artificial viscosity contour integrals have been neglected altogether.

Newton's method can be introduced by setting:

$$\Delta U=U^{n+1}-U^{n}$$

where $U = (\rho,u,v)$ represents the unknowns. The finite element discretization is introduced by setting:

$$U=\sum_{j=1}^{df}N_j\hat{U}_j \qquad (2.5)$$

where N_j are the shape or interpolation functions, identical to the weight functions, W_i, which for bilinear elements are:

$$N_j=\frac{1}{4}(1+\xi\xi_j)(1+\eta\eta_j) \qquad j=1,\ldots,4$$

where the non-dimensional coordinates ξ_j and η_j are the locations of the nodes given as $\xi_j,\eta_j=\pm 1$ as shown in Fig. 1.

The derivatives of the shape functions with respect to the x,y coordinates can be expressed in terms of the local derivatives. Let:

$$\begin{bmatrix}\dfrac{\partial N_i}{\partial \xi}\\[2ex] \dfrac{\partial N_i}{\partial \eta}\end{bmatrix}^{(e)}=\begin{bmatrix}\dfrac{\partial x}{\partial \xi} & \dfrac{\partial y}{\partial \xi}\\[2ex] \dfrac{\partial x}{\partial \eta} & \dfrac{\partial y}{\partial \eta}\end{bmatrix}^{(e)}\begin{bmatrix}\dfrac{\partial N_i}{\partial x}\\[2ex] \dfrac{\partial N_i}{\partial y}\end{bmatrix}^{(e)}=[J]^{(e)}\begin{bmatrix}\dfrac{\partial N_i}{\partial x}\\[2ex] \dfrac{\partial N_i}{\partial y}\end{bmatrix}^{(e)}$$

The elements of the Jacobian of the coordinate transformation can be obtained directly from the local shape functions. For example:

$$\frac{\partial x}{\partial \xi}=\sum_{j=1}^{df}\frac{\partial N_i}{\partial \xi}\hat{x}_j \tag{6}$$

The global derivatives can then be calculated by setting:

$$\begin{bmatrix}\dfrac{\partial N_i}{\partial x}\\[2ex] \dfrac{\partial N_i}{\partial y}\end{bmatrix}^{(e)}=[J]^{-1}\begin{bmatrix}\dfrac{\partial N_i}{\partial \xi}\\[2ex] \dfrac{\partial N_i}{\partial \eta}\end{bmatrix}^{(e)}$$

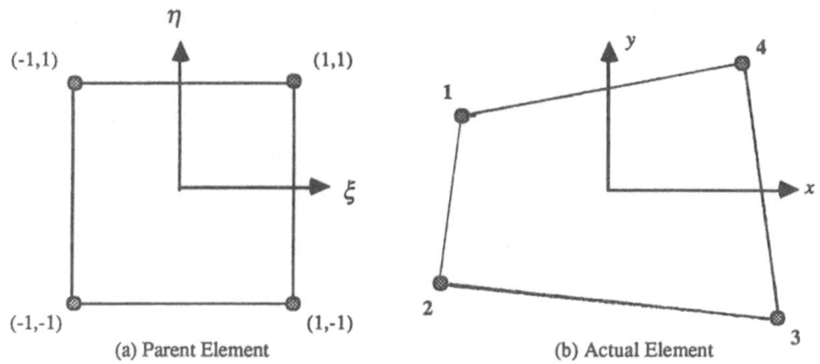

Fig. 1 Isoparametric linear element

The Newton-Galerkin form of the continuity equation can be written as:

$$\sum_{j=1}^{df}[K_{i,j}^{\rho}]\,\Delta\hat{\rho}_j+\sum_{j=1}^{df}[K_{i,j}^{u}]_{\rho}\Delta\hat{u}_j+\sum_{j=1}^{df}[K_{i,j}^{v}]_{\rho}\Delta\hat{v}_j=-\{R_i\}_{\rho}+\int_S W_i\,\rho(\bar{q}\cdot\bar{n})dS \tag{2.7a}$$

where:

$$[K_{i,j}^{\rho}]_{\rho}=\int_A\!\!\int N_j\left(u^n\frac{\partial W_i}{\partial x}+v^n\frac{\partial W_i}{\partial y}\right)dA-\upsilon_1\int_A\!\!\int\left(\frac{\partial W_i}{\partial x}\frac{\partial N_j}{\partial x}+\frac{\partial W_i}{\partial y}\frac{\partial N_j}{\partial y}\right)dA$$

$$[K_{i,j}^{u}]_{\rho}=\int_A\!\!\int \rho^n N_j\frac{\partial W_i}{\partial x}\,dA$$

$$[K_{i,j}^{v}]_{\rho}=\int_A\!\!\int \rho^n N_j\frac{\partial W_i}{\partial y}\,dA$$

$$\{R_i\}_\rho = \iint_A \left[(\rho u)^n \frac{\partial W_i}{\partial x} + (\rho v)^n \frac{\partial W_i}{\partial y} \right] dA - \upsilon_1 \iint_A \left(\frac{\partial W_i}{\partial x}\frac{\partial \rho^n}{\partial x} + \frac{\partial W_i}{\partial y}\frac{\partial \rho^n}{\partial y} \right) dA$$

The x-momentum equation becomes:

$$\sum_{j=1}^{df} [K^\rho_{i,j}]_u \Delta\hat{\rho}_j + \sum_{j=1}^{df} [K^u_{i,j}]_u \Delta\hat{u}_j + \sum_{j=1}^{df} [K^v_{i,j}]_u \Delta\hat{v}_j = -\{R_i\}_u$$

$$+ \int_S W_i \, \rho u (\bar{q} \cdot \bar{n}) dS + \int_S W_i \, p \frac{\partial y}{\partial S} dS \tag{2.7b}$$

where:

$$[K^\rho_{i,j}]_u = \iint_A N_j \left[\left(\frac{\gamma+1}{2\gamma} (u^2)^n - \frac{\gamma-1}{2\gamma} (v^2)^n + \frac{\gamma-1}{\gamma} H_\infty \right) \frac{\partial W_i}{\partial x} + (uv)^n \frac{\partial W_i}{\partial y} \right] dA$$

$$[K^u_{i,j}]_u = \iint_A N_j \left[\frac{\gamma+1}{\gamma} (\rho u)^n \frac{\partial W_i}{\partial x} + (\rho v)^n \frac{\partial W_i}{\partial y} \right] dA - \upsilon_2 \iint_A \left(\frac{\partial W_i}{\partial x}\frac{\partial N_j}{\partial x} + \frac{\partial W_i}{\partial y}\frac{\partial N_j}{\partial y} \right) dA$$

$$[K^v_{i,j}]_u = -\iint_A N_j \left[\frac{\gamma-1}{\gamma} (\rho v)^n \frac{\partial W_i}{\partial x} - (\rho u)^n \frac{\partial W_i}{\partial y} \right] dA$$

$$\{R_i\}_u = \iint_A \left[\left(\frac{\gamma+1}{2\gamma} (\rho u^2)^n - \frac{\gamma-1}{2\gamma} (\rho v^2)^n + \frac{\gamma-1}{\gamma} \rho H_\infty \right) \frac{\partial W_i}{\partial x} + (\rho u v)^n \frac{\partial W_i}{\partial y} \right] dA$$

$$- \upsilon_2 \iint_A \left(\frac{\partial W_i}{\partial x}\frac{\partial u^n}{\partial x} + \frac{\partial W_i}{\partial y}\frac{\partial u^n}{\partial y} \right) dA$$

The y-momentum equation becomes:

$$\sum_{j=1}^{df} [K^\rho_{i,j}]_v \Delta\hat{\rho}_j + \sum_{j=1}^{df} [K^u_{i,j}]_v \Delta\hat{u}_j + \sum_{j=1}^{df} [K^v_{i,j}]_v \Delta\hat{v}_j = -\{R_i\}_v$$

$$+ \int_S W_i \, \rho v (\bar{q} \cdot \bar{n}) dS - \int_S W_i \, p \frac{\partial x}{\partial S} dS \tag{2.7c}$$

where:

$$[K^\rho_{i,j}]_v = \iint_A N_j \left[(uv)^n \frac{\partial W_i}{\partial x} + \left(\frac{\gamma+1}{2\gamma} (v^2)^n - \frac{\gamma-1}{2\gamma} (u^2)^n + \frac{\gamma-1}{\gamma} H_\infty \right) \frac{\partial W_i}{\partial y} \right] dA$$

$$[K^{u}_{i,j}]_v=\iint_A N_j\left[(\rho v)^n\frac{\partial W_i}{\partial x}-\frac{\gamma-1}{\gamma}(\rho u)^n\frac{\partial W_i}{\partial y}\right]dA$$

$$[K^{v}_{ij}]_v=\iint_A N_j\left[(\rho u)^n\frac{\partial W_i}{\partial x}+\frac{\gamma+1}{\gamma}(\rho v)^n\frac{\partial W_i}{\partial y}\right]dA-\upsilon_2\iint_A\left(\frac{\partial W_i}{\partial x}\frac{\partial N_j}{\partial x}+\frac{\partial W_i}{\partial y}\frac{\partial N_j}{\partial y}\right)dA$$

$$\{R_i\}_v=\iint_A\left[(\rho uv)^n\frac{\partial W_i}{\partial x}+\left(\frac{\gamma+1}{2\gamma}(\rho v^2)^n-\frac{\gamma-1}{2\gamma}(\rho u^2)^n+\frac{\gamma-1}{\gamma}\rho H_\infty\right)\frac{\partial W_i}{\partial y}\right]dA$$

$$-\upsilon_2\iint_A\left(\frac{\partial W_i}{\partial x}\frac{\partial v^n}{\partial x}+\frac{\partial W_i}{\partial y}\frac{\partial v^n}{\partial y}\right)dA$$

The unsteady Euler system contains the time varying terms and the complete energy equation must be added :

$$\frac{\partial}{\partial t}(\rho)+\frac{\partial}{\partial x}(\rho u)+\frac{\partial}{\partial y}(\rho v)=\upsilon_1\left(\frac{\partial^2\rho}{\partial x^2}+\frac{\partial^2\rho}{\partial y^2}\right) \tag{2.8a}$$

$$\frac{\partial}{\partial t}(\rho u)+\frac{\partial}{\partial x}(\rho u^2+p)+\frac{\partial}{\partial y}(\rho uv)=\upsilon_2\left(\frac{\partial^2 u}{\partial x^2}+\frac{\partial^2 u}{\partial y^2}\right) \tag{2.8b}$$

$$\frac{\partial}{\partial t}(\rho v)+\frac{\partial}{\partial x}(\rho uv)+\frac{\partial}{\partial y}(\rho v^2+p)=\upsilon_2\left(\frac{\partial^2 v}{\partial x^2}+\frac{\partial^2 v}{\partial y^2}\right) \tag{2.8c}$$

$$\frac{\partial}{\partial t}(\rho e)+\frac{\partial}{\partial x}[u(\rho e+p)]+\frac{\partial}{\partial y}[v(\rho e+p)]=\upsilon_2\left(\frac{\partial^2 e}{\partial x^2}+\frac{\partial^2 e}{\partial y^2}\right) \tag{2.8d}$$

$$p=(\gamma-1)\rho\left(e-\frac{u^2+v^2}{2}\right) \tag{2.8e}$$

The equations are recast in the Galerkin integral form in the same fashion as the steady equations and the time terms are discretized by backward Euler differences.

The integrals are evaluated numerically by Gauss-Legendre quadrature using 3x3 integration points per element after transforming each of their terms into local coordinates as follows:

$$K_{i,j}=\int_{-1}^{1}\int_{-1}^{1}K_{i,j}[x(\xi,\eta),y(\xi,\eta)]\,|J|\,d\xi\,d\eta$$

3. Numerical Results

The nozzle in Fig. 2 has a 30% area restriction and is discretized using 100x24 isoparametric bilinear elements. The Mach number at the inlet of the nozzle is 2.2 and the pressure ratio is 7.1:1. Converged solutions, to machine accuracy, are obtained in less than 5 iterations in all cases considered. The results of figures 3 demonstrate the non-unique solution of the 2-D steady Euler equations. In Fig. 3a, a shock appears in the convergent part of the nozzle while Fig. 3b shows that a shock in the divergent part is also an acceptable solution, for the same back pressure and inlet conditions.

The stability of both shocks is then examined using the time dependent form of the equations with a backward Euler finite difference discretization of the time dependent terms. The shock in the divergent part is proved to be stable, i.e. if the shock is misplaced to the right or left of the steady-state position it returns back to it. The shock in the convergent part, however, is unstable and the shock motion is demonstrated in the sequence of Figs. 4 and 5. Note that if the shock is initially placed to the right side of the steady state it is eventually swallowed by the throat where the flow is sonic and the shock has zero strength and disappears. On the other hand, if the shock is placed on the left side of the steady state position it continuously moves towards the exit. More details of these calculations can be found in [5].

References

1. Steinhoff, J. and Jameson, A., "Multiple Solutions of the Transonic Potential Flow Equation", AIAA Paper 81-1019, Proceedings of the AIAA 5th Computational Fluid Dynamics Conference, Palo Alto, California, 1981.

2. Salas, M.D., Jameson, A. and Melnick, R.E., "A Comparative Study of the Nonuniqueness Problem of the Potential Equation", AIAA Paper 83-1888, Danvers, Massachusetts, 1983.

3. Hafez, M.M., Palaniswamy, S. and Mariani, P., "Calculation of Transonic Flows with Shocks Using Newton's Method and Direct Solver. Part II", AIAA Paper 88-0226, Reno, Nevada, 1988.

4. Yee, H.C., Warming, R.F. and Harten, A., "On a Class of TVD Schemes for Gas Dynamic Calculations", in Numerical Methods for the Euler Equations of Gas Dynamics, Angrand et al. (eds), SIAM, Philadelphia, 1985.

5. Baruzzi, G., "Implicit Finite Element Formulations of the Euler Equations", Concordia University, Montreal, Canada, M. Eng. Thesis, 1989.

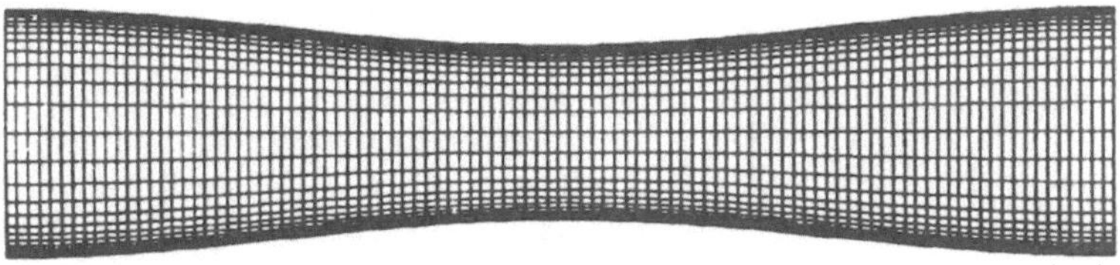

Fig.2 Finite Element grid and shape of nozzle

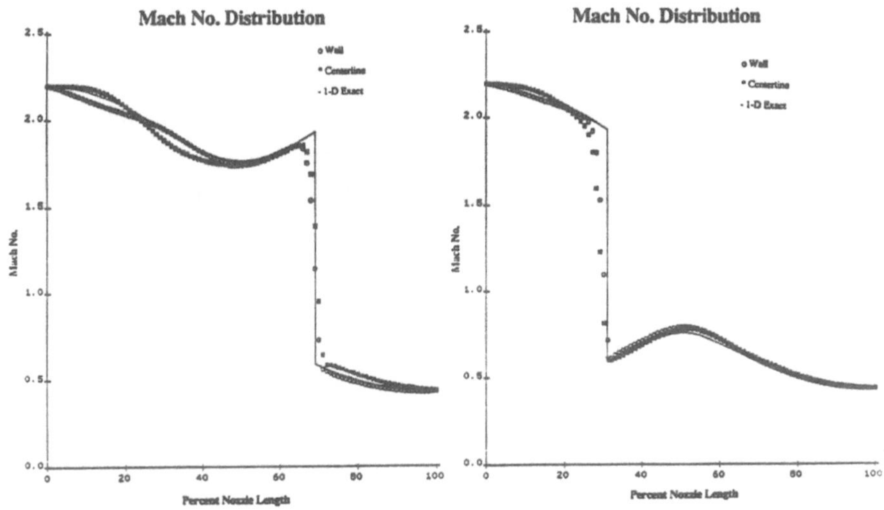

Fig.3 2-D results in a nozzle: shock in convergent and divergent parts

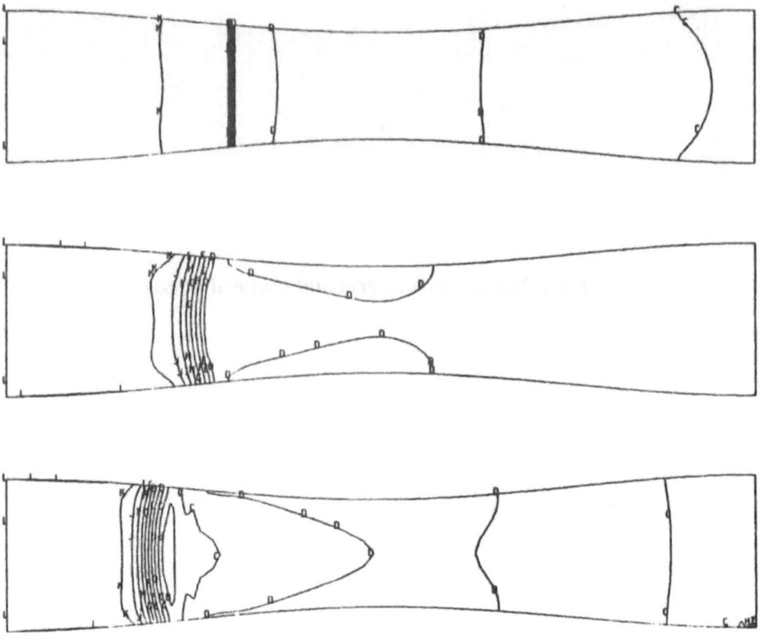

Fig.4 Mach contours vs time step for the unstable solution

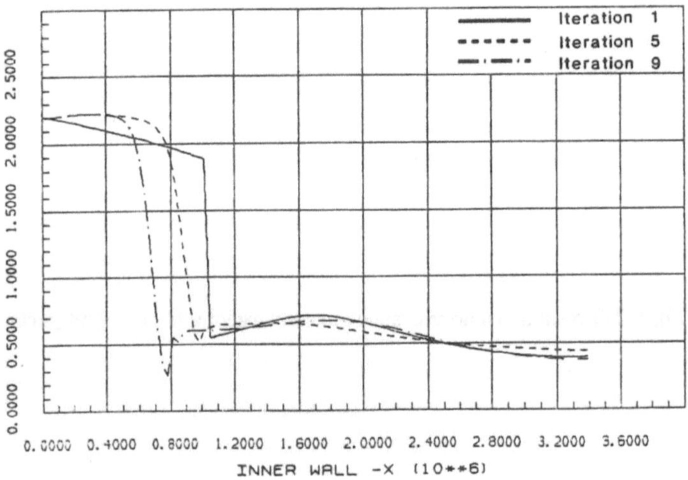

Fig.5 Mach number distribution over surface vs time for the unstable solution

Navier-Stokes Computations of Three-Dimensional Laminar Flows with Buoyancy in a Channel with Wing-Type Vortex Generators

Gautam Biswas, Nimai Kumar Mitra, and Martin Fiebig

1. Introduction

Longitudinal vortices embedded in shear flows can substantially increase the heat transfer between the fluid and the channel wall by inducing azimuthal velocity which mixes continuously hot and cold fluids [1]. This technique of heat transfer enhancement is of special interest in channel flows where the heat transfer between the fluid and channel walls detiorates as the boundary layer grows on the channel walls and the flow tends to become fully developed. Longitudinal vortices can be generated in the channel flows by mounting vortex generators on the channel wall. Alternatively part of the channel wall can be used as vortex generators. To this purpose small rectangular or triangular pieces can be punched out of the channel wall. These pieces will remain attached to the channel wall at the base and project into the flow with an angle of attack β with the main flow direction. The longitudinal vortices are generated along the side edge of the vortex generator due to the pressure difference between front surface facing the flow and the back surface of the vortex generator.

Because of their form these vortex generator can be termed wing-type vortex generator and in short they are called delta wing when the vortex generator has a triangular shape.

In contrast to the flow around a delta wing in an infinite medium, the flow in a channel with an attached delta wing will show no trailing edge vortex. At the juncture of the wing and the channel wall both at the pressure and suction side separation bubbles may appear.

Finally, because of the favorable pressure gradient in the channel, the longitudinal vortices will be stable and their spiraling motion will persist to a length which is many times the length of the delta wing.

An area of application of heat transfer enhancement through vortex generator can be the flow in gas side of a compact plate-fin heat exchanger where vortex generators in form of slender delta wings can be punched out of the plate-fin. Typically the Reynolds and Mach numbers in such applications lie below 2000 and 0.02 respectively. So the gas flow can be treated as laminar and the density variation with respect to pressure will be negligible. However, the variation of density with respect to the temperature may not be negligible in a heat exchanger. A heat exchanger working with air between 300 K and 330 K will induce a density variation of 10 %. Along with the density variation, the thermal buoyancy will influence the flow.

Theoretical and experimental investigations show that counter rotating longitudinal vortices can be generated in a horizontal channel without a built-in obstacle by combined forced and free convection [2, 3]. This mixed convection can result in local heat transfer enhancement of 400 % over heat transfer in a channel with forced convection only. Computation of flow fields and heat transfer in a channel with built-in delta wing and delta winglets have been presented elsewhere [4]. These computations neglected the effect of buoyancy. The computational results showed that longitutudinal vortices behind the delta wing can locally increase the heat transfer by factor of 2. The purpose of the present work is numerical investigations of mixed convection flow structure and heat transfer in a rectangular channel with a built-in slender delta wing.

The base of the wing is fixed on the bottom plate and the apex faces the incoming flow with an angle of attack. For computational simplicity, the wing will be assumed to have zero thickness. The flow medium will be treated as incompressible and the effect of free convection will be modeled by Boussinesq approximation.

2. Basic Equations and Method of Solution

Figure 1 shows the flow configuration consisting of a rectangular channel of length L, width B and height H, with a built-in delta wing of aspect ratio Λ and angle of attack β. The wing has zero thickness. The governing equations are unsteady incompressible Navier-Stokes and energy equations with Boussinsq approximation for free convection. In non-dimensional form these equations in cartesian coordinates are

$$\frac{\delta u_i}{\delta x_i} = 0 \tag{1}$$

$$\frac{\delta u_i}{\delta t} + u_j \frac{\delta u_i}{\delta x_j} + \frac{\delta p}{\delta x_i} = \frac{\nabla^2 u_i}{Re} + k \frac{Gr}{Re^2} T \tag{2}$$

$$\frac{\delta T}{\delta t} + u_j \frac{\delta T}{\delta x_j} = \frac{\nabla^2 T}{Re\ Pr} \tag{3}$$

Here: Re = Reynolds number = $\bar{u}_{av}\ H / \nu$
Gr = Grashof number = $g\ (\bar{T}_\infty - \bar{T}_w)\ H^3 / \nu^2\ \bar{T}_\infty$
Pr = Prandtl number = $\mu\ c_p / \lambda$
k = 1 for y - momentum equation
k = 0 for x - and z - momentum equation
λ = heat conductivity
μ = viscosity
$\nu = \mu / \rho$
ρ = density

The velocity components are nondimensionalized by the average velocity in the channel, inlet $\bar{u}_{av}$, the nondimensional temperature is $T = (\bar{T} - \bar{T}_\infty) / (\bar{T}_w - \bar{T}_\infty)$ and the nondimensional pressure is $p = (\bar{p} - \bar{p}_\infty) / \rho \, \bar{u}_{av}^2$ where subscripts ∞ and w stand for inlet and wall conditions respectively. The lengths are nondimensionelized with the channel height H.

The top and bottom walls have temperatures $\bar{T}_w$ and side walls are adiabatic. No-slip boundary conditions are used on top and bottom walls and symmetry conditions on side walls. At the channel inlet the flow is hydrodynamically fully developed and with a uniform temperature $\bar{T}_\infty$ ($\bar{T}_\infty < \bar{T}_w$). At the channel exit second deriatives of the flow variables are set equal to zero.

A modified version of Marker And Cell (MAC) technique has been used to solve the basic equations [5]. The computational domain is divided into a set of cartesian cells (Fig. 2a). Staggered grids are used in which velocity components are defined at cell faces to which they are normal (Fig. 2b). The pressure and temperature are defined at the cell center. MAC is a semi-implicit finite difference technique consisting of three steps. In the first step the momentum equations are solved explicitly in order to obtain a new (provisional) velocity field from a given velocity, temperature and pressure field. In the second step the continuity equation is implicitly solved by a pressurevelocity iteration until the velocity divergence at each cell becomes negligible. This iteration process amounts to an SOR-solution of the Poisson eyuation for pressure [6]. In the third step the energy equation is solved by an SOR-method. This solution technique is continued until a steady state solution is obtained.

Second upwind differences are used for convective terms and central differences are used for viscous terms. The computational grids are 36 * 15 * 26 in x, y and z directions. Computations with these grids in a channel with thermally developing flow for Re = 500 give at the channel exit a Nusselt number which lies within 5 % of the exact value [7].

3. Results and Discussion

A large number of computations have been performed with different Reynolds numbers (Re = 500, 1000 and 1500) and Grashof numbers (Gr / Re^2 = 1, 2 and 4) for flows in a channel with a delta wing of aspect ratio Λ = 1 and with angle of attack β = 15° and 26°. Figure 3 shows the typical generation, transport of longitudinal vortices and the flattening of their cross section as they move along the channel. In these computations, the effect of free convection has not been considered. Figure 4 compares the cross stream velocity vectors at a distance of x = 3.8 from the inlet for different Grashof numbers. The Reynolds number for these cases is 500. The cross stream velocity vectors become somewhat larger (upto 10 %) when free convection is considered than for the case with Gr = 0. The increase in cross stream velocity vectors is dependent on Grashof number. For all the cases of Grashof number 8, the vortices diverge towards side walls. However, this shift towards the side walls becomes more prominent as the Gr increases.

Velocity vector plots at x = 2.5 (not presented here) show little influence of the Grashof number. The effect of free convection becomes more and more prominent as the flow moves further downstream of the channel inlet. It has already been mentioned that mixed convection flows in a channel will give rise to longitudinal vortices. For the temperature boundary conditions used in this case, the orientation of the longitudinal vortices due to mixed convection in the channel without vortex generator will be apparently the same as that of vortices due to delta wing in the channel with only forced convection.

Figure 5 shows velocity vectors on x-y plane at a distance z = 0.38 from the midplane for different Grashof numbers. The location of the maximum of the axial velocity profile shifts towards the bottom wall as the Grashof number increases.

Figure 6 compares the isolines of Nusselt numbers on the bottom plate for different Gr. Free convection increases Nusselt number directly downstream of the wing. This can be seen by comparing the isoline 9. Similar Nusselt number plots on the top wall (not shown here) show an increase of area of smaller Nusselt number as the Grashof number increases.

Figure 7 shows the spanwise average Nusselt number on the bottom plate. The behavior of the Nusselt number at and downstream of the channel inlet is, as expected, typical of channel flow. It falls down very quickly. Downstream of the apex of the delta wing (at $x = 1$) it starts increasing due to the swirling motion associated with longitudinal vortices until it reaches a maximum of approximately of 10. Then it starts again falling down to a second minimum at the junction of the wing and the plate where a small separation bubble appears. Farther downstream, the Nusselt number again starts increasing. This increase is due to the persistence of the longitudinal vortices. Upto a distance of $x = 2.7$, the effect of free convection on the Nusselt number is not prominent. The peak value of Nusselt number at $x = 2$ increases approximately 4 % as the Gr increases from 0 to $5 * 10^5$. However after $x = 2.7$, the effect of free convection dominates and the three curves for three Gr diverge. With $Gr = 0$, the Nusselt number levels off at $x = 3.5$ and then starts decreasing slowly. This is expected since the longitudinal vortices are losing strength because of the viscous dissipation. In contrast the Nusselt number with nonzero Gr starts increasing at $x = 2.7$. The work of Incropera et al [2, 3] showed that the longitudinal vortices due to mixed convection in the channel appear far downstream from the entrance.

It can be conjectured that in the present case the longitudinal vortices due to mixed convection are augmenting the longitudinal vortices from wing at $x \geq 2.7$. The Nusselt numbers at the exit ($x = 4.5$) are 14.5, 12.5 and 10 for $Gr = 5 * 10^5$, $2.5 * 10^5$ and 0 respectively. Because the computation has been terminated at $x = 4.5$, it is not known where the Nusselt number for mixed convection peaks up.

The spanwise average Nusselt numbers on the top plate show that increasing Gr results in decrease in Nu. This reduction of heat transfer is caused by the upward drift of the hot fluid due to buoyancy.

Figure 8 compares the combined spanwise average Nusselt numbers of the top and bottom plates of the channel. At $x \geq 3.15$ the Nu increases with Gr. At the exit, Nu for the larges Gr is 22.5 % larger than that for Gr = 0. The solid curve shows the Nu in the channel without a wing and free convection. The wing even without free convection can the increase the exit Nusselt numbers by 78 % over the flow in channel without wing.

4. Conclusion

Heat transfer can be increased significantly in channel flows by generating longitudinal vortices with wing-type vortex generators. Free convection causes a farther augmentation of heat transfer. In the present investigations the lacations of heat transfer enhancement due to vortex generator and free convection have been different since the effect of free convection manifested itself near the exit of the channel. It will be of interest to investigate the optimum location of the vortex generator.

References

1. M. Fiebig, P. Kallweit, N.K. Mitra: In Proc. 8th International Heat Transfer Conf. in San Francisco, ed. by C.L. Tien, V.P. Carey, J.K. Ferell, Hemisphere Pub. Cop.
2. F.P. Incropera, J.A. Schutt: Numerical Heat Transfer 8, 707 (1985)
3. F.P. Incropera, A.L. Knox, J.R. Maughan: Jour. Heat Transfer (ASME) 109, 434 (1987)
4. M. Fiebig, U. Brockmeier, N.K. Mitra, T. G)ntermann: Numerical Heat Transfer, to appear
5. C.W. Hirt, B.D. Nichols, N.C. Romero: Los Alamos Scientific Lab. Rep. LA - 5652 (1975)
6. A. Brandt, J.E. Dendy, H. Ruppel: J. Comp. Phys. 34, 348 (1980)
7. R.K. Shoh, A.L. London: Advances in Heat Transfer, Supp. 1, Academ (1978)

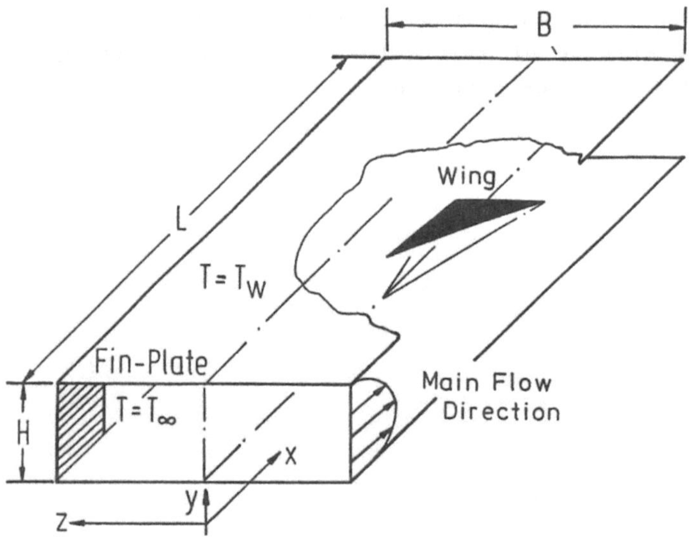

Fig.1 Flow configuration and computational domain

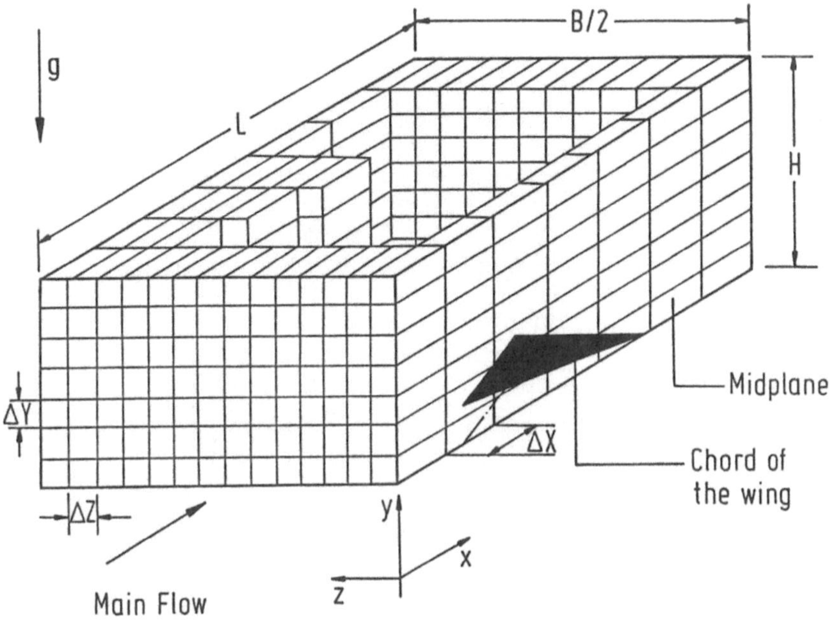

a

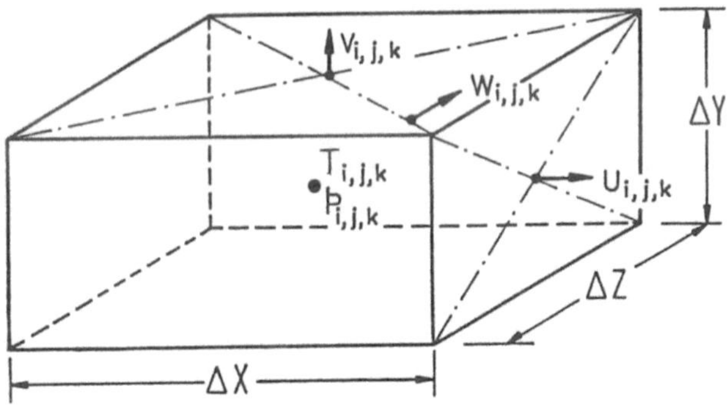

b

Fig.2 Schematic of three dimensional rectangular grids (a) and staggered grid (b)

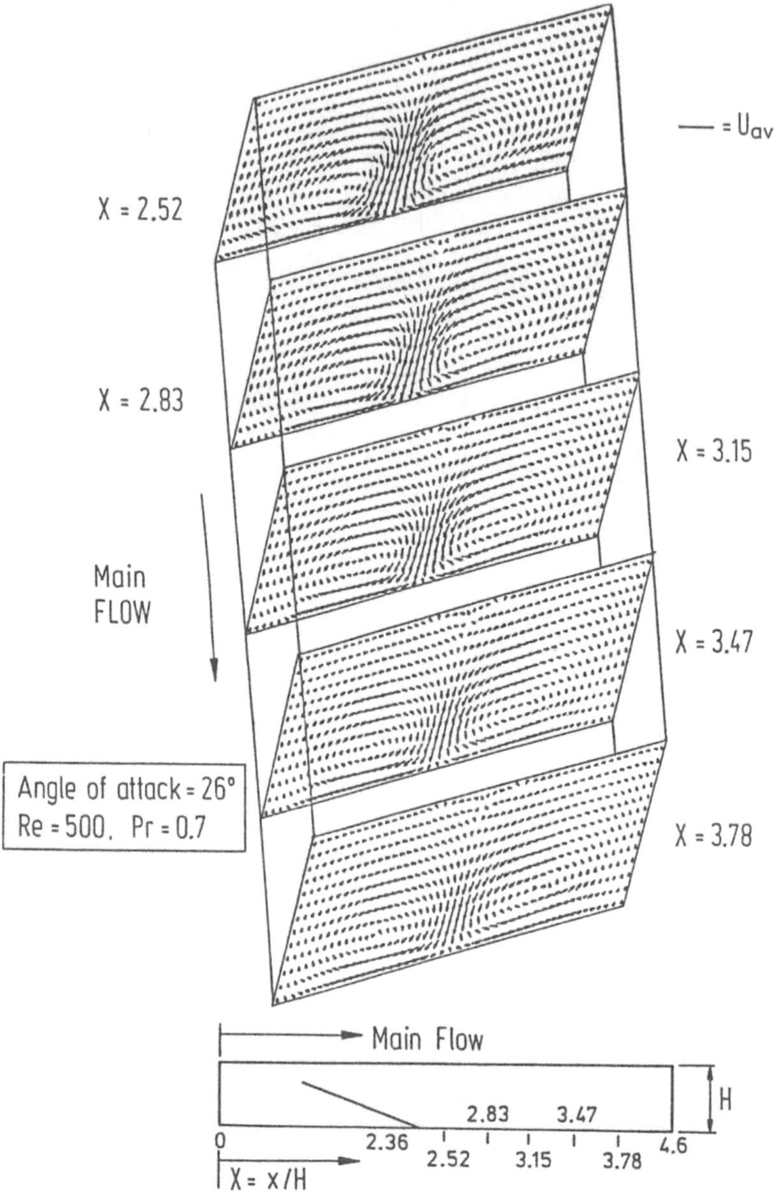

Fig.3 Velocity vectors at different (y-z)-cross sections for a flow with Re = 500, Pr = 0.7, Gr = 0, $\beta = 26^0$, $\Lambda = 1$, $\alpha = B/H = 2$

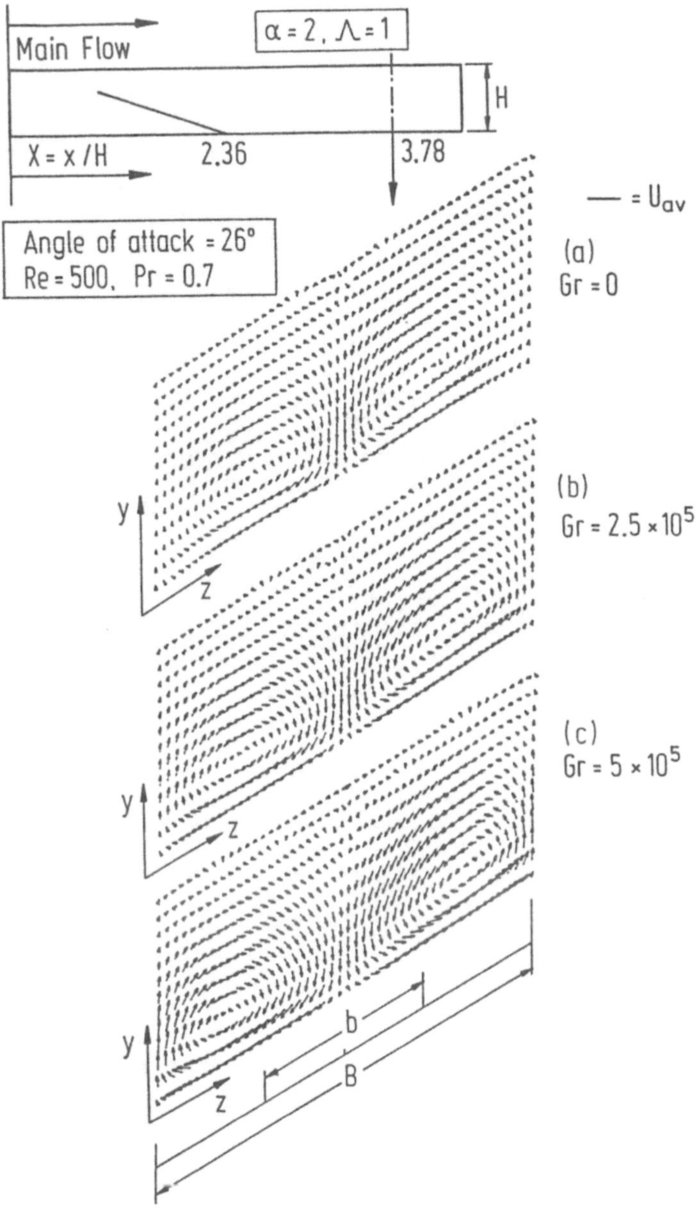

Fig.4 Velocity vectors for different Grashof numbers at (y-z)-cross section ($x = 3.78$) with $Re = 500$, $Pr = 0.7$, $\beta = 26^0$, $\Lambda = 1$, $\alpha = B/H = 2$

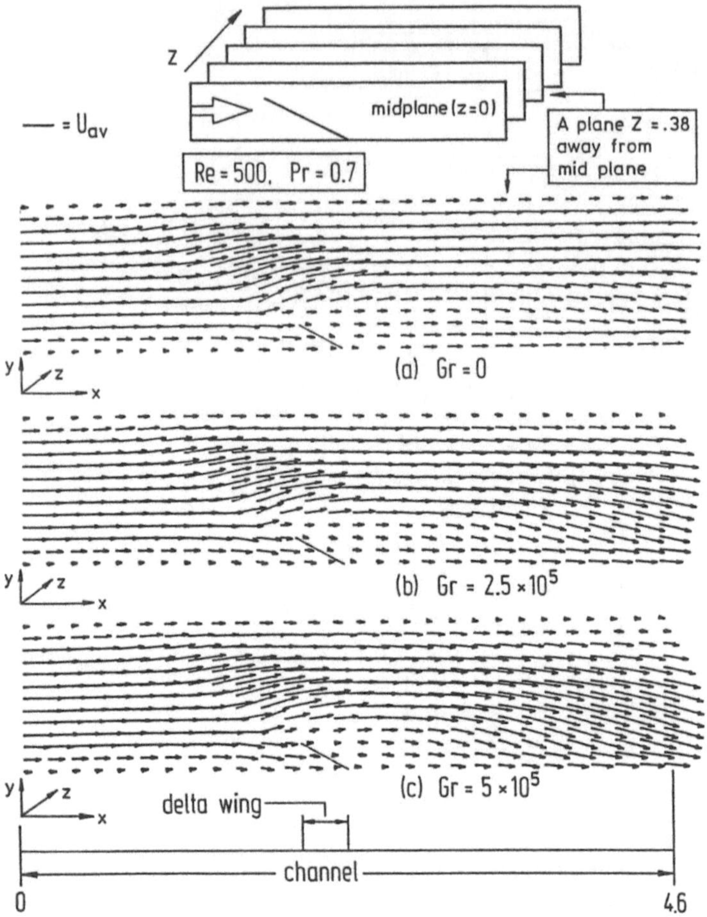

Fig.5 Velocity vectors for different Grashof numbers at a (x-y)-cross section (z = 0.38) for a flow with Re = 500, Pr = 0.7, Λ = 1, α =2

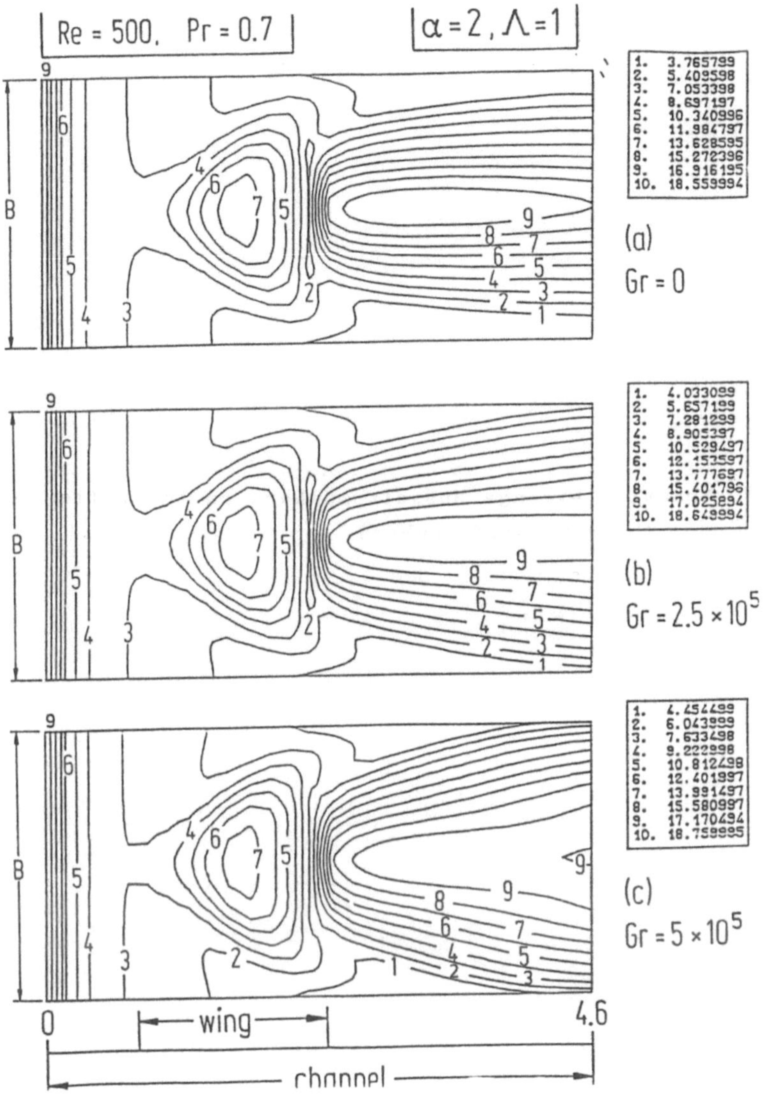

Fig.6 Iso-lines for Nusselt numbers on the bottom plate for different Grashof numbers for a flow with Re = 500, Pr = 0.7, $\beta = 26^0$, $\Lambda = 1$, $\alpha = 2$

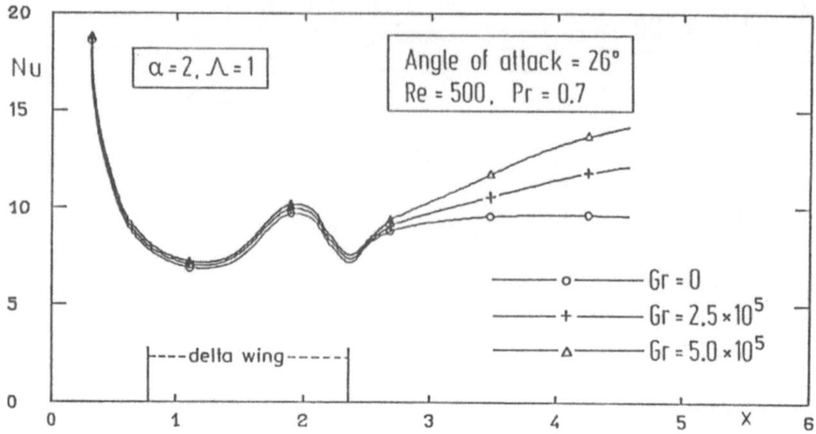

Fig.7 Spanwise average Nusselt numbers on the bottom plate for three Grashof numbers

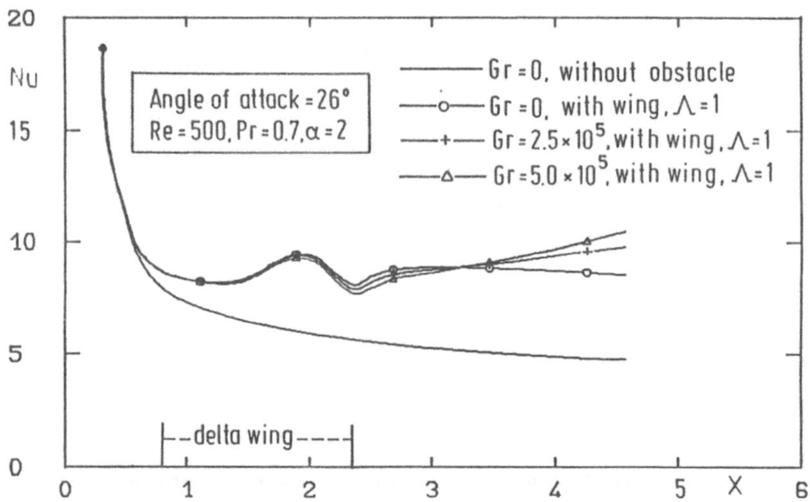

Fig.8 Spanwise average Nusselt numbers in the channel for three Grashof numbers

The Numerical Solution of the Biharmonic Equation, Using a Spectral Multigrid Method

R.W. de Vries and P.J. Zandbergen

Introduction

Spectral methods can be adequate to solve problems in fluid dynamics numerically. A comprehensive treatment of these methods can be found in the book of Canuto, Hussaini, Quarteroni and Zang [1].

In [2] among others some experiences are reported with spectral methods for a flow with a re-entrant corner. In 1982 Zang, Wong and Hussaini introduced a spectral-multigrid method [3], [4]. In [5] we reported our own experiences with a spectral multigrid method. For spectral methods preconditioning is essential and the preconditioning matrix must be a good approximation of the differential operators involved. On a non equidistant grid the preconditioning matrix is usually non-symmetric [3], [4], which has several disadvantages.

In this paper we use a symmetric preconditioning matrix and we apply a non stationary Richardson relaxation process and an I.C.C.G. method to solve the resulting set of equations for the corrections on the grids. To obtain an efficient spectral-multigrid Poisson solver, fast fourier transform techniques are used to transform from the physical plane to the phase plane and vice versa. This Poisson solver is used to solve the biharmonic equation. As is known the biharmonic equation emerges in the stream function vorticity formulation in the Stokes approximation of the Navier-Stokes equations. To solve the biharmonic equation, it will be split into two Poisson problems, one for the stream function and one for the vorticity. A non trivial problem is the treatment of the boundary conditions for the vorticity. This problem can be solved by the Kleiser-Schumann method [6], [7], [2] or by the more general variational setting of Glowinsky and Pironneau [8], where Green's theorem plays an important role and so it does in the paper of Dennis and Quartapella [9].

In this paper we also use Green's theorem extensively both for the derivation of the symmetric preconditioning matrix and for the solution of the boundary condtions for the vorticity.
In section I a description is given of the spectral multigrid method, section II describes the solution of the biharmonic equation and in section III numerical results are presented.

I Description of the spectral-multigrid method

Although spectral multigrid methods are described in different papers [1], [3], [4] we will give a short description for completeness. One wishes to solve the following problem; $Lv = f$, v given on Γ, the boundary of the square $[-1,1] \times [-1,1]$. L is an elliptic differential operator and $f(x,y)$ a given function. As usual we expand v in a Chebyshev series:

$$v(x,y) = \sum_{k=0}^{N}{}'' \; b_k(y)\, T_k(x) \text{ and } b_k(y) = \sum_{j=0}^{M}{}'' \; a_{jk}\, T_j(y) \qquad (1)$$

In (1) $T_k(x)$ and $T_j(y)$ are Chebyshev polynomials of order k and j and the double prime " denotes that the terms with $k = 0$ or N or $j = 0$ or M are taken with factor $\frac{1}{2}$. As is seen from (1) a two-dimensional problem can be treated as a repeated one-dimensional problem. For the description of the method we can restrict ourselves to the description in one dimension. So

$$v(x) = \sum_{j=0}^{N}{}'' \; a_j T_j(x) \text{ with } T_j(x_k) = \cos\,(j \arccos x_k) \text{ and}$$

if the collocation points $x_k = \cos \frac{\pi k}{N}$, $k = 0,1,\dots.N$ are used $v(x)$ reduces to:

$$v(x_k) = \sum_{j=0}^{N}{}'' \; a_j \cos j \frac{\pi k}{N}, \quad k = 0,1,\dots N \qquad (2)$$

The coefficients a_j are given by [10]:

$$a_j = \frac{2}{N} \sum_{k=0}^{N}{}'' \; v(x_k)\, T_j(x_k) = \frac{2}{N} \sum_{k=0}^{N}{}'' \; v_k \cos \frac{jk\pi}{N} \qquad (3)$$

where for $v(x_k)$ we write v_k.
Because the series in (2) and (3) are Fourier-cosine series the fast fourier transform can be used to evaluate these sums.

The F.F.T. routine of the NAG library [11] gives the discrete fourier transform of the function v(x) which

is given by $\hat{Z}_k = \frac{1}{\sqrt{N}} \sum_{j=0}^{N-1} v_j \exp\left(-\frac{2\pi ijk}{N}\right)$; $i = \sqrt{-1}$ and $k = 0,1,\ldots$

An even continuation of v(x) in such a way that $v_j = v_{2N-j}$ gives after some simple algebra: $\hat{Z}_k = \frac{1}{\sqrt{2N}} a_k$.

If L_{sp} is the spectral representation of L, then with a_k known

$$L_{sp}\, v = \sum_{j=0}^{N-2}{}'' \; b_j T_j(x) \text{ and}$$

$$b_j = \sum_{\substack{i=j+2 \\ i+j \text{ even}}}^{N-1} i(i^2-j^2)\, a_i + \frac{1}{2}(N^2-j^2)\, Na_N, \text{ see [12]}.$$

For the transormation from the phase plane to the physical plane $\sum_{j=0}^{n-2}{}'' \; b_j T_j(x)$ has to be evaluated, which gives $\frac{d^2v}{dx^2}$ in the physical plane.

For this evaluation the recurrence relation of Clenshaw is used. This recurrence relation is stable [10] and gives an efficient way for summing the series $\sum_{r=0}^{N} C_r \varphi_r(x)$ where φ obeys the recursion

$$\varphi_{r+1}(x) + \alpha_r(x)\varphi_r(x) + \beta_r(x)\varphi_{r-1}(x) = 0$$

and Clenshaw's recurrence relation yields:

$$\sum_{r=0}^{N} C_r\varphi_r(x) = d_0(x)\varphi_0(x) + d_1(x)\{\varphi_1(x) + \alpha_0(x)\varphi_0(x)\}$$

with $C_r = d_r(x) + \alpha_r(x)d_{r+1}(x) + \beta_{r+1}d_{r+2}(x)$ and $d_{N+1} = d_{N+2} = 0$, $r = N, N-1,\ldots 0$. In the Chebyshev case we have $\alpha_r(x) = -2x$, $\beta_r = 1$ and the sum of Chebyshev polynomials is given by:

$$\sum_{n=0}^{N}{}'' \; C_n T_n(x) = \frac{1}{2}(d_o(x)-d_2(x)) - \frac{1}{2} d_N T_N(x) \;;$$

$$d_r = C_r + 2xd_{r+1} - d_{r+2};\; d_{N+1} = d_{N+2} = 0.$$

So we have efficient and fast transform methods for the transformation from the physical to the phase plane and vice versa.

For the numerical solution of the problem $Lv = f$ a non stationary Richardson relaxation process is used:

$$v^{(k+1)} = v^{(k)} + \omega_k A^{-1}(L_{sp}v^{(k)}-f) \qquad (4)$$

In (4) A^{-1} is a semi inverse of L and A preferably symmetric. The parameters $1/\omega_k$ are the zero's of the Chebyshev polynomial $T_k(\frac{b+a-2x}{b-a})$ where [a,b] is the interval of the eigenvalues of $A^{-1}L_{sp}$ [13]. In our case k = 4.

A symmetric preconditioning matrix A can be obtained in the following way [14].

Consider the problem $-\Delta u = f$ on $[-1,1] \times [-1,1]$ and a non equidistant mesh is given by:

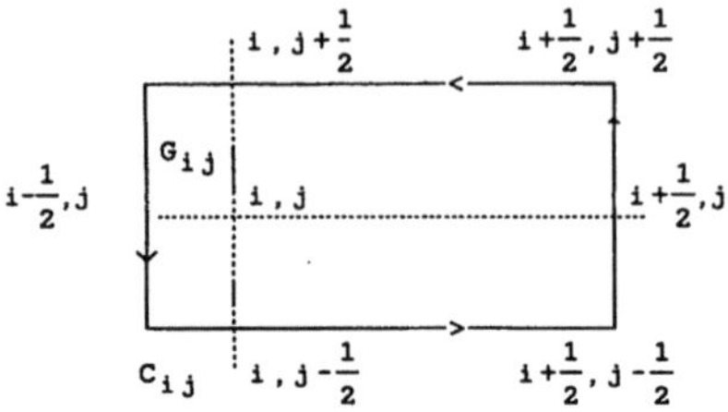

Fig. 1

$h_i = x_{i+1} - x_i$, $k_j = y_{j+1} - y_j$.

Integration of $-\Delta u = f$ over G_{ij} (fig. 1) gives:

$$-\iint_{G_{ij}} (u_{xx}+u_{yy})\,dxdy = \iint_{G_{ij}} f\,dxdy$$

Green's identity: $\iint_{G_{ij}} (S_x-T_y)\,dxdy = \oint_{C_{ij}} (Tdx+Sdy)$ yields:

$$-\iint_{G_{ij}} (u_{xx}+u_{yy})\,dxdy = -\oint_{C_{ij}} (u_x dy-u_y dx)$$

An approximation for the line-integral is given by:

$$-\int_{i+\frac{1}{2},j-\frac{1}{2}}^{i+\frac{1}{2},j+\frac{1}{2}} u_x\,dy = -u_x(x_{i+\frac{1}{2}},y_j)(k_j+k_{j-1})/2$$

For u_x an approximation is given by central differences: $u_x \simeq (u_{i+1,j}-u_{i,j})/h_i$. In the same way $\int u_y dx$ can be approximated and so we arrive at the following approximation scheme for the integrated Poisson problem:

$$-\frac{(k_{j-1}+k_j)}{2h_{i-1}} U_{i-1,j} + \left\{\frac{(k_{j-1}+k_j)}{2}\left(\frac{1}{h_i}+\frac{1}{h_{i-1}}\right) + \frac{h_{i-1}+h_i}{2}\left(\frac{1}{k_j}+\frac{1}{k_{j-1}}\right)\right\}U_{i,j}$$

$$-\frac{(k_{j-1}+k_j)}{2h_i} U_{i+1,j} - \frac{(h_{i-1}+h_i)}{2k_j} U_{i,j+1} - \frac{(h_{i-1}+h_i)}{2k_{j-1}} U_{i,j-1} = \iint_{G_{ij}} f\,dxdy \tag{5}$$

The left-handside of (5) yields a symmetric matrix. The right-handside will be discretized by $\iint_{G_{ij}} f\,dxdy \simeq f_{i,j}\ (h_i-h_{i-1})(k_j-k_{j-1})/4$ (6)

The discretized form of the problem $-\Delta u = f$, including the boundary conditions, can be written as AU = F, with A a symmetric matrix and $F = \iint_{G_{ij}} f\,dxdy$.

In this case the Richardson relaxation scheme as given by eq.4 becomes $A\underline{W}^{(k)} = \omega_k r^{(k)}$, with $\underline{W}^{(k)} = v^{(k+1)} - v^{(k)}$, the corrections and $\underline{r}^{(k)} = L_{sp} v^{(k)} - \underline{F}$, the residuals. (7)

To solve the problem (7) an ICCG method as proposed by Kershaw [15] is used. Although fast and efficient methods are used for the transformations and the solution of the set of equations (7), the Richardson relaxation process on a fixed grid converges slowly. As pointed out by A. Brandt [16] multigrid speeds up the convergence of a relaxation process considerably. We use two grids a fine grid with most 16 Chebyshev polynomials and a coarse grid with 8 polynomials.

On each grid 4 iterations are performed. Restriction and prolongation operators are given by Zang et al [3], [4]. In this way we cycle from fine to coarse grid and back untill the preset accuracy is reached on the fine grid.

II Numerical solution of the biharmonic equation

Introducing the streamfunction and vorticity into the Navier-Stokes equation and neglecting the convective terms gives rise to a biharmonic problem. $\Delta\Delta\psi = f$, $\psi = 0$ on the boundary, $\frac{\partial\psi}{\partial n}$ known on the boundary. To solve this problem we write the biharmonic equation in its coupled form:

$$\left.\begin{array}{ll} \Delta\psi = \varsigma & \psi = 0 \text{ on the boundary} \\ \Delta\varsigma = f & \frac{\partial\psi}{\partial n} \text{ known on the boundary} \end{array}\right\} \tag{8}$$

The Poisson problems in (8) can be solved if the boundary condition for ς as a function in ψ en $\frac{\partial\psi}{\partial n}$ on the boundary is known.

To retrieve the boundary condition for ς a function $e_m(x,y)$ is introduced and from Green's theorem it follows that [9]:

$$\iint_o (e_m\Delta\psi - \psi\Delta e_m)\, dxdy = \oint_{\partial o} \left[e_m \frac{\partial\psi}{\partial n} - \psi \frac{\partial e_m}{\partial n} \right] d\tau = \oint_{\partial o} e_m \frac{\partial\psi}{\partial n}\, d\tau \tag{9}$$

If $\Delta e_m = 0$ and $e_m(x)$ given on the boundary

$$\iint_o e_m \varsigma\, dxdy = \oint_{\partial o} e_m \frac{\partial\psi}{\partial n}\, d\tau \tag{10}$$

where $\psi = 0$ on the boundary is used.

ς is represented by the following series:

$$\varsigma = \varsigma_0 + \sum_{k=1}^{4(N-1)} a_k \varsigma_k \tag{11}$$

In this series ς_0 is obtained from:
$\Delta\varsigma_0 = f$ and $\varsigma_0 = 0$ on the boundary and ς_k follows from $\Delta\varsigma_k = 0$ and ς_k given on the boundary.
The set ς_k on the boundary can be an arbitrary set of independent functions.* Here the boundary condition $\varsigma_k(x_1) = \delta_{k1}$ is used, thus the coefficients a_k in (11) are the boundary conditions for ς. To calculate ς_o and ς_k, one Poisson problem and 4(N-1) Laplace problems must be solved. For our problem symmetry considerations reduce the number of Laplace problems to N/2.
From (10) and (11) the set of equations:

$$\sum_{k=1}^{4(N-1)} a_k \iint_O e_m \varsigma_k \, dxdy = \oint_{\partial O} e_m \frac{\partial\psi}{\partial n} d\tau \quad , \; m = 1,2,\ldots 4(N-1) \text{ is derived.}$$

If $B_{pq} = \iint_O e_p \varsigma_q dxdy$ then the matrix B is called the influence matrix. The set $\{e_p\}$ is chosen to be the same set as $\{\varsigma_p\}$.
To calculate the elements of the influence matrix B we here consider $B_{pq} = \int_{-1}^{1}\int_{-1}^{1} \varsigma_p \varsigma_q \, dxdy$, and write for $e_p\varsigma_q$ the spectral representation:

$$\varsigma_p \varsigma_q = \sum_{k=0}^{N}{}'' \sum_{m=0}^{N}{}'' a_{km}^{(pq)} T_k(x)\, T_m(y) \tag{12}$$

With (12) we find:

$$B_{pq} = \sum_{k=0}^{N}{}'' \sum_{m=0}^{N}{}'' a_{km}^{(pq)} \int_{-1}^{1}\int_{-1}^{1} T_k(x)\, T_m(y) dxdy$$

After some algebra we find:

$$\int_{-1}^{1} T_m(y)dy = \begin{cases} \dfrac{(-1)^{m+1}}{2}\left(\dfrac{1}{m+1} - \dfrac{1}{m-1}\right) - \dfrac{1}{2}\left(\dfrac{1}{m+1} - \dfrac{1}{m-1}\right) & m \neq 1 \\ 0 & m = 1 \end{cases}$$

and

$$B_{pq} = \sum_{i=0}^{N/2}{}'' \sum_{j=0}^{N/2}{}'' a_{2i2j}^{(p,q)} \frac{4}{(4i^2-1)(4j^2-1)} \tag{13}$$

* Numerical experiments with a set of modified Chebyshev polynomials show some advantages over the normally used boundary condition $\varsigma_k(x_m) = \delta_{km}$ [8], [9].

Also the elements of B can be calculated by a mid-point integration rule. In this case the overall accuracy is less compared with the calculation of B_{pq} by (13).

III Numerical results

In this section we give the results which are obtained for:

- the influence functions with a Lagrangian Polynomial representing the 1 on the boundary with 16 and 32 Chebyshev polynomials.
- the comparison of the calculated values of the problem $\Delta\Delta\psi = 8$, with $\psi = (1-x^2)(1-y^2)$, which is the exact solution.
- the so called "driven cavity" problem.

If the relaxation process is convergent, the quantity

$$d = \left\{ \frac{\sum_{i=1}^{N-1} \sum_{j=1}^{N-1} \left[F_{ij} - (Lv)_{ij} \right]^2}{(N-1)(M-1)} \right\}^{\frac{1}{2}}$$

is used to stop the iteration. (N and M are the number of Chebyshev polynomials in x- and y-direction)

In all calculations the relaxation process is ended if $d < 10^{-8}$, on the finest grid.

Table I gives the iteration history for the influence function with 1 on $x = \cos \frac{\pi}{N}$ and $y = -1$. V is the number of visists of the finest grid.

V	d	
1	486635	
2	1580	
3	10	
4	9.10^{-2}	
5	9.20^{-4}	
6	$1.3.10^{-5}$	
7	$1.6.10^{-7}$	
8	$3.0.10^{-9}$	

Table I

To get an idea of the accuracy of the influence functions tables II and III give some values of influence functions with 1 on $x = \cos \frac{\pi}{N}$ and $y = -1$. These values are obtained by representing the 1 by a 16th degree, Lagrangian polynomial and solving the problem with 16 and 32 Chebyshev polynomials.

y	x = 0.9807853		x = 0	
	16 polynomials	32 polynomials	16 polynomials	32 polynomials
- 1	1	1	1	1
- 0.98	0.36511	0.37488	0.86018	0.85989
- 0.92	0.02738	0.03884	0.57591	0.57513
- 0.83	0.00480	0.00357	0.34139	0.34067
- 0.78	0.00010	0.00070	0.20375	0.20350
- 0.56	0.00050	0.00021	0.12918	0.12911

Table II | Table III

As can be expected, the accuracy near the corners of the domain is less than in the middle of the domain. In table IV the biharmonic problem $\Delta\psi=\varsigma$, $\Delta\varsigma=8$ with solution $\psi = (1-x^2)(1-y^2)$ is compared with the numerical solution.

x \ y=0.981	ς_{ex}	ς_{num}	ψ_{ex}	ψ_{num}
0.98	- 0.076120	- 0.08236	0.00145	0.00145
0.92	- 0.29289	- 0.28906	0.00557	0.00557
0.83	- 0.617317	- 0.61989	0.011748	0.011748
0.707	- 1.0	- 0.99797	0.019030	0.019030
0.555	- 1.38268	- 1.38447	0.026312	0.026313
0.383	- 1.70711	- 1.70545	0.032486	0.032486
0.195	- 1.92388	- 1.92553	0.036612	0.036612
0.0	- 2	- 1.99821	0.03806	0.03806

Table IV

Table IV shows that ψ_{num} is in excellent agreement with its analytical value ψ_{ex}.

As our last application of the spectral multigrid method we study the driven cavity problem, which describes the viscous flow in a square with a sliding wall at zero Reynoldsnumber. If this problem is formulated in the vorticity-stream fucntion formulation, a singularity is present in the vorticity and its normal derivative. To compare the results of the spectral multigrid method with the modified integral equation method of Kelmanson [17] we have to rescale our problem. Instead of $\frac{\partial\psi}{\partial n} = -1$ on the sliding wall, we have $\frac{\partial\psi}{\partial n} = -\frac{1}{2}$ and so we arrive at:

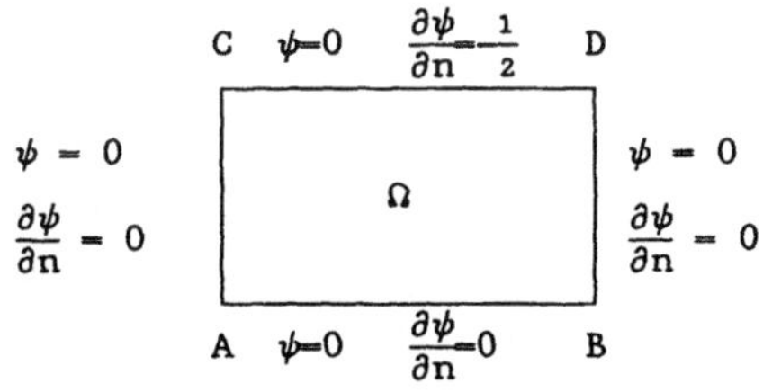

Fig. 2

$\Delta\Delta\psi = 0$ in Ω, $\psi = \frac{\partial\psi}{\partial n} = 0$ on CABD (fig. 2), $\psi = 0$ $\frac{\partial\psi}{\partial n} = -\frac{1}{2}$ on DC. Figure 3 gives the results for the vorticity and stream function. Comparing these results with [17] show a reasonable agreement.

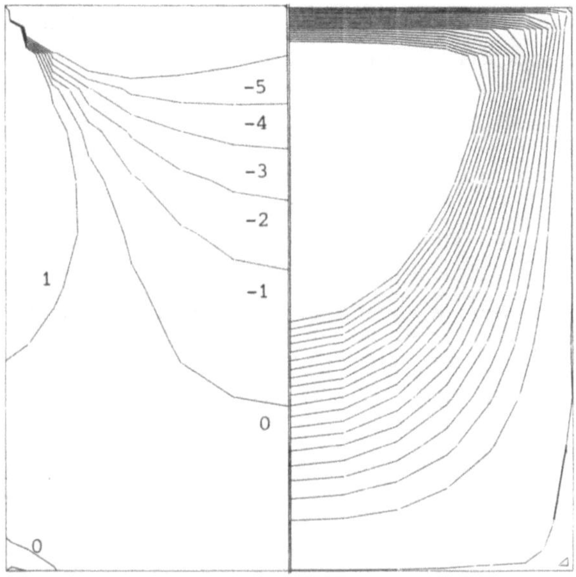

Fig. 3
Equivorticity lines and Streamfunction pattern

The results of Kelmanson who compares his results with those of Burggraf are:

Primary vortex center $(x_{vc}, y_{vc}) = (0.5, 0.76)$, $\psi_{vc} = 0.0998$ and $\varsigma_{vc} = -3.201$. With spectral multigrid we find $(x_{vc}, y_{vc}) = (0.5, 0.78)$, $\psi_{vc} = 0.09975$ and $\varsigma_{vc} = -3.36$. Although we do not subtract out the analytic behaviour of the flow in the neighbourhood of the singularities, the results are in reasonable agreement with those of Kelmanson.

Conclusion

By using a symmetric preconditioning matrix an efficient spectral multigrid method is obtained. It is possible to use fast solvers and most of the CPU time gets in the calculation of restriction and prolongation.
On squares even non smooth problems can be treated in a direct way.
For biharmonic problems the use of Green's identity is a natural way to retrieve the "missing boundary" conditions. The calculation of the influence matrix is time (CPU) consuming and so its worth while to modify the boundary conditions for the influence functions in such a way that the structure of the influence matrix will be improved.

References

1. C. Canuto, Hussaini, M.Y., Quarteroni, A. Zang, T.A. : Spectral methods in fluid dynamics. Springer-Verlag 1988.

2. R. Verstappen, Ten Thye Boonkamp, J., de Vries, R., Zandbergen, P.J. : Lecture Notes in Physics 264. Proceedings of the 10th ICNMFD, Beijing 1986.

3. T.A. Zang, Wong, Y.S., Hussaini, M.Y. : Spectral multigrid methods for elliptic equations. Jrnl. of Comp. Physics 48, 485-501 (1982).

4. id. : Spectral multigrid methods for elliptic equations II, Jrnl. of Comp. Physics 54, 489-507 (1983).

5. A.J. Renkema, Verstappen, R., de Vries, R.W., Zandbergen, P.J. : Some experiences with spectral methods; Notes on numerical fluid Mechanics, Vol. 17, p. 101-115 (1987).

6. L. Kleiser : Numerische Simulationen zum laminar-turbulenten Umschlagsprozess der ebenen Poissenille Strömung. Karlsruhe, april 1982.

7. L. Kleiser, Schumann, U. : Treatment of incompressibility and boundary conditions in 3-d numerical simulations of Plane channel flows. Proc. Third GAMM conf. 1980.

8. R. Glowinsky, Pironneau, O. : Numerical methods for the first Biharmonic equation and for the 2-dimensional Stokes problem. Siam Sirev Vol. 21 No.2 april 1979.

9. S.C.R. Dennis, Quartapella, L. : Direct solution of the Vorticity - Stream Function ordinary differential equations by a Chebyshev Approximation. Jrnl. of Comp. Ph. 52, 448-463 (1983).

10. L. Fox, I.B. Parker : Chebyshev polynomials in numerical analysis, London, Oxford Univ. Press 1968.

11. Numerical Algorithm Group (1984) : Fortran library manual, Mark 11, Vol 1.

12. D. Gottlieb, S.A., Orszag : Numerical analysis of spectral methods: Theory and applications. Soc. Ind. Appl. Math. Philadelphia (1977)

13. G.E. Forsythe, W.R., Wasow : Finite difference methods for partial differential equations. John Wiley and Sons 1967.

14. R.S. Varga : Matrix iterative Analysis. Prentice-Hall Inc. 1962.

15. D.S. Kershaw : The Incomplete Cholesky-Conjugate Gradient method for the Iterative Solution of systems of linear Equations. Jrnl. of Comp. Ph. 26, 43-65 (1978).

16. A. Brandt : Multi-Level Adaptive Solutions to Boundary-Value problems, Math. of Comp vol. 31, nr. 138 (1977).

17. M.A. Kelmanson : Modified Integral equation solution of Viscous flows near sherp corners. Computers and Fluids Vol. 11, p. 307-304 (1983).

Finite Volume Shape and Placement for Efficiency, Accuracy, and Physics

Harry A. Dwyer

Introduction

The purpose of this article is to demonstrate that the shape and placement of finite volume cells can have a profound influence on the algebraic equations which result from the application of the basic physical laws. As an example of the use of this concept three-dimensional results will be presented with a "singular" point in the mesh (such as the center of a cylindrical or spherical like geometry), which has been remove by the use of special **"triangular"** shaped cells. These special cells are essentially hexilateral volumes with one or more zero area sides which are placed at the singular point. With the zero area sides placed at the singular point all coupling of mesh points is removed, and the use of factored or line type implicit solvers, References [1] through [3], can proceed in an efficient manner.

Another subject addressed in the paper is the influence of the placement of the cell center or cell area at the physical boundaries of the computational domain, and the resulting consequences on the ease of application of boundary conditions. Examples will be given from flows with both mass and heat addition, as well as for the pressure boundary conditions for low Mach Number flows. The final topic to be covered in the paper is the convergence of the algebraic equations at the cell center or the cell area in a finite volume formulation. The choice of the proper location for convergence can strongly influence the properties of an algorithm.

Singular Meshes

When a mesh is generated about a point in space or a plane, there can be a problem of excessive independent variable coupling due to the mesh points converging to a single point or so called "singular" point. A schematic of the situation is shown in figure (1) where typical cell centers and volumes are shown, and this figure will be utilized to illustrate the problem and a solution. The general problem is that all the independent variables become coupled to the mesh singularity, and it is more difficult to form efficient implicit finite difference or finite volumes equations from the basic laws. Also, the use of the popular factored methods becomes very difficult, and one is forced to use explicit techniques which can be inefficient for the solution of the Navier-Stokes equations.

The solution to the problem is illustrated in figure (1), and this solution consists of **not** placing a grid point or cell center at the singular point. What is placed at the singular point is the cell boundary, and this cell boundary is chosen to have zero area. With a zero area boundary some of the cell areas then become triangular rather than quadrilaterals, and the typical cell volumes are shown at the bottom of figure (1). The zero area boundaries completely remove the excessive coupling and this can be seen from the basic equations of continuity and momentum, which are given in integral form below

$$\text{Continuity} \quad \partial/\partial t\, [\textstyle\int\int\int_V \rho\, d\mathcal{V}] + \int\int_S \rho\, \mathbf{V}\bullet\mathbf{dA} = 0 \quad (1)$$

$$\text{Momentum} \quad \partial/\partial t\, [\textstyle\int\int\int_V \rho\mathbf{V}\, d\mathcal{V}] + \int\int_S \rho\mathbf{VV}\bullet\mathbf{dA} = \quad (2)$$

$$-\textstyle\int\int_S p\mathbf{dA} + \int\int_S \tau''\bullet\mathbf{dA}$$

where ρ = density, $\mathbf{V}$ = fluid velocity, $d\mathcal{V}$ = cell volume, $\mathbf{dA}$ = cell area, p = pressure, and τ'' = stress tensor.

It is seen from the equations given above that all contributions to the area integrals from the zero area boundary are null, and that all coupling of variables is avoided at the singular point. Of course there remains coupling along all the other boundaries of the cell, and the stress gradients cause coupling through derivative operations across the finite area boundaries. For the cells shown in figure (1) the accuracy of the gradients across the other cell boundaries are determined by the mesh spacing and the order of the approximation, and the influence of the zero area boundary does not negatively influence the overall accuracy of the approximation. The only mathematical operation that is influenced by the cell shape is a divergence or gradient operation at the center of cell volume, since it must be evaluated in a one sided way (Note: These type of mathematical operations are not needed in equations (1) and (2)).

Physical Boundaries

Another type of problem occurs when one is confronted with the placement of the cell boundary or cell center at the physical boundary of a computational region. Some of the typical options are shown in figure (2), and these will be discussed presently. The placement of the cell center on the boundary is shown in the top part of figure (2), and this type of placement favors Dirchlet or known value type of boundary conditions, where the value of the variable does not have to be calculated. However, when Neumann or derivative boundary conditions are necessary, there is a requirement of a half cell analysis of the physical equation or the use of nonconservative mathematical boundary conditions.

The problem of Neumann conditions with this type of grid placement is the most difficult when mass is injected into the boundary. With mass injected the flux and cell center are at the same location, and this violates the basic control volume formulation. A much more "natural" formulation for derivative boundary conditions is the placement of the cell boundary at the physical boundary, as shown in the second portion of figure (2). With this situation a boundary condition is not needed and the flux is treated as a given quantity.

In complex problems both types of boundary conditions are usually encountered, and the resulting situation does not favor one formulation over the other. In order to apply Dirchlet boundary conditions, when the cell boundary and the physical boundary are at the same location, it is necessary to introduce a fictitious cell "inside" the physical boundary, as shown in figure (2). Also, in some circumstances there is sometimes a need to apply fictitious boundary conditions in order to produce the desired influence of the boundary condition. An example of a widely used fictitious boundary condition are the so called reflection conditions, which easily create the effect of a solid wall.

It should be mentioned that the use of highly expanded or "Navier-Stokes" grids near boundaries makes the use of first order approximations more attractive. This result is caused by the fact that the cells are very fine near the boundary, and the resulting truncation error is not a serious problem. Although first order approximations are not to be encourage, there are situations where other algorithm considerations are more dominant in the formulation of a problem. Therefore, it is seen that there is not one way of applying boundary conditions, and the investigator has a choice of conditions which best fits the algorithm or the physics of a particular problem.

Location of Variable Convergence

During the development of numerical methods for problems in fluid mechanics the use of multiple or staggered meshes have played a useful role in many formulations. The basic reason that multiple meshes have been useful is caused by the appearance of both volume and area terms in the basic equations. Volumes terms favor a convergence at the cell center, while surface terms would like to converge at the cell boundaries (This fact is more important for lower order methods). If a duel set of grid points can be generated, then convergence can be obtained at both the cell center and the cell boundary. However, for complex grids it is very cumbersome to employ two sets of geometric quantities, and there is need to develop other methods to accomplish the same result.

An important problem area where multiple meshes have been utilized effectively is low Mach number flows where the pressure filed is determined from the continuity equation, References [4] through [6]. For incompressible flows the continuity equation, eq. (1), involves only the values of the velocity on the cell boundary, and it is necessary to interpolate from the cell centers. On a rough mesh this interpolation can lead to serious error in most pressure correction methods, and staggered mesh schemes have shown themselves to have advantages. A good alternative to a staggered grid is two variables, one at the cell boundary and one at the cell center.

The basic two variable method can be illustrated with use of figure (3), where a schematic of the pressure/velocity correction method is shown. The method determines a pressure correction α, which corrects the velocities across the cell boundaries, so that the continuity equation is satisfied. The resulting velocity correction is applied at the cell boundary and the cell center, but only the velocities at the cell boundaries are capable of converging the continuity equation to machine zero. The velocity field at the cell center always has truncation error which violates the continuity equation, but the truncation error is not a problem for the volume or stress area terms in the equations. Therefore, the terms in the momentum equation are evaluated in the following way

$\partial/\partial t\,[\iiint_V \rho \mathbf{V}\, dV]$	Cell Center Velocities
$\iint_S \rho \mathbf{V}\mathbf{V}\bullet d\mathbf{A}$	Exact Boundary Velocities
$\iint_S p d\mathbf{A}$	Boundary Average of the Cell Center Pressures
$\iint_S \tau''\bullet d\mathbf{A}$	Cell Center Velocities

The two variable method does add complexity, but it is far less than two meshes for complex geometries, and it does treat the sensitive pressure correction in an accurate way. The important aspect of this discussion for control volume analysis is that physical insight into the problem is very clear, and the control volume formulation suggests the proper type of analysis.

Results

The methods that have been discussed is this paper have been applied to a variety of applications which will be briefly described. The discussion will begin with the flow over an ellipse at angle of attack where a three-dimensional mesh is generated by rotating the two-dimensional mesh shown in figure (4) about the north/south line given in the figure. The mesh "singularity" at the north/south line is removed by placing a zero area side of a cell along the line, and some typical results are given in figure (5).

Figure (5) shows the isotherm patterns about the ellipse at angle of attack at a intermediate Reynolds number of thirty. This calculation does not contain any artificial viscosity, and the viscous stresses and thermal conductivity terms are significant. Under these conditions any problems caused by the mesh system would be clearly evident in the isotherm patterns. As can be seen from the results given in figure (5) the solution is very smooth, and the mesh singularity has effectively been removed. It should also be mentioned that the solution method employed consisted of a predictor/corrector technique, References [7] and [8], which solves the basic laws with a line implicit solver.

An example in which the two variable pressure-velocity correction has been employed has been the burning of a octane fuel droplet in a hot air environment. This problem involves time varying density, which is a volume term in the continuity equation, and the introduction of fuel at the gas liquid interface. With the use of the two variable velocity scheme the continuity equation is satisfied to machine zero at every time step, and the calculation is completely conservative. A "Navier-Stokes" grid has been employed at the boundaries, and a first order conservative mass flux condition has been employed with the boundary cell center located at the physical boundary.

The isotherms at a given instant around the droplet are shown in figure (6), and the solution has shown itself to have very good convergence and accuracy properties. A complete discussion of this problem is beyond the scope of the present article, and the reader is referred to a more complete discussion found in Reference [8]. It has been found that the use of a control volume formulation greatly helps in the evaluation of transport properties, and the formulation of complex mass transfer boundary conditions.

The major conclusions of the present article are the following:

1. The use of a control volume formulation allows the investigator to apply knowledge of fluid physics to overcome problems in computational fluid mechanics.
2. The removal of mesh "singular" points can be easily accomplished with special shaped cells, which do not negatively damage the band width of implicit solvers.
3. The pressure-correction algorithm for low speed flows can be improved with the use of two variable methods, which are less complex than multiple grids.

References

1. Steger, J.L., Implicit Finite Difference Simulation of Flow About Arbitrary Two-Dimensional Geometries", **AIAA J.**, Vol. 16, No.7, July 1978, pp. 679-686.
2. Beam, R.M. and Warming, R.F., "An Implicit Factored Scheme for the Compressible Navier-Stokes Equations", AIAA J., Vol.16, No. 4 (1978) pp.393-402.
3. Vinokur, M., "An Analysis of Finite-Difference and Finite-Volume Formulation of Conservation Laws", Sterling Software, Palo Alto, Ca. 1986.
4. Patankar, **Numerical Heat Transfer and Fluid Flow**, Hemisphere Publishing Corporation, New York, 1980.

5. Anderson, D.A., Tannehill, J.C., and Pletcher, R.H., **Computational Fluid Mechanics and Heat Transfer**, Hemisphere Publishing Corporation, New York, 1984.
6. Roache, P., **Computational Fluid Dynamics**, Hermosa Publishers, Albuquerque, New Mexico, 1979.
7. Dwyer, H.A. and Sanders, B.R., "Unsteady Influences in Droplet Dynamics and Combustion", To appear in a Special Issue of Combustion and Flame, Editor R.M. Soo, Also ASME AMD-Vol.81, Dec. 1986.
8. Dwyer, H.A., "Calculations of Droplet Dynamics in High Temperature Environments", To Appear In Combustion Science and Technology, Spring 1989.

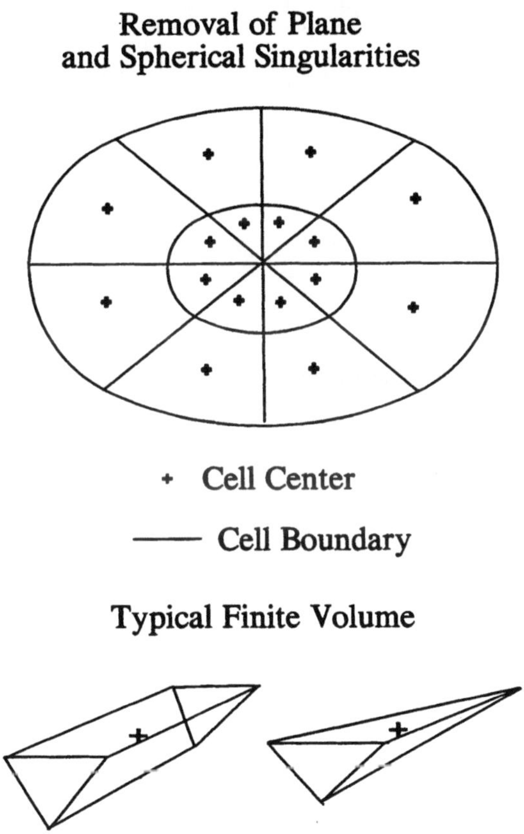

Figure (1) Singular Mesh with Zero Area Cells

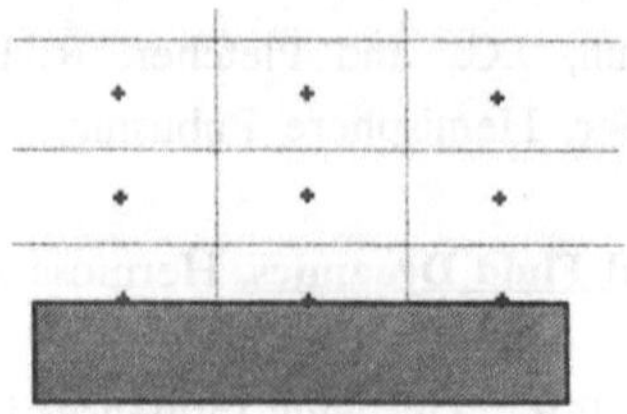

Cell Center on Boundary

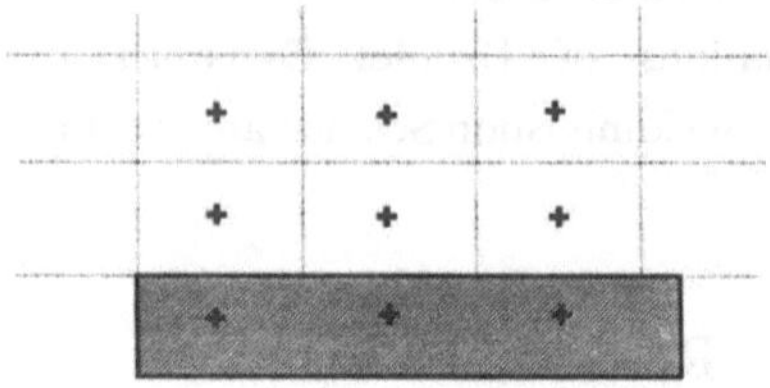

Cell Boundary on Boundary

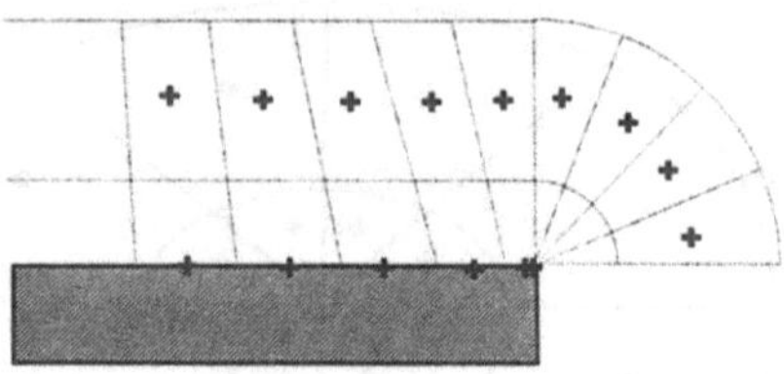

Cell Fan

Figure (2) Placement of Cell Center on Physical Boundaries

$$\iint \vec{V} \bullet d\vec{A} = \iint \nabla \alpha \bullet d\vec{A}$$

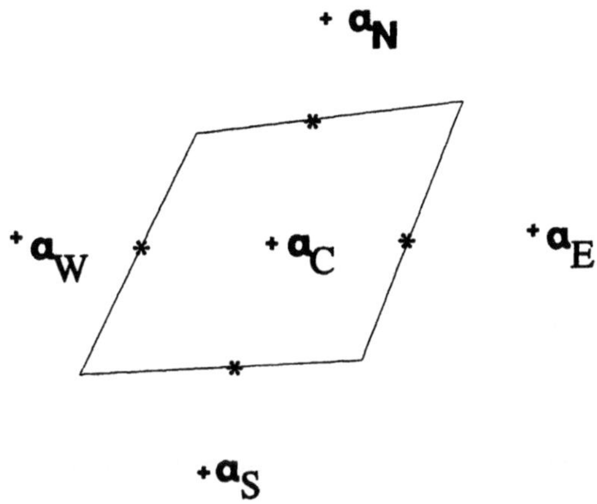

Figure (3) Pressure Correction Illustration

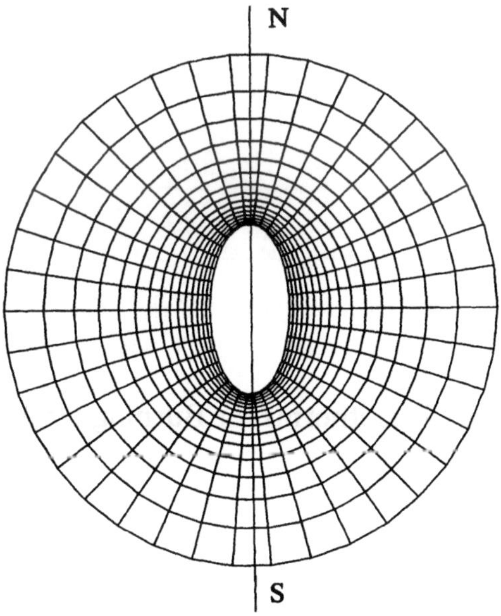

Figure (4) Coordinate System for Ellipse

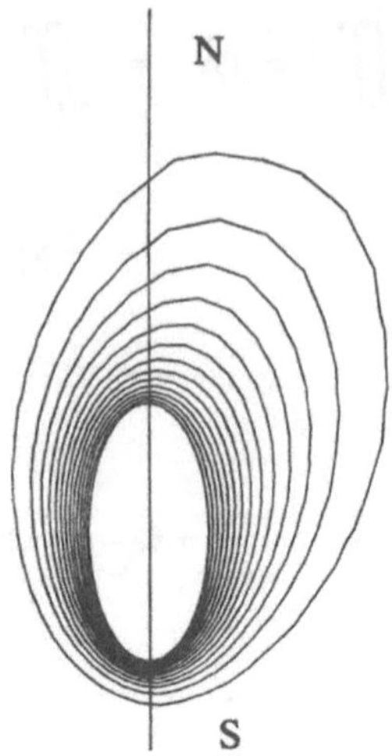

Figure (5) Isotherms for Flow over an Ellipse

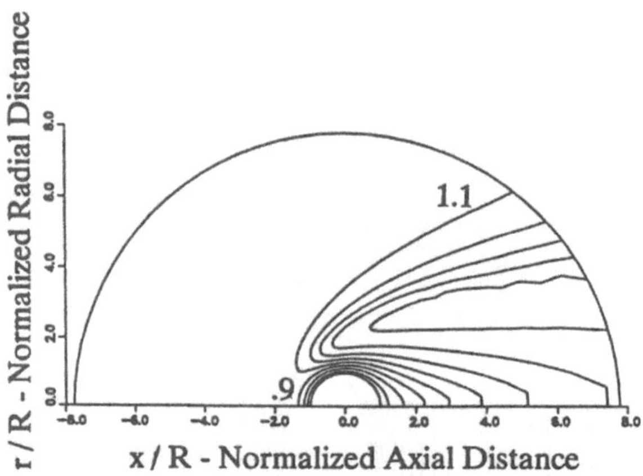

Figure (6) Isotherms for Burning Droplet

NND Schemes and their Application to Numerical Simulation of Complex Plume Flow Problems

Zhuang Fenggan

1. Introduction

In the calculation of complex flow fields containing shock waves, much attention has been paid to the shock capturing method. In order to capture shock waves smoothly without spurious oscillations near or in the shock regions, mixed dissipative schemes (containing free parameters) with first order accuracy near shocks and second order schemes elsewhere have been widely used [1],[2],[3]. There are inherent disadvantages in employing these schemes: first, the free parameters are basically determined through numerical experiments, and second, the resolution of shock is not very satisfactory. Naturally the development of high resolution non-oscillatory dissipative schemes (NND) containing no free parameters has been much emphasized recently, such as TVD schemes [4]-[8] and ENO schemes [9].

In [10], ZHANG found with a linearized analysis of one dimensional Navier-Stokes equations, a proper manipulation of the coefficients of third order derivatives ν_3, the spurious oscillation both upstream and downstream of shocks can be totally suppressed. The choice of ν_3 must be in accordance with the second law of thermodynamics. The numerical experiments with one dimensional non-linear Navier-Stokes equations confirm that the conclusion reached by linearized analysis is valid in general. The importance of the role played by ν_3 was not realized before. While it was usually thought that the third order dispersion terms in the modified differential equations would not play an essential role in a linear problem e.g. in the stability analysis etc, they do contribute through non-linear interactions to the unwanted wave structure and even lead to chaos, if they are not properly treated.

Based on the above physical concepts, we suggest a new NND scheme which is non-oscillatory, dissipative and contains no free parameters. The details are described in the following section. In order to test the effectiveness of the scheme, we have carried out numerical simulations of plume flows with the inviscid Euler equations and the results are given in the paper. Finally some concluding remarks are drawn and suggestions for future work are indicated.

2. A Formulation of NND Schemes

We consider the one-dimensional Navier-Stokes equations without heat conduction and modified by the addition of dispersion terms with a variable coefficient. The equations are:

$$\frac{\partial \rho}{\partial t} + \frac{\partial \rho u}{\partial x} = 0$$

$$\rho\frac{\partial u}{\partial t} + \rho u\frac{\partial u}{\partial x} + \frac{\partial p}{\partial x} = \frac{\partial}{\partial x}\left(\frac{4}{3}\nu\frac{\partial u}{\partial x}\right) \tag{1}$$

$$\rho\frac{\partial H}{\partial t} - \frac{\partial p}{\partial t} + \rho u\frac{\partial H}{\partial x} = \frac{\partial}{\partial x}\left(\frac{4}{3}\mu\frac{\partial H}{\partial x}\right) + \frac{\partial}{\partial x}\left(\nu_3\frac{\partial^2 H}{\partial x^2}\right)$$

where usual notations are adopted and the Prandtl number is assumed to be 3/4. From equations (1) we may obtain an equation for entropy S

$$\rho T\frac{DS}{Dt} = \frac{4}{3}\mu\left(\frac{\partial u}{\partial x}\right)^2 + 3\nu_3\frac{\partial u}{\partial x}\frac{\partial^2 u}{\partial x^2}. \tag{2}$$

Here DS/Dt is the substantial derivative of entropy. For a true physical shock, we have upstream of the shock

$$\frac{\partial u}{\partial x} < 0, \quad \frac{\partial^2 u}{\partial x^2} < 0 \tag{3}$$

hence

$$\left(\frac{\partial u}{\partial x}\right)\left(\frac{\partial^2 u}{\partial x^2}\right) > 0 \tag{4}$$

and downstream of the shock

$$\frac{\partial u}{\partial x} < 0, \quad \frac{\partial^2 u}{\partial x^2} > 0 \tag{5}$$

hence

$$\left(\frac{\partial u}{\partial x}\right)\left(\frac{\partial^2 u}{\partial x^2}\right) < 0. \tag{6}$$

For inviscid flow or flow at high Reynolds numbers, we may disregard the viscous contribution to the entropy, therefore we may observe that if $\nu_3 > 0$ in the whole shock region, then the entropy increase condition is met but not in the downstream region and this is associated with the appearance of downstream spurious oscillations; on the other hand if $\nu_3 < 0$ in the whole region, then the entropy increase condition is not met in the upstream and upstream spurious oscillations occur upstream. Now it is obvious that if we can keep $\nu_3 > 0$ upstream and $\nu_3 < 0$ downstream, we may have a smooth shock transition. Numerical results verify the above conclusion. In passing we note that there might be individual points in the shock region where $(\partial u/\partial x)(\partial^2 u/\partial x^2) = 0$ and hence when $\mu = 0$, the entropy increase condition is not satisfied at these points.

To construct a NND scheme we start with a one dimensional scalar equation

$$\frac{\partial u}{\partial t} + \frac{\partial f(u)}{\partial x} = 0 \tag{7}$$

where $f = au$, and $a = \partial f/\partial u$, and a is the characteristic speed. We may write

$$a = a^+ + a^- \tag{8}$$

where

$$a^+ = (a + |a|)/2, \quad a^- = (a - |a|)/2 \tag{9}$$

Define

$$f^+ = a^+ u \quad \text{and} \quad f^- = a^- u \tag{10}$$

so that

$$f = f^+ + f^-. \tag{11}$$

Equation (7) may be rewritten as

$$\frac{\partial u}{\partial t} + \frac{\partial f^+}{\partial x} + \frac{\partial f^-}{\partial x} = 0. \tag{12}$$

It is easy to verify that if the second order upwind difference scheme is used to evaluate the term $\partial f^+/\partial x$, the coefficient of the third order derivative in the right hand side of the modified equation is positive and the coefficient of the fourth order dissipative term is negative; if the second order central difference is used, the coefficient of the third order derivative is negative and the coefficient of the fourth order dissipative term is zero. Similarly if the second order upwind scheme is used to evaluate $\partial f^-/\partial x$, the corresponding coefficient of third order derivative is negative, while we still have the coefficient of fourth order dissipative term negative; if the second order central difference is used, the coefficient of third order derivative is negative, while we still have the coefficient of fourth order dissipative term negative; if the second order central difference is used, the coefficient of third order derivative is positive and the coefficient of the fourth order dissipative term remains zero. Based on the above findings, we may employ the following difference schemes to evaluate space derivatives:
In the region upstream of a shock, use

1. second order upwind difference for $\partial f^+/\partial x$,
2. second order central difference for $\partial f^-/\partial x$.

In the region downstream of a shock, use

1. second order central difference for $\partial f^+/\partial x$,

2. second order upwind difference for $\partial f^-/\partial x$.

In so doing, we have made the proper choice of the sign of the coefficient of third order derivative in the modified differential equation i.e. ν_3 is positive upstream of a shock and ν_3 is negative downstream of the shock, and this provides us with a sharp shock transition without spurious oscillations both upstream and downstream of shocks. At the same time we notice that in the entire region, the coefficient of fourth order dissipative derivative is negative, which helps suppress odd-even decoupling in smooth regions of physical flows.
Now we write down the semi-discrete difference form of (12) as follows:

In the region upstream of a shock,

$$\left(\frac{\partial u}{\partial t}\right)_j^n = -\frac{3f_j^{+n} - 4f_{j-1}^{+n} + f_{j-2}^{+n}}{2\Delta x} - \frac{f_{j+1}^{-n} - f_{j-1}^{-n}}{2\Delta x} \tag{13}$$

In the region downstream of a shock,

$$\left(\frac{\partial u}{\partial t}\right)_j^n = -\frac{-3f_j^{-n} + 4f_{j+1}^{-n} - f_{j+2}^{-n}}{2\Delta x} - \frac{f_{j+1}^{+n} - f_{j-1}^{+n}}{2\Delta x} \tag{14}$$

Equations (13) and (14) can be combined together using the expressions for numerical flux functions to form

$$\left(\frac{\partial u}{\partial t}\right)_j^n = -\frac{1}{\Delta x}\left(h_{j+\frac{1}{2}}^n - h_{j-\frac{1}{2}}^n\right) \tag{15}$$

where

$$h_{j+\frac{1}{2}} = f_{j+\frac{1}{2}L}^+ + f_{j+\frac{1}{2}R}^- \tag{16}$$

and

$$f_{j+\frac{1}{2}L} = \begin{matrix} f_j^+ + \frac{1}{2}\Delta f_{j-\frac{1}{2}}^+ & \text{(upstream)} \\ f_j^+ + \frac{1}{2}\Delta f_{j+\frac{1}{2}}^+ & \text{(downstream)} \end{matrix} \tag{17}$$

$$f_{j+\frac{1}{2}R} = \begin{matrix} f_{j+1}^- - \frac{1}{2}\Delta f_{j+\frac{1}{2}}^- & \text{(upstream)} \\ f_{j+1}^- - \frac{1}{2}\Delta f_{j+\frac{3}{2}}^- & \text{(downstream)} \end{matrix} \tag{18}$$

$$\Delta f_{j+\frac{1}{2}}^{\pm} = f_{j+1}^{\pm} - f_j^{\pm} \tag{19}$$

It is clear from equation (17) we have to choose between $\Delta f_{j-\frac{1}{2}}^+$ and $\Delta f_{j+\frac{1}{2}}^+$ for the upstream and downstream regions. For a shock wave with a monotonic u profile it is easy to convince ourselves that in general (up to second order accuracy) that $\Delta f_{j-\frac{1}{2}}^+$ and $\Delta f_{j+\frac{1}{2}}^+$ have the same algebraic sign and that in the upstream region the absolute value of $\Delta f_{j-\frac{1}{2}}^+$ is smaller than $\Delta f_{j+\frac{1}{2}}^+$, and that the opposite is true in the downstream region. Similarly $\Delta f_{j+\frac{1}{2}}^-$ and $\Delta f_{j+\frac{3}{2}}^-$ also have the same sign and the absolute value of $\Delta f^-_{j+\frac{1}{2}}$ is smaller than that of $\Delta f^-_{j+\frac{3}{2}}$ in the upstream while the opposite is true in the downstream region. Therefore if we introduce a min mod function, such that min mod (x, y) is equal to the argument with the smallest absolute value when x and y have the same algebraic sign, then (17) and (18) can be rewritten as

$$f^+_{j+\frac{1}{2}L} = f^+_j + \frac{1}{2}\min\operatorname{mod}(\Delta f^+_{j-\frac{1}{2}}, \Delta f^+_{j+\frac{1}{2}}) \tag{20}$$

$$f^-_{j+\frac{1}{2}R} = f^-_{j+1} - \frac{1}{2}\min\operatorname{mod}(\Delta f_{j+\frac{1}{2}}, \Delta f^-_{j+\frac{3}{2}}) \tag{21}$$

We have remarked that there might exist individual points where the entropy increase condition is not satisfied and the finite difference solution might possess undesired oscillations, so we impose, that min mod $(x, y) = 0$ when x and y are of different algebraic signs. The finite difference scheme is then reduced to first order accuracy.

The equations (15), (16), (20), and (21) define the so called NND schemes. The schemes are second order accurate except possibly near some points where $(\partial u/\partial x)(\partial^2 u/\partial x^2) = 0$. Therefore when compared against other mixed schemes which are first order accurate in the shock region and second order accurate elsewhere, the present scheme would possess the capability of high resolution of shock waves.

It is not difficult to extend formally the scheme for scalar equation (10) to the Euler equations

$$\frac{\partial U}{\partial t} + \frac{\partial F(U)}{\partial x} = 0 \tag{22}$$

where U is a vector and F is a function of vector U. The NND scheme reads in semi-discrete form

$$\left(\frac{\partial U}{\partial t}\right)^n_j = -\frac{1}{\Delta x}\left(H^n_{j+\frac{1}{2}} - H^n_{j-\frac{1}{2}}\right) \tag{23}$$

where

$$H_{j+\frac{1}{2}} = F^+_{j+\frac{1}{2}L} + F^-_{j+\frac{1}{2}R} \tag{24}$$

$$F^+_{j+\frac{1}{2}L} = F^+_j + \frac{1}{2}\min\operatorname{mod}(\Delta F^+_{j-\frac{1}{2}}, \Delta F^+_{j+\frac{1}{2}}) \tag{25}$$

$$F^-_{j+\frac{1}{2}R} = F^-_{j+1} - \frac{1}{2}\min\operatorname{mod}(\Delta F^-_{j+\frac{1}{2}}, \Delta F^-_{j+\frac{3}{2}}) \tag{26}$$

For one dimensional Navier-Stokes equations

$$\frac{\partial U}{\partial t} + \frac{\partial F}{\partial x} = \frac{\partial F_\nu}{\partial x} \tag{27}$$

where F_ν is due to the viscous contribution. For the convective part, NND scheme is used and for the viscous diffusive part, the usual central difference schemes can be employed.

For three dimensional flow, the well known time splitting method may be easily adapted, the problem is reduced to three one-dimensional problems and the entropy increase condition is preserved.

It was shown in [10], NND schemes possess TVD property. The scheme developed here is explicit. The linearized stability analysis suggests for first order time discretization that the maximum allowable Courant number is equal to 2/3 and for second order time discretization it is equal to 1.

3. Applications to solutions of Euler Equations

To test the effectiveness of NND scheme, we calculated two dimensional and axial symmetric plume flows with Euler equations. The equations can be written as follows:

$$\frac{\partial U}{\partial t} + \frac{\partial F}{\partial x} + \frac{\partial G}{\partial y} + \epsilon \frac{\tilde{H}}{y^\epsilon} = 0 \tag{28}$$

where

$$\begin{aligned} U &= (\rho, \rho u, \rho v, e)^T \\ F &= [\rho u, \rho u^2 + p, \rho u v, (e+p)u]^T \\ G &= [\rho v, \rho u v, \rho v^2 + p, (e+p)v]^T \\ \tilde{H} &= [\rho v, \rho u v, \rho v^2, (e+p)v]^T \end{aligned} \tag{29}$$

The usual notations are used. For two dimensional flow $\epsilon = 0$, and for axial symmetric flow $\epsilon = 1$.

With a time splitting technique, the problem is reduced to solving the following two equations:

$$\frac{\partial U}{\partial t} + \frac{\partial F}{\partial x} = 0 \tag{30}$$

$$\frac{\partial U}{\partial t} + \frac{\partial G}{\partial y} + \epsilon \frac{\tilde{H}}{y^\epsilon} = 0 \tag{31}$$

Define operators $\mathcal{L}_x$ and $\mathcal{L}_y$

$$\mathcal{L}_x U_{ij}^n = U_{ij}^n - \frac{\Delta t}{\Delta x}(H^x_{i+\frac{1}{2},j} - H^x_{i-\frac{1}{2},j})^n \tag{32}$$

$$\mathcal{L}_y U_{ij}^n = U_{ij}^n - \frac{\Delta t}{\Delta x}(H^y_{i,j+\frac{1}{2}} - H^y_{i,j-\frac{1}{2}})^n - \Delta t \epsilon \frac{\tilde{H}_{ij}^n}{y_j^\epsilon} \tag{33}$$

where

$$H^x_{i+\frac{1}{2},j} = F^+_{i+\frac{1}{2},jL} + f^-_{i+\frac{1}{2},jR} \tag{34}$$

$$H^y_{i,j+\frac{1}{2}} = G^+_{i,j+\frac{1}{2},L} + G^-_{i,j+\frac{1}{2}R} \tag{35}$$

then we have

$$U^{n+1} = \mathcal{L}(\Delta t)U^n \tag{36}$$

where

$$\mathcal{L}(\Delta t) = \mathcal{L}_x(\frac{\Delta t}{2})\mathcal{L}_y(\frac{\Delta t}{2})\mathcal{L}_y(\frac{\Delta t}{2})\mathcal{L}_x(\frac{\Delta t}{2}) \tag{37}$$

or

$$\mathcal{L}(\Delta t) = \mathcal{L}_y(\frac{\Delta t}{2})\mathcal{L}_x(\frac{\Delta t}{2})\mathcal{L}_x(\frac{\Delta t}{2})\mathcal{L}_y(\frac{\Delta t}{2}) \tag{38}$$

In the above expression

$$\Delta t = \min(\Delta t_x, \Delta t_y) \tag{39}$$

where

$$\Delta t_x = CFL_{max}\Delta x / |\lambda_A|_{max} \tag{40}$$

$$\Delta t_y = CFL_{max}\Delta y / |\lambda_B|_{max} \tag{41}$$

$|\lambda_A|_{max}$ is the maximum absolute eigenvalue of matrix $A = \partial F/\partial U$ and $|\lambda_B|_{max}$ is the corresponding value of matrix $B = \partial G/\partial U$.
Three cases are studied in [11].

1. Axial symmetric free jet.
   The ratio of exit total pressure P_j to external pressure P_∞ is 50, exit Mach number $M = 1.0$. The external stream is at rest ($M_\infty = 0$). The total enthalpy is the same for both external and internal streams ($H_j = H_\infty$).

2. Axial symmetric jet with external supersonic stream.
   Here we have the following parameters: $P_{oj}/P_\infty = 60, M_j = 1.0, M_\infty = 3.42, H_j = H_\infty$.

We assume that in all the cases the gas is perfect and $\gamma_\infty = \gamma_j = 1.4$. The distinguished feature of all these numerical calculations is the capability of capturing shocks and other contact discontinuities as exemplified by the graphical representations showing pressure density and temperature contours. Figures 1, a, c, and d respectively give the density, pressure, and temperature contours for the axial symmetric free jet. The existence of drum shock and Mach disc is clearly visible, and are the temperature and density jumps across the contact discontinuity. Figure 1b is the flow pattern obtained from experiment using laser induced fluorescence method in [12], where the location of the Mach disc nearly coincides with present calculation. Figure 2 gives a comparison of our calculated axial pressure, density, temperature and Mach number distributions with the PLM scheme in reference (13). The agreement is good upstream of the Mach disc; behind the Mach disc there appears slight oscillations in the temperature profile [13], while there are none in our calculation. Figures 3 a, b, and c give the pressure, density and Mach number contours for the axial symmetric jet with supersonic external stream. The appearance of oblique shock in the external stream and the regular reflection in the jet is evident. Figures 3 d, e, and f give the axial distributions of Mach number, density, and temperature. The results of calculations for the case 3 are given in Figure 4 in terms of pressure and density contours. Again the existence of strong and weak shocks is clearly visible.

4. Concluding Remarks

Through these preliminary calculations, we have gained confidence on the use of the NND scheme suggested here. The scheme possesses good stability characteristics and converged accuracy and this is essential to any high shock resolution scheme. Besides the min mod operator is applied to flux i.e. we use flux limiter, and in the present form it seems to be the simplest, thus the amount of numerical work is much reduced in comparison with some other high resolution TVD schemes. The present one also suffers in that it is of first order accuracy around individual points and when all the min mod functions at these points are zero, it simply reduces to a first order upwind scheme.

The NND schemes have been applied to the numerical simulation of the Navier-Stokes equations with some successes. Its full potential is yet to be explored. In some cases implicit NND schemes should be developed in order to obtain steady state solutions and the work is now underway.

References

1. Pulliam, T. H.: AIAA Paper 85-0438, 1985.
2. Jameson, A. et al: AIAA Paper 85-0293, 1985.
3. Zhang, H.X. et al: Applied Math. and Mech. 4, 1, 1985.
4. Harten, A.: SIAM J. Num. Anal., 21, 1-23, 1984.
5. Van Leer, B.: J. Comp. Phys. Vol. 32, 101-136, 1979.
6. Chakravarthy, S.R. and Osher, S.: AIAA Paper 85-0363, 1985.
7. Davis, S. F.: ICASE Report 84-20, 1984.
8. Yee, H. C.: NASA TM 89464, 1987.
9. Harten, A. et al: ICASE Report 86-18, 1986.
10. Zhang, H. X.: CARDC Report 87-3019, 1987.
11. Zheng, M.: CARDC Report 87-3005, 1987.
12. Teshima, K. et al: J. Japan Soc. Aero. Space Sci., 32, 1984.
13. Saito, T. et al: Trans. Japan Soc. Aero. Space Sci., 28, 1986.

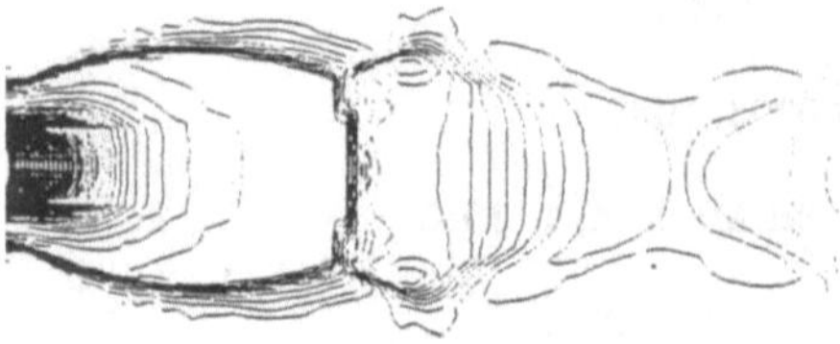

(a) pressure contours

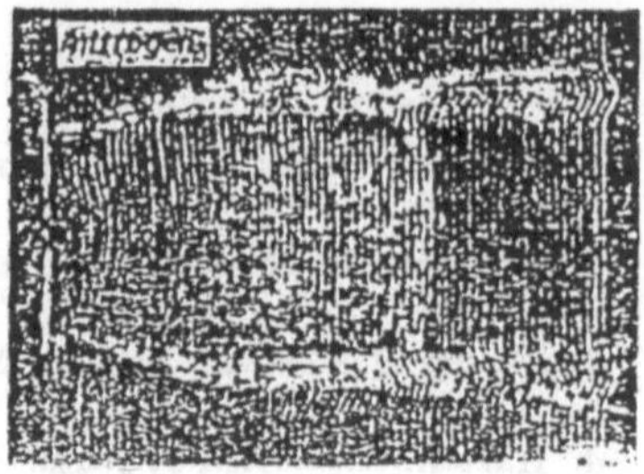

(b) visualization picture

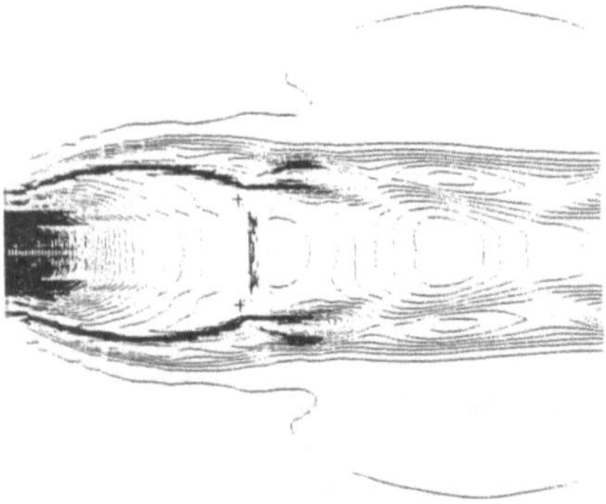

(c) density contours

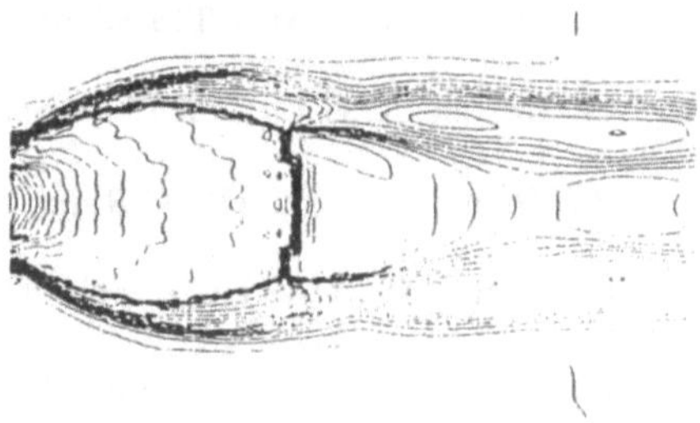

(d) tempertature contours

Fig. 1 Pressure, density and temperature contours of an axial free jet, also its LIF visualization picture

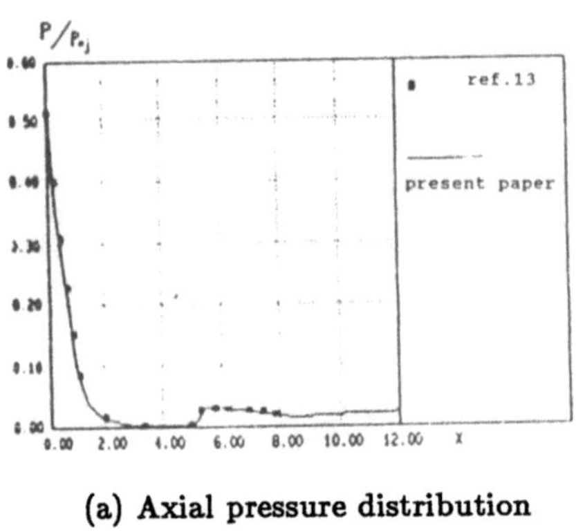

(a) Axial pressure distribution

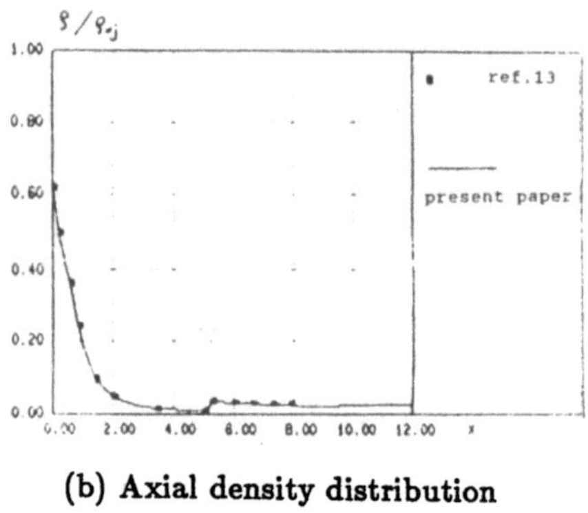

(b) Axial density distribution

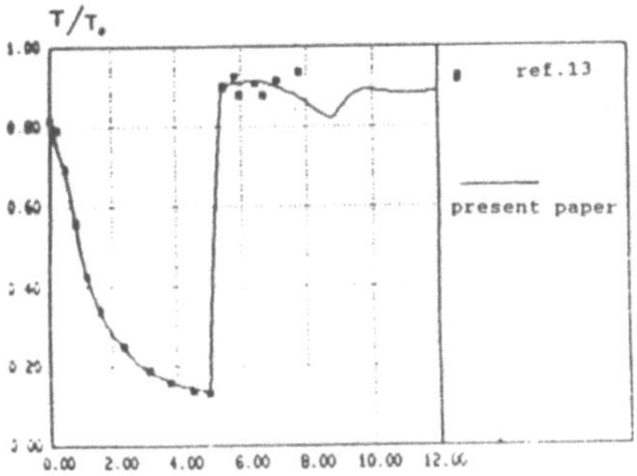

(c) axial temperature distribution

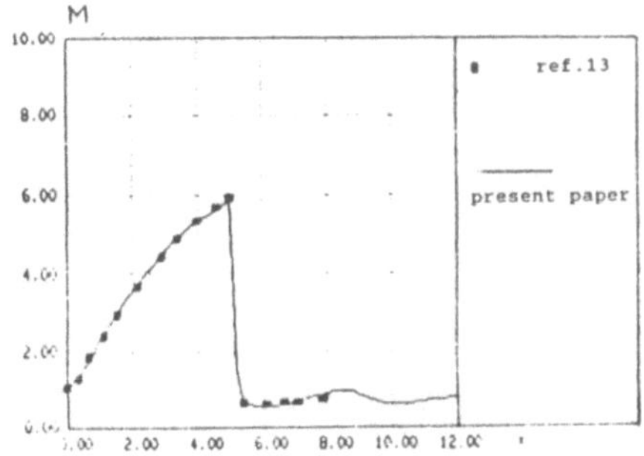

(d) axial mach number distribution

Fig.2 Comparison between the present calculated results of axial pressure, density, temperature and Mach number distributions and those obtained in ref. 13

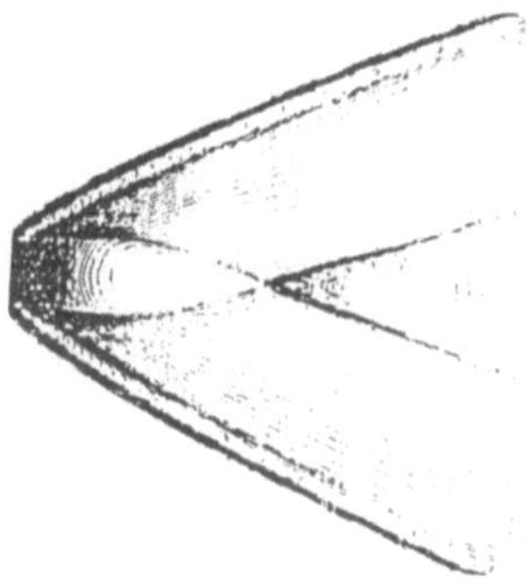

(a) pressure contours

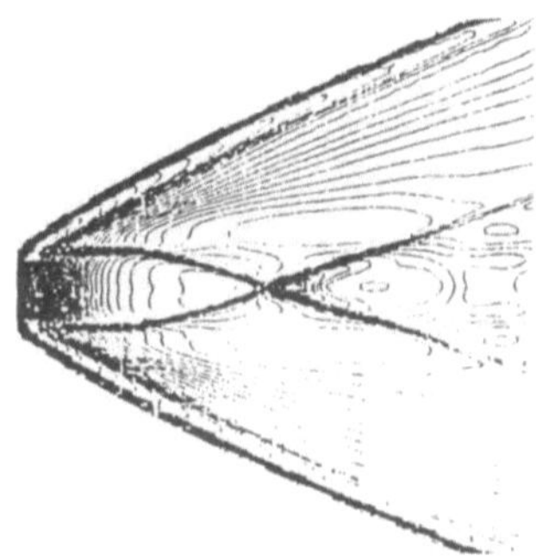

(b) density contours

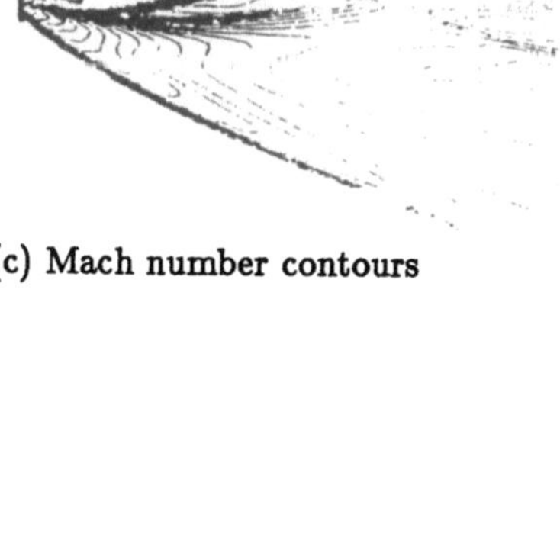

(c) Mach number contours

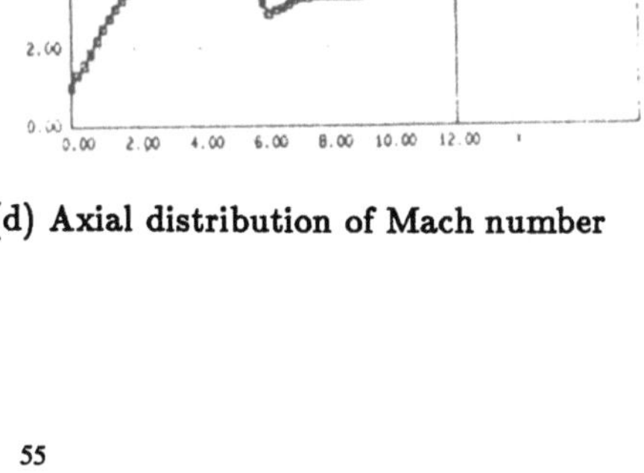

(d) Axial distribution of Mach number

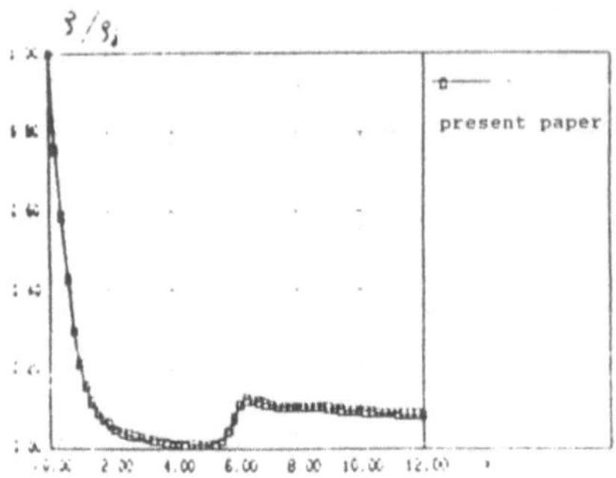

(e) axial density distribution

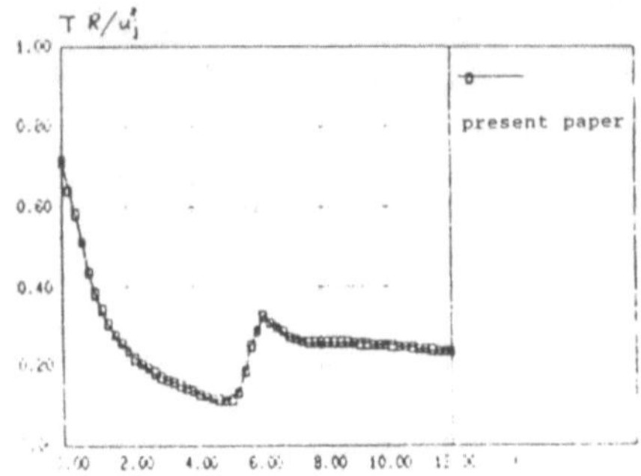

(f) axial temperature distribution

Fig. 3 The results for an axial symmetric jet with external supersonic stream

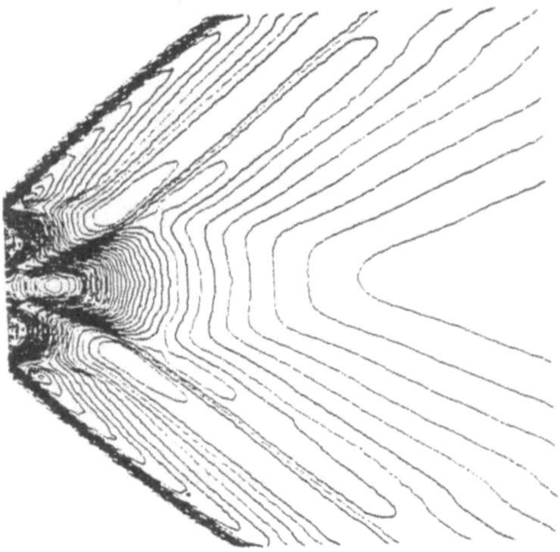

(a) pressure contours

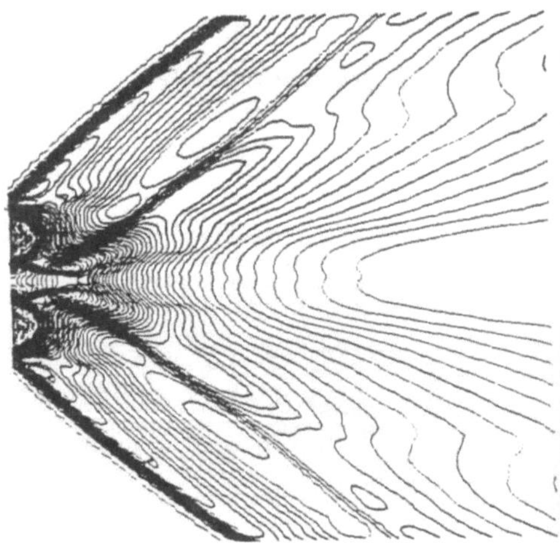

(b) density contours

Figure 4 The results of interaction of two two-dimensional jet

Symbolic Analysis and Computational Algorithm Construction

Clive A.J. Fletcher

1. Introduction

In this paper an analysis is presented that, in the Maurice Holt tradition, provides the means of obtaining specific information about the system of governing equations that can be exploited to construct more efficient computational algorithms. Symbolic analysis is very easy to apply, robust and usually produces considerable insight into the likely behaviour of the solution of the governing equations.

Symbolic analysis is an a priori analysis, based on taking a Fourier transform of a suitably linearised version of the governing equations, that will determine if the governing equations are elliptic or non–elliptic. As with characteristics analysis an eigenvalue problem is solved as an intermediate step. Symbolic analysis is more effective than characteristics analysis because:

a) It is able to take into account all terms in the governing equations, not just the highest derivatives.

b) The analysis is able to identify precisely which terms in the governing equations are responsible for the elliptic behaviour.

c) It is more robust in that degenerate systems are avoided.

For viscous flows with a dominant flow direction it is often possible to reduce the full Navier–Stokes (FNS) equations to a simpler form (RNS equations) through neglect of specific terms on an order–of–magnitude basis. For steady flows this raises the possibility that the mathematical character of the RNS equations may be non–elliptic with respect to the dominant flow direction. If this occurs stable computational algorithms can be constructed that permit the solution to be obtained in a *single* downstream march. Symbolic analysis is very effective in determining the ellipticity (or not) of the RNS equations.

For steady external subsonic/transonic flows with a dominant flow direction even reduced forms of the Navier–Stokes equations are generally elliptic. Symbolic analysis indicates that the addition of a transverse pressure gradient perturbation to the main flow direction momentum equation generates a non–elliptic system of equations (FLETCHER and ARMFIELD [6]); consequently stable *single–march* computational algorithms can be constructed. Numerical experiments (FLETCHER [4]) confirm the symbolic analysis and indicate that the single–march computational algorithm forms the basis of a very efficient iterative algorithm for the unperturbed Navier–Stokes equations.

Symbolic analysis is described in Section 2 of this paper in relation to a Vigneron RNS formulation. It is shown in Section 3 that when combined with a von Neumann stability analysis the restriction on the marching stepsize is related directly to the ellipticity of the governing equations. Two single–march algorithms for internal flow are examined using symbolic analysis in Section 4.

2. External Flow RNS Equations

The technique will be illustrated here for an external two–dimensional steady subsonic viscous flow in which a Vigneron strategy (FLETCHER, [5]) is used to control the elliptic influence of the pressure gradient term in the streamwise momentum equation.

The governing equations are written as

$$\rho\partial u/\partial x + \rho\partial v/\partial y + u\partial\rho/\partial x + v\partial\rho/\partial y = 0 \qquad (1)$$

$$\rho u\partial u/\partial x + \rho v\partial u/\partial y + \omega\partial p/\partial x - (1/\mathrm{Re})\partial^2 u/\partial y^2 = 0 \qquad (2)$$

$$\rho u\partial v/\partial x + \rho v\partial v/\partial y + \partial p/\partial y - (1/\mathrm{Re})\partial^2 v/\partial y^2 = 0, \qquad (3)$$

and a constant total enthalpy condition,

$$1 + \gamma M_\infty^2 p = \rho\{1 + 0.5(\gamma-1)M_\infty^2[1 - (u^2+v^2)]\}\ , \qquad (4)$$

is used to eliminate p from eqs. (1) to (3). The parameter, ω, multiplying $\partial p/\partial x$ in eq. (2) was introduced by VIGNERON et al. [12] as a means of obtaining an approximate solution in a single downstream march. Symbolic analysis will be used to determine a limiting value of ω to make the above equation system non–elliptic with respect to the x direction. Consequently a stable computational solution can be obtained from a single downstream march.

Eliminating p from eqs. (2) and (3) gives,

$$(\omega a^2/\gamma)\partial\rho/\partial x + \rho u\partial u/\partial x + \rho v\partial u/\partial y - (\omega(\gamma-1)/\gamma)(\rho u\partial u/\partial x + \rho v\partial v/\partial x) - (1/Re)\partial^2 u/\partial y^2 = 0 \qquad (5)$$

$$(a^2/\gamma)\partial\rho/\partial y - ((\gamma-1)/\gamma)\rho u\partial u/\partial y + \rho u\partial v/\partial x + (\rho v/\gamma)\partial v/\partial y - (1/Re)\partial^2 v/\partial y^2 = 0, \qquad (6)$$

where the nondimensional sound speed is defined by

$$a^2 = 1/(\rho M_\infty^2) + \gamma p/\rho. \qquad (7)$$

To apply symbolic analysis to the system, (1), (5) and (6), all undifferentiated terms are frozen and complex Fourier series are introduced for ρ, u and v as

$$\rho = (\hat{\rho}/4\pi)\exp(i\sigma_x x)\exp(i\sigma_y y) \qquad (8)$$

$$u = (\hat{u}/4\pi)\exp(i\sigma_x x)\exp(i\sigma_y y) \qquad (9)$$

$$v = (\hat{v}/4\pi)\exp(i\sigma_x x)\exp(i\sigma_y y), \qquad (10)$$

where only one mode in a general Fourier representation has been retained due to the linear nature of the frozen coefficient governing equations. The amplitudes $\hat{\rho}$, $\hat{u}$ and $\hat{v}$ can also be interpreted as the Fourier transforms of ρ, u and v. Equations (8) to (10) allow the qualitative behaviour of ρ, u and v as a function of x and y to be determined once the relationship between σ_x and σ_y is known.

This relationship is obtained by substituting (8) to (10) into (1), (5) and (6),

$$\begin{bmatrix} i\Lambda & i\rho\sigma_x & i\rho\sigma_y \\ i(\omega a^2/\gamma)\sigma_x & i\rho\Lambda - i\rho u\omega G\sigma_x + \sigma_y^2/Re & -i\rho v\omega G\sigma_x \\ i(a^2/\gamma)\sigma_y & -i\rho u G\sigma_y & i\rho u\sigma_x + i\rho v\sigma_y/\gamma + \sigma_y^2/Re \end{bmatrix} \begin{bmatrix} \hat{\rho} \\ \hat{u} \\ \hat{v} \end{bmatrix} = 0, \qquad (11)$$

where $\Lambda = u\sigma_x + v\sigma_y$ and $G = (\gamma-1)/\gamma$. Imposing det [ ] = 0 generates the following polynomial in σ_x,

$$[(\sigma_x/\sigma_y)^2\{u^2[1-\omega G] - \omega a^2/\gamma\} + (\sigma_x/\sigma_y)\{uv[H-\omega G] - iu\sigma_y/\rho Re\}$$
$$+\{(v^2-a^2)/\gamma - iv\sigma_y/\rho Re\}][\rho(u\sigma_x + v\sigma_y) - i\sigma_y^2/Re] = 0, \qquad (12)$$

where $H = (\gamma+1)/\gamma$.

The second factor, which is associated with the convection diffusion operator, has the solution,

$$\sigma_x = - v\sigma_y/u + i\sigma_y^2/(\rho uRe). \qquad (13)$$

For any real choice of σ_y substitution of eq. (13) into eqs. (8) to (10) indicates the solution behaviour in the x direction. If u is positive it is clear that the solution may have a component that is oscillatory in x (σ_x real) and a component that demonstrates an exponential decay in x (σ_x positive imaginary).

If σ_x is negative imaginary, exponential growth in the x direction is indicated. It can be shown (ARMFIELD and FLETCHER [1]) that this corresponds to the equation system being elliptic with respect to the x direction. The exponential growth also implies that a spatial march in the x direction would not generate a stable, and hence meaningful, solution. Thus negative imaginary solutions for σ_x are to be avoided if efficient computational algorithms are to be constructed based on spatial marching.

A consideration of the first factor in eq. (12) with $v = 0$ and $Re = \infty$ provides the following solution,

$$\sigma_x/\sigma_y = \pm\, a/\{u^2[1-\omega G] - \omega a_a^2/\gamma\}^{1/2}. \qquad (14)$$

To avoid a negative imaginary value for σ_x it is necessary that

$$\omega \leq \gamma M_x^2/\{1 + (\gamma-1)M_x^2\}. \qquad (15)$$

This is the same result as obtained by VIGNERON et al. [12] via a simplified characteristics analysis. If $Re = \infty$ but $v \neq 0$ a limiting value of ω can still be found, for which the equation system, (1) to (4), is non–elliptic. However if Re is large, but not infinite, eq. (12) indicates that there will be a small elliptic effect associated with the viscous terms. In relation to eq. (2), if $M_x = u/a$ is supersonic, eq. (15) indicates that a stable single–march solution can be obtained with $\omega = 1.0$. But if $M_x < 1.0$ then $\omega < 1$ is required in eq. (2) to obtain a stable solution.

3. Connection with von Neumann Stability Analysis

The Fourier analysis of the governing equations, (1), (5) and (6), is conceptually similar to the von Neumann stability analysis of the discretised equations. Thus it is of interest to determine if the non–ellipticity condition, (15), corresponds to a marching stepsize restriction in the algorithm formed from the discretised equivalent of eqs. (1), (5) and (6).

If v is locally zero a fully implicit discretisation of the inviscid form of (1), (5) and (6) is

$$\frac{u_j^n(\rho_j^{n+1} - \rho_j^n)}{\Delta x} + \frac{\rho_j^n(u_j^{n+1} - u_j^n)}{\Delta x} + \rho_j^n L_y v_j^{n+1} = 0 \; , \tag{16}$$

$$\left[\frac{\omega a^2}{\gamma}\right]_j^n \frac{(\rho_j^{n+1} - \rho_j^n)}{\Delta x} + (\rho u')_j^n \frac{u_j^{n+1} - u_j^n}{\Delta x} = 0 \; , \tag{17}$$

$$\left[\frac{a^2}{\gamma}\right]_j^n L_y \rho_j^{n+1} - (\rho' u)_j^n \; L_y u_j^{n+1} + (\rho u)_j^n \frac{\nu_j^{n+1} - \nu_j^n}{\Delta x} = 0 \; , \tag{18}$$

where $u' = u[1 - \omega(\gamma - 1)/\gamma]$, $\rho' = \rho(\gamma - 1)/\gamma$, $L_y \equiv \{1,0,-1\}^T/2\Delta y$ and indices n and j indicate the increasing x and y directions respectively. To apply a Fourier analysis to (16) to (18) the equations are linearised by freezing the terms multiplying the difference expressions. Terms in the difference expressions are given a complex Fourier representation in y only. Thus

$$\rho_j^{n+1} \sim \hat{\rho}^{n+1} e^{ij\theta} \; , \tag{19}$$

where $\theta = m\pi\Delta y$. With similar representations for u and ν the following matrix equation is obtained

$$\underline{A}\hat{q}^{n+1} = \underline{B}\hat{q}^n \; , \tag{20}$$

where $\hat{q} \equiv \{\hat{u}, \hat{\nu}, \hat{\rho}\}^T$.

To ensure that (20) is stable it is necessary that all eigenvalues

$$\lambda_{\underline{A}^{-1}\underline{B}} \leq 1.0 \,, \tag{21}$$

where

$$\underline{A}^{-1}\underline{B} = \begin{bmatrix} 1 - \omega'\sigma^2/\beta & \dfrac{(\gamma-1)\rho M_x^2 \omega' \sigma^2}{u\beta} & \dfrac{\gamma M_x^2 \rho \omega' \sigma}{u\beta} \\ \dfrac{u\sigma^2\omega}{\gamma\rho M_x^2 \beta} & 1 - \left[\dfrac{\gamma-1}{\gamma}\right]\dfrac{\omega\sigma^2}{\beta} & -\dfrac{\omega\sigma}{\beta} \\ -\dfrac{u\sigma\omega\omega_1}{\rho\beta} & (\gamma-1)M_x^2 \dfrac{\sigma}{\beta}\omega\omega_1 & \dfrac{\gamma M_x^2 \omega\omega_1}{\beta} \end{bmatrix}, \tag{22}$$

with $\sigma = i(\Delta x/\Delta y)\sin 2\theta$, $\omega' = 1-\omega(\gamma-1)/\gamma$, $\beta = \sigma^2+\gamma M_x^2\omega\omega_1$, $\omega_1 = (1/\omega_{cr}-1/\omega)$ and $\omega_{cr} = \gamma M_x^2/[1+(\gamma-1)M_x^2]$, i.e. from (15). Introducing $\tau = 1-\lambda$ leads to the following equation for the eigenvalues of $\underline{A}^{-1}\underline{B}$,

$$\tau(\beta\tau^2 - 2\sigma^2\tau + \sigma^2) = 0 \,. \tag{23}$$

Substituting for β and introducing $\epsilon^2 = -\sigma^2 = (\Delta x \sin 2\theta/\Delta y)^2$ gives

$$[1 + (\gamma-1)M_x^2](\omega-\omega_{cr})\tau^2 = \epsilon^2(\tau-1)^2 \,. \tag{24}$$

The two cases $\omega < \omega_{cr}$ and $\omega > \omega_{cr}$ are of interest since they correspond to the inviscid form of (1),(5) and (6) with $\nu = 0$, being non–elliptic and elliptic, respectively, in relation to the proposed marching direction, x. For the non–elliptic case, $\omega < \omega_{cr}$, there is no restriction on Δx.

For $\omega > \omega_{cr}$, (24) gives

$$\mp \alpha^{1/2}(1-\lambda) = \epsilon\lambda \,, \tag{25}$$

where $\alpha = [1 + (\gamma-1)M_x^2](\omega - \omega_{cr})$. Therefore λ is given by

$$\lambda = \mp \frac{\alpha^{1/2}}{(\mp\alpha^{1/2} + \epsilon)} \,. \tag{26}$$

The condition $\lambda < 1.0$ leads to $\epsilon > \alpha^{1/2}$. The condition $\lambda > -1.0$ leads to $\epsilon > 2\alpha^{1/2}$ or

$$\frac{\Delta x}{\Delta y} \sin\theta\cos\theta > \alpha^{1/2} .$$

On a finite grid, $\theta_{min} = \Delta y \pi / y_{max} \simeq \sin\theta$ and $\cos\theta \approx 1$. Thus

$$\Delta x > \alpha^{1/2} \frac{y_{max}}{\pi} \text{ or}$$

$$\Delta x > \{[1 + (\gamma-1)M_x^2](\omega - \omega_{cr})\}^{1/2} \frac{y_{max}}{\pi} , \tag{27}$$

where y_{max} is the transverse extent of the computational domain. Equation (27) indicates that when the governing equations are elliptic there is a step–size restriction of the form $\Delta x > (\Delta x)_{min}$, where $(\Delta x)_{min}$ is proportional to the degree of ellipticity. If $(\Delta x)_{min}$ is too large then accurate solutions will not be possible with a single spatial march.

The dependence of $(\Delta x)_{min}$ on y_{max} is similar to that found by LUBARD and HELLIWELL [10] for the subsonic sublayer in an otherwise supersonic flow. Also RUBIN and LIN [11] have obtained a restriction of the form $\Delta x > k y_{max}$ for incompressible flow.

It is apparent that symbolic analysis used in conjunction with the von Neumann stability analysis provides insight into the link between ellipticity and the marching step size restriction. It is expected that such a combined analysis would be useful in conjunction with applying Telenin's method and the Method of Lines (HOLT [7]) to elliptic problems.

4. Internal Flow RNS Equations

Steady internal flows lend themselves to spatial marching techniques because the condition of constant mass flow implies that the average downstream velocity component must be positive. That is, any reversal in the axial flow can only be local. In addition the constant mass flow condition provides an extra piece of information that can be exploited in constructing the marching algorithm.

In this section two single–march internal flow algorithms will be investigated using symbolic analysis to demonstrate the essential features responsible for the non–elliptic behaviour in the downstream direction. The first algorithm is appropriate for duct flows with small axis curvature; the second algorithm for duct flows with large axis curvature.

For incompressible laminar flow in a straight rectangular duct the nondimensional reduced Navier–Stokes equations can be written as

$$\partial u/\partial x + \partial v/\partial y + \partial w/\partial z = 0 \tag{28}$$
$$L^c u + \partial p^{inv}/\partial x + \partial p^v/\partial x - L^d u = 0 \tag{29}$$
$$L^c v + \partial p^{inv}/\partial y + \partial p^v/\partial y - L^d v = 0 \tag{30}$$
$$L^c w + \partial p^{inv}/\partial z + \partial p^v/\partial z - L^d w = 0, \tag{31}$$

where $L^c u \equiv u\partial u/\partial x + v\partial u/\partial y + w\partial u/\partial z$ and $L^d u \equiv (1/Re)(\partial^2 u/\partial y^2 + \partial^2 u/\partial z^2)$.

For the above equation system positive x corresponds to the downstream direction. It is assumed that a preliminary inviscid solution is available which provides p^{inv}. Consequently the momentum equations, (29) to (31), contain a "viscous" pressure correction which is obtained as part of the solution, whereas the p^{inv} terms are treated as known source terms. For uniform inviscid flow in a straight duct the gradients of p^{inv} in eqs (29) to (31) are zero and will be dropped. The viscous pressure is split into two parts,

$$p^v(x,y,z) = \bar{p}^v(x) + p^{v,c}(x,y,z) \tag{32}$$

where $\bar{p}^v$ is the mean flow and is determined from the mass flow constraint,

$$\int\int \partial u/\partial x \, dydz = 0, \tag{33}$$

after integrating the axial momentum equation over the cross–sectional area. Clearly, after substituting eq. (32), $\bar{p}^v$ will only appear in eq. (29). The term, $p^{v,c}$, is a correction to the mean pressure to account for transverse pressure behaviour. Using the above information eqs. (28) to (31) become

$$\partial u/\partial x + \partial v/\partial y + \partial w/\partial z = 0 \tag{34}$$
$$(L^c - L^d)u + \beta\partial p^{v,c}/\partial x = -\partial\bar{p}/\partial x \tag{35}$$
$$(L^c - L^d)v + \partial p^{v,c}/\partial y = 0 \tag{36}$$
$$(L^c - L^d)w + \partial p^{v,c}/\partial z = 0. \tag{37}$$

The coefficient β is introduced into eq. (35) to study the effect of dropping $\partial p^{v,c}/\partial x$ (i.e. $\beta = 0$) from the axial momentum equation. For ducts with small axis curvature this does not introduce significant error.

Applying the symbolic analysis of Section 2 to eqs (34) to (37) produces the following polynomial for σ_x,

$$\{u\sigma_x + v\sigma_y + w\sigma_z - i(\sigma_y^2 + \sigma_z^2)/Re\}^2\{\beta\sigma_x^2 + \sigma_y^2 + \sigma_z^2\} = 0 \tag{38}$$

The first factor arises from the convection diffusion operator in the momentum equations and has no negative imaginary root for σ_x as long as u is positive. However the second factor does produce a negative imaginary root if β is non–zero. The appearance of the $\beta\sigma_x^2$ term in (38) is associated with the interaction of $\partial p^{v,c}/\partial x$ in (35) with $\partial u/\partial x$ in (34). Consequently to render equation system, (34) to (37), non–elliptic it is necessary to set $\beta = 0$. In turn this permits stable solutions to be obtained in a single downstream march, using the algorithm described by BRILEY [2].

For flows with large curvature of the duct axis the transverse velocity components v and w can be of the same order as the streamwise velocity component u. Consequently significant transverse pressure variation occurs and the pressure splitting introduced in eq. (32) is inappropriate as a means of making the system of equations non–elliptic, since it will not lead to accurate solutions in a single downstream march.

An alternative non–elliptic approach, which is suitable for flows with large curvature, is given by BRILEY and McDONALD [3]. In this formulation an inviscid solution for the same duct geometry provides a preliminary approximate solution U^i,V^i,W^i and p^i, which is modified by the subsequent solution of the RNS equations to predict the viscous flow behaviour in the duct.

The transverse velocity field is split as

$$v = V^i + v^\varphi + v^\phi\ ,\ w = W^i + w^\varphi + w^\phi\ , \tag{39}$$

where the transverse rotational velocity field (v^φ, w^φ) is generated by streamwise vorticity which, in turn, is caused by axis curvature. The transverse potential velocity field (v^{ϕ}, w^{ϕ}) is generated by the streamwise velocity gradient, $\partial u/\partial x$, and is necessary to ensure that the continuity equation is satisfied.

An order–of–magnitude analysis indicates that the potential velocity field is small compared with the rotational velocity field. Consequently the governing equations can be reduced. For incompressible flow in Cartesian coordinates they take the form

$$\partial u/\partial x + \partial v^{\phi}/\partial y + \partial w^{\phi}/\partial z = 0 \tag{40}$$

$$L^{c}u + \partial p/\partial x - L^{d}u = 0 \tag{41}$$

$$L^{c}(\beta v^{\phi} + v^{\varphi}) + \partial p/\partial y - L^{d}v^{\varphi} = 0 \tag{42}$$

$$L^{c}(\beta w^{\phi} + w^{\varphi}) + \partial p/\partial z - L^{d}w^{\varphi} = 0 \tag{43}$$

$$\partial v^{\varphi}/\partial y + \partial w^{\varphi}/\partial z = 0 \tag{44}$$

$$\partial v^{\phi}/\partial z - \partial w^{\phi}/\partial y = 0 \tag{45}$$

where L^c and L^d are as above.

The transverse momentum equations, (42) and (43), contain terms, βv^{ϕ} and βw^{ϕ}. On an order–of–magnitude basis it is appropriate to set $\beta = 0$. However for the moment, these terms will be retained to examine their impact on the character of the equation system.

Equations (44) and (45) follow from the transverse streamfunction, φ, and potential, ϕ, on which the transverse rotational and potential velocity fields are based. It may be noted that the full pressure field is retained in eqs. (41) to (43).

Symbolic analysis is applied to eqs (40) to (45) after linearisation of the convective operator, L^c, and introduction of complex Fourier series, like eq. (8), for $u, v^{\phi}, w^{\phi}, v^{\varphi}, w^{\varphi}$ and p. The resulting polynomial for σ_x takes the form,

$$(\sigma_y^2 + \sigma_z^2)\Lambda^t\{\beta\, \sigma_x^2\, \Lambda^c + (\sigma_y^2 + \sigma_z^2)\Lambda^d\} = 0 \tag{46}$$

where $\Lambda^t = \Lambda^c - i\Lambda^d$,

$\Lambda^c = u\sigma_x + v\sigma_y + w\sigma_z$ and $\Lambda^d = (\sigma_y^2 + \sigma_z^2)/Re$.

Thus Λ^t coincides with the first factor in eq. (38) and $\Lambda^t = 0$ does not generate a negative imaginary value for σ_x unless u is negative. Consequently, for positive u, this term does not introduce any elliptic behaviour in the x direction. However, if β is non–zero, the third factor in (46) is cubic in σ_x and will have a negative imaginary solution, implying elliptic behaviour with respect to the streamwise direction, x. Thus to obtain the solution via a single downstream march it is necessary that $\beta = 0$, i.e. that the transverse potential velocity field does not appear in the differentiated terms in the momentum equations. The σ_x^2 term multiplying β in (46) comes from the interaction of $\partial p/\partial x$ in (41) with $\partial u/\partial x$ in (40). Thus the underlying elliptic mechanism is the same as for the equation system, (34) to (37).

The algorithm based on the present formulation does not use eqs. (40), (42) and (43) directly. Instead eqs. (42) and (43) are re–expressed in terms of ψ and the streamwise vorticity, $\Omega = \partial w^{\varphi}/\partial y - \partial v^{\varphi}/\partial z$, and eq. (40) is replaced by a Poisson equation for φ (FLETCHER [5]). Applications of the present formulation are provided by BRILEY and MacDONALD [3] and by PLATFOOT and FLETCHER [8,9].

5. Conclusion

The effectiveness of symbolic analysis has been demonstrated for determining the elliptic character of equation systems governing both external (Section 2) and internal flows (Section 4). The analysis is particularly useful in constructing spatial marching algorithms. As an example it has been shown that the marching stepsize restriction can be related to the degree of ellipticity when using a Vigneron RNS strategy. Symbolic analysis is described in greater detail and, in a wider context, by FLETCHER [5].

6. References

1. S.W. Armfield and C.A.J. Fletcher: "Symbolic Analysis of Partial Differential Equations" Comp. Math. Applic., to appear, 1989.
2. W.R. Briley: J. Comput. Physics, 14, 8–28, 1974.
3. W.R. Briley and H. McDonald: J. Fluid Mechanics, 144, 47–77, 1984.
4. C.A.J. Fletcher: "Computational Exploitation of the Reduced Navier–Stokes Equations", Proc. ISCFD–Sydney (eds G. de Vahl Davis and C.A.J. Fletcher), North–Holland, Amsterdam, 407–416, 1988.
5. C.A.J. Fletcher: *Computational Techniques for Fluid Dynamics*, Springer–Verlag, Heidelberg, Vols 1 and 2, 1988.
6. C.A.J. Fletcher and S.W. Armfield: Comm. Appl. Numerical Methods, 4, 219–226, 1988.

7. M. Holt: *Numerical Methods in Fluid Dynamics*, Springer–Verlag, Heidelberg, 1984.

8. R.A. Platfoot and C.A.J. Fletcher: "Curvilinear Coordinate System–fitted to the Three–Dimensional Navier–Stokes Equations for Flows in Ducts" Proc. CTAC–85 (eds. B.J. Noye and R.L. May), North–Holland, Amsterdam, 312–325, 1986.

9. R.A. Platfoot and C.A.J. Fletcher: "Prediction of Three–Dimensional Turbulent Flow in a Circular–Arc Duct using Transformed Equations", Proc. ISCFD – Sydney (eds G. de Vahl Davis and C.A.J. Fletcher), North–Holland, Amsterdam, 599–608, 1988.

10. S.C. Lubard and W.S. Helliwell: AIAA J, 12, 965–974, 1974.

11. S.G. Rubin and A. Lin: Israel J. Technology, 18, 21–31, 1980.

12. Y.C. Vigneron, J.V. Rackich and J.C. Tannehill: NASA Tech Memo 78500, 1978.

Velocity and Mass Flux Distribution Measurements of Spherical Glass Beads in Air Flow in a 90° Vertical-to-Horizontal Bend

Yannis Kliafas

Abstract

The mean streamwise and radial velocities and the associated Reynolds stresses have been measured simultaneously in a square-sectioned (10 cm x 10 cm), 90° vertical-to-horizontal bend using a 2-component Laser-Doppler velocimeter. 50 μm and 100 μm diameter glass-beads were used. The air velocity measurements were performed in the absence of the glass-spheres.

Mass-flux was defined as the number of solid particles detected by the probe-volume of the LDA per second per unit area and was measured along the center-line at the following angles to the horizontal 0°, 15°, 30°, 60° and 75°.

The radius ratio of the bend was 1.76. The measurements were taken for two different Reynolds numbers 2.2×10^5 and 3.47×10^5 corresponding to mass ratios of 1.5×10^{-4} and 9.5×10^{-5} respectively. In this paper the mean streamwise and radial velocities will be presented in detail. The glass-spheres velocity profiles were very flat. Higher negative slip-velocities were observed for the 100 μm particles than those for the 50 μm particles. At angular displacements $\theta = 0°$ the radial velocity is directed towards the inner wall for both sizes of glass beads. Most of the particle wall collisions occur between the $\theta = 30°$ and 60° stations.

List of symbols

D	hydraulic diameter (100 mm)
De	Dean number, $D_e = (D/2R)^{1/2} R_e$
m	mass flow rate
N	number of particles per second (detected by the probe volume)
r	radial coordinate direction
r_i	radius of curvature of the inner wall

r_o	radius of curvature of the outer wall
r^*	normalized radial coordinate, $r^*=(r-r_o)/(ri-r_o)$
R	mean radius of curvature
R_e	Reynolds number , $R_e=U_bD/\nu$
R_r	radius ratio, $R_r=R/D$
U_θ, U_z	mean streamwise velocity
U_r, U_y	mean radial velocity
U_b	bulk velocity
$\tilde{u}_\theta, \tilde{u}_z$	rms fluctuating streamwise velocity
$\tilde{u}_r, \tilde{u}_y$	rms fluctuating radial velocity
$\overline{u_\theta u_r}$	θ - r shear stress component
$\overline{u_z u_y}$	z - y shear stress component
x	spanwise coordinate direction
x^*	normalized spanwise coordinate, $x^*=x/(D/2)$
y	radial coordinate direction in straight ducts
y*	normalized radial coordinate in straight ducts, $y^*=y/D$
z	streamwise coordinate direction in straight ducts
z*	normalized streamwise coordinate in straight ducts, $z^*=z/D$

Greek symbols

θ	streamwise coordinate direction
ν	kinematic viscosity of air

1. Introduction

The dispersion of solid and liquid particles in turbulent flow is important in a number of industrial applications such as rocket-nozzle performance analysis, coal-fired furnaces, pneumatic conveying systems, aerosol studies, studies on deposition of solid particles or liquid droplets from a turbulent flow to channel wall etc. The blockage or rapid erosion of the conventional measuring devices (Pitot tube, hot-wire anemometer, etc.) were responsible for the slow progress of experimentation on two-phase flows over the years. These were eliminated with the development of the non-intrusive laser Doppler velocimetry (LDV) technique. The basic concern of almost all experimentalists using LDV in two-phase flows is the differentiation between the signals coming from the small (seeding) and large particles. A number of methods have been suggested which relate characteristics of the anemometer signal to particle size, including the maximum amplitude of the pedestal signal (e.g. LEVY and LOCKWOOD [1], MODARRESS and TAN [2]) and the signal modulation depth ('visibility'), (e.g.

FARMER [3], DURST [4], BACHALO [5]). The relationships between these signal characteristics and particle diameter can be found from theory. Since the underlying principles of these two methods are based on intensity, these relationships may not hold in practice because of the random, time-dependent attenuation of the laser light due to the partial interruption of the beams as these pass through the two-phase flow.

KLIAFAS, TAYLOR and WHITELAW [6], [7], quantify the influence of the particle size, concentration and depth of field on the accuracy of particle sizing based on measurements of either the pedestal amplitude or the visibility. It is recommended that for smooth-surfaced, transparent or opaque particles the phase (interferometric) technique (BACHALO and HAUSER [8]) must be used, which alleviates the problems posed by flow turbidity. For other particles, the likely errors using intensity-based methods can be estimated from the information given in the above mentioned papers.

No experimental work on turbulent two-phase flows in curved ducts has been done thus far. The lack of detailed experimental studies and the complexity of the mechanisms of particle and flow interactions have thus far prevented the development of a theoretical model to treat two-phase flows. Most of the existing numerical schemes use numerous simplifying assumptions in order to allow solutions for the flow-field variables.

YEUNG [9] and POURAHMADI and HUMPHREY [10] developed computational models capable of treating two-phase gas flows in curved ducts. They considered dilute suspensions so that the gas flow field is not significantly affected by the presence of particles. YEUNG [9] referred to the Stokes number, $St = W_i T_m / R$, where W_i is the entry velocity, R the mean radius of curvature of the curved pipe and T_m the momentum equilibration time. In terms of St he found that:

(a) for $St \gg 3$ the particle trajectories are straight lines;
(b) for $St \leqslant 0.1$ some particles leave the pipe without making an impact with the outer wall;
(c) only the outer wall is impinged by particles.

CRANE and EVANS, [11], examined the inertial deposition of particles in a pipe bend. They found that the secondary flow has a significant effect on particle motion and plays an important role in determining the origin, in the bend inlet plane, of deposited particles. The secondary flow was shown to have a relatively small effect on bend deposition rate. The impaction efficiency in a 90° bend was not found to be strongly dependent on the ratio of bend radius to pipe radius or on the particle Reynolds number.

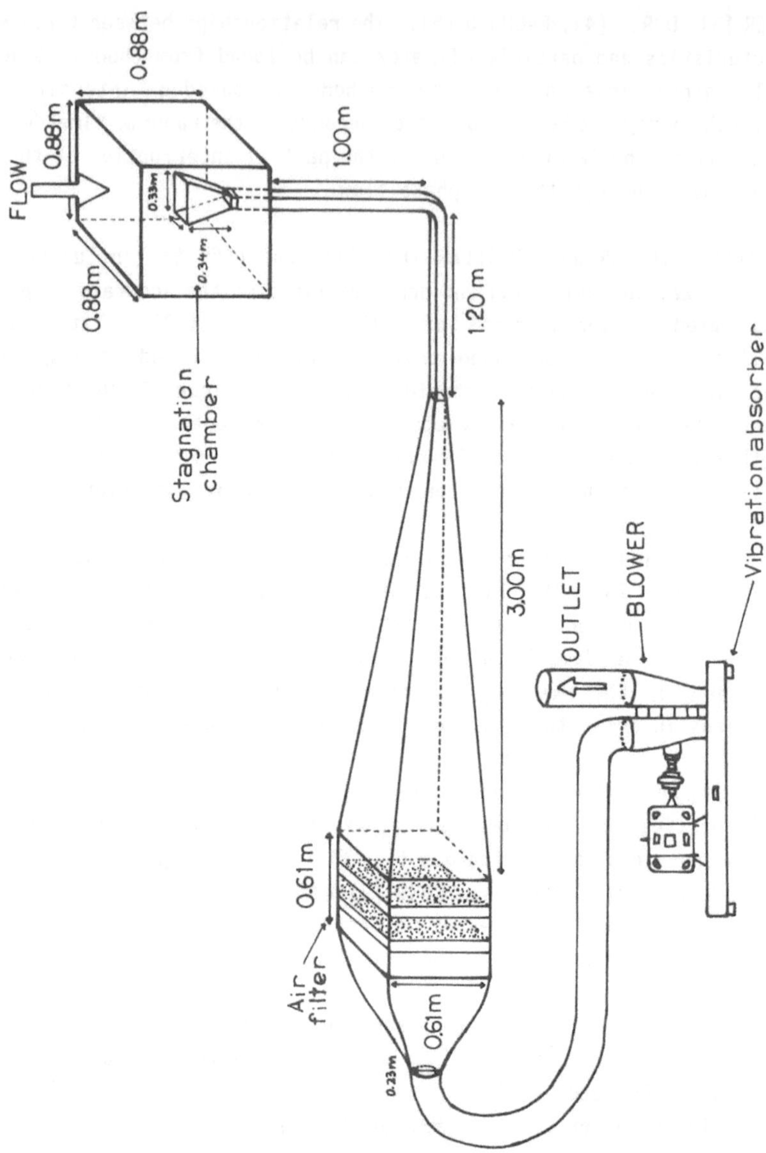

Fig.1. General flow system

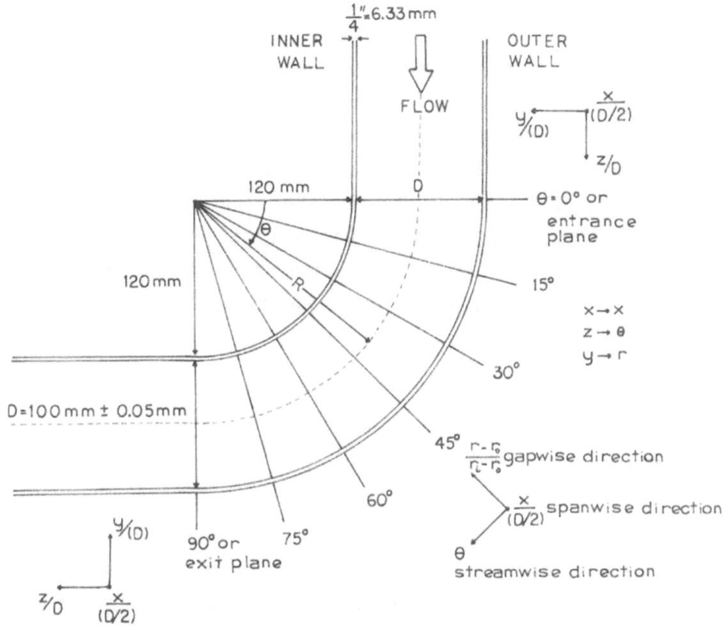

Fig.2. Geometry of the curved square duct

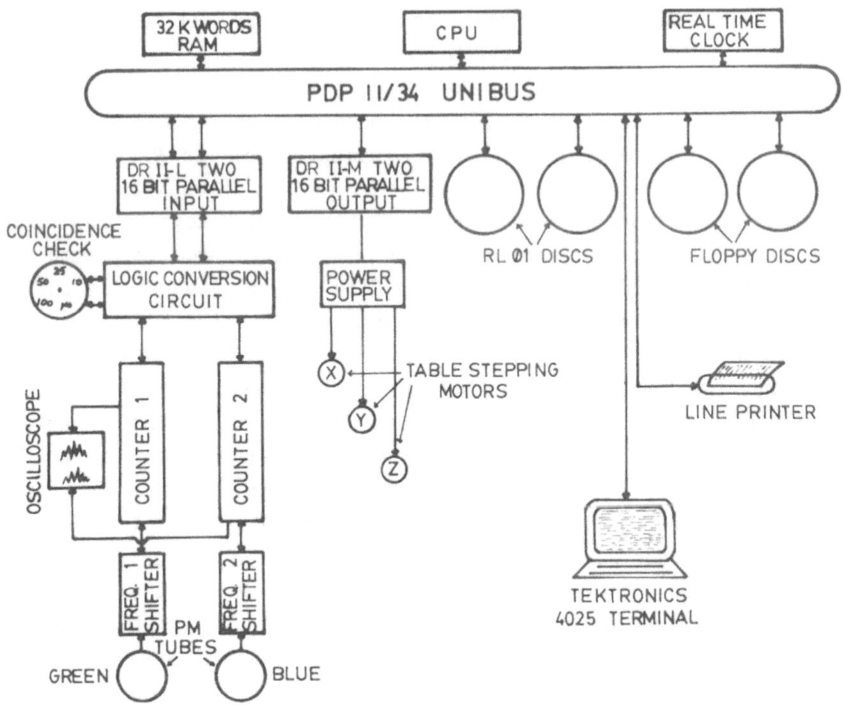

Fig.3. LDV electronic components and coincidence check hardware.

The direct application of the present study would be to the development of large coal gasification plants. Detailed measurements of the mean air velocities and the corresponding Reynolds stresses at 15 different planes were also obtained. To improve the understanding of turbulent high speed air flows in bends of strong curvature we have provided as much information as possible about the mean and turbulent quantities of the flow. These results appear in KLIAFAS [12] and KLIAFAS, BACHALO and HOLT [13]. As a continuation of the present study the velocity profiles of air in the presence of solids should be obtained. The phase method will be applied to have simultaneous measurements of gas velocity, solid particle velocity and diameter.

2. Experimental apparatus

2.1 Air flow

Figure 1 shows the general flow system. The air flow was driven by a four-stage Lamson blower-exhauster. A schematic of the test section and the coordinate systems used is shown in Fig. 2. The test section was manufactured from clear plexiglass, 6.3 mm thick, and consisted of a 90° bend (vertical to horizontal) and two straight ducts of square cross-section (100 ± 0.05 x 100 ± 0.05 mm). The upstream and downstream duct lengths are 1 and 1.2 m, respectively (or 10 and 12 hydraulic diameters respectively). The straight sections of the duct were matched very accurately (0.05 mm) to the curved section. The ratio R_r , of the bend mean radius of curvature R to the hydraulic diameter D (called the radius ratio) was $R_r = R/D = 1.76$. The room-temperature air entered the test section through a diffuser with an area contraction ratio of 1:10.9. A hexagonal honeycomb (50 mm thick with 3 mm cell diameter) followed by a fine mesh screen was placed inside the inlet diffuser. A uniform steady flow was obtained at the outlet of the contraction. Four jet atomizers symmetrically placed with respect to the entrance diffuser were employed to generate a fine mist of mineral oil which was the seeding of the flow for the LDV measurements. A pitot tube was fitted near the end of the downstream section. The pitot tube centerline velocity profiles were compared with the LDV measurements to verify the accuracy of the velocity measurements. Leaving the test section, the air proceeded through a diffuser connected to the end of the plexiglass duct. The diffuser angle was 4°. In turn, the other end of the exit diffuser was attached to the filter box. With careful positioning of the filters, high quality filtration of the air was obtained on its way to the blower. The advantage of the suction tunnel (with a blower at the exit) which was employed in the present study, is that air coming from the room is less disturbed than that moving from the blower.

2.2 Solid phase

The measurements of the solid and gas phase were performed separately. The same experimental apparatus was used for the solid phase measurements as for the gas phase measurements with the following modifications:

(a) The mineral oil jet atomizers were removed from the stagnation chamber and replaced by the solid particle feeder/shaker assembly.

(b) A cyclone particle separator was placed 2.5 m downstream of the bend exit to collect the glass beads entrained in the flow.

(c) The side walls in the bend were made of glass in this part of the experiment.

Two sizes of spherical glass beads were used in the present investigation: The nominal 50 μm particles (distribution in the range 48.5 μm ± 9%) and the 100 μm diameter ones (distribution in the range 97 μm ± 8.8%). Their density and refractive index were 2,990 kg/m^3 and 1.63, respectively. The glass spheres were stored in a hopper of conical shape. A 9 cm long drill was used to transport the particles and is referred to as the screw feeder. The glass beads left the screw feeder via another hopper and met a vibrating screen used to ensure uniform distribution of particles in the test section. The particle feeder/shaker assembly was calibrated for different r.p.m. of the motor. In the present study, the mass flow rate was kept constant at 3.6 g/min, for both the 50 and 100 μm diameter beads.

The cyclone particle separator was found to collect 97% of the glass beads. The rest were stopped by the air filters.

The use of glass side-walls proved to be beneficial since no scratches were noticed on the glass surfaces during the course of the experiment.

2.3 LDV arrangement and signal-processing system

Simultaneous two-component measurements of the mean velocity and turbulent stress were made using the LDV technique in the forward-scatter mode. A Lexel 2-watt Argon-ion water-cooled laser was used in this study. The optics were of the DISA 55X modular series. A DISA 55N10, two-channel frequency shifter was used for the electronic frequency shifting. In our case, since highly fluctuating and reversing flow was expected, a combination of both optical and electronic shifting was used. Data acquisition was accomplished with the use of two DISA 55L90 counters. The LDV optical system parameters are shown in Table 1. For two-component measurements, signals validated on the counters were checked by a coincidence logic to ensure that the signal in each counter came from the same particle. To achieve this, a 10 μs coincidence window was allowed between validated signals from the two counters. When the signals satisfied the coincidence check, the data acquisition software package was initiated and retrieved the Doppler information stored in the output register.

Table 1. LDV optical system parameters

Parameters	Laser beam	
	Green	Blue
Beam diameter (mm)	1.25	1.25
Focal length (mm)	310	310
Wavelength of laser line (nm)	514.5	488
Beam separation (mm)	53.48	53.48
Intersection angle [deg]	9.79	9.79
Beam waist diameter (mm)	0.08	0.08
Probe volume length (mm)	0.98	0.9
Measuring volume diameter (mm)	0.09	0.09
Fringe space (μm)	3.02	2.86
Fringe number	28	28
Calibration factor (ms^{-1}/mHz)	3.02	2.86

The laser, mirror stage assembly and optics were mounted on a thick aluminum plate, which was itself firmly bolted to an X-Y-Z traversing mechanism. Three stepping motors, with linear encoders, monitored by the PDP 11/34a minicomputer served to position the traversing mechanism during data acquisition. The PDP 11/34a minicomputer functioned as the central data acquisition and reduction controller, and processor.

Values of the mean velocity and turbulent stress components were determined from population samples consisting of 2,000 - 3,000 data points in groups of 1,000 data points for both phases. Data rates of about 1 kHz, with a validation percentage of around 50% were obtained for the air measurements. The data rate and the validation percentage were much smaller for the solid phase than the ones observed for the air measurements. In the present work McLAUGHLIN and TIEDERMAN's [14] two dimensional velocity correction was used to reduce the velocity bias effect.

The ratio of particle diameter to fringe spacing was around 11 for the

50 μm glass beads and around 22 for the 100 μm beads. The Doppler frequency modulation of the signals was very satisfactory. The system operated with signal-to-noise ratio around 20 (measured on the oscilloscope). Another usual problem in two-phase flows is the presence of cross-talks between signals from small and large particles in the measurements. This problem was not encountered in the present study since the solid-phase measurements were performed after the mineral oil jet atomizers (used for artificial seeding of the air) were removed.

Table 2a presents the total estimated systematic and random errors affecting the LDV measurements in the present study. For the conditions of this study some of the largest errors appear only in the near wall regions of the flow. Table 2b is a test on experimental reproducibility. It contains the range of errors observed for the mean velocities and turbulent stresses. The measurements were performed twice or three times for identical experimental conditions. The data showed that good experimental reproducibility can be achieved.

Table 2.a Estimated maximum measurement errors; b Experimental reproducibility test for air measurements

Variable	Total estimated error (systematic and random) a	Range of errors (% difference between different runs) b
U	-2 to 4%	0 to 2%
U_r	-2 to 4%	0 to 2%
$u\theta$	-3 to 5%	0 to 2%
u_r	-3 to 5%	0 to 4%
$u_\theta u_r$	-4 to 10%	0 to 9%
U_B	0 to 4%	

3. Results and discussion

The mean streamwise and radial velocities together with some "mass flux" distributions will be presented in this paper. The velocity measurements were performed along the center-line at four different stations $\theta = 0°$, 15°, 30°, and 45°, where θ is the angular displacement in the center plane. The gas and the solid phases were measured separately while maintaining the same flow conditions. 50 μm and 100 μm glass spheres were used to represent the solid phase. The bulk velocities, U_b, corresponding to the two different flow rates, were 33.09 and 52.19 m/s. These will be referred to as the low- and high- speed flow, respectively.

The present work is restricted to a very dilute suspension so that the gas flow field is negligibly affected by the presence of particles. The mass ratio was 1.5 x 10^{-4} and the volumetric ratio was 6 x 10^{-8} for the low-speed flow. For the high-speed flow, the mass and volumetric ratios were 9.5 x 10^{-5} and 3.8 x 10^{-8}, respectively. The mass and volumetric ratios are defined as the mass or volume of glass, respectively, divided by the mass or volume of air that enters the test section in one second. Detailed measurements of the gas phase were performed at 16 different planes but these results appear in KLIAFAS [12] and KLIAFAS, BACHALO and HOLT [13]. The gas flow field was obviously three-dimensional. However the streamwise and cross-stream velocities were of interest in this study. The "mass flux" measurements were performed with LDV at the following stations: θ=0°, 15°, 30°, 45°, 60° and 75°.

3.1 Mean velocities

The polar coordinate system (r, θ) is shown in Fig.2. The radii of the inner and outer walls are r_i and r_o, respectively. The normalized radial coordinate is $r^* = (r-r_o)/(r_i-r_o)$. The mean streamwise velocity component is U_θ (or U_z) while the mean radial component is U_r (or U_y). The bulk velocity is U_b and Reynolds number $R_e = (U_b D)/\nu$.

Figures 4 and 5 present plots of U_θ/U_b against r^* for the high speed flow (U_b = 52.19 m/s). For the most part, the 50 µm particles were found to lag behind the gas (positive slip-velocity). The velocity profiles for the gas and solid phase crossed over near the outer wall, where the glass beads seemed to maintain a constant finite velocity. The length of the negative slip velocity region was 0.23 of the duct diameter. The solid velocity profile was found to be considerably flat.

The U_θ profiles for the 50 µm glass beads were not affected by the streamwise pressure gradient which caused the displacement of the maximum gas velocities towards the inner wall. The air velocity profile is due to the favourable and unfavourable streamwise pressure gradient which exists near the inner and outer walls respectively. It appears that the air flow behaves in a manner consistent with potential flow solutions. At θ = 0° the maximum streamwise slip-velocity, 0.23 x U_b, appears near the inner wall.

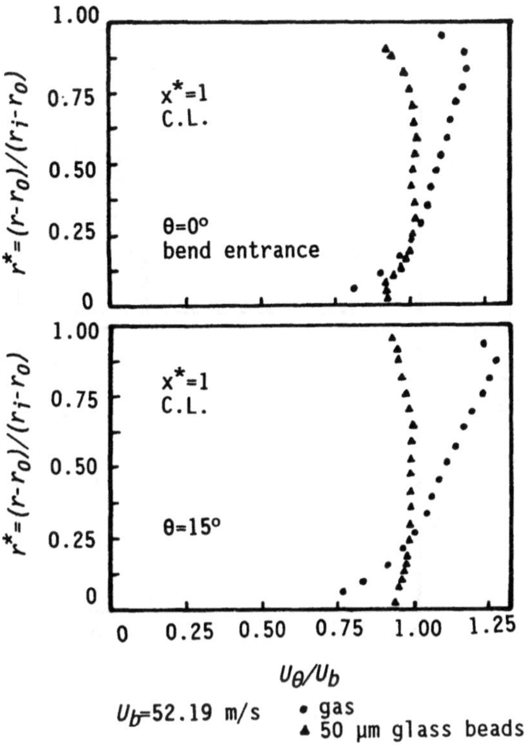

Fig. 4 Mean streamwise velocity profiles at θ=0° and 15° stations (along the center line), U_b = 52.19 m/s

The particle streamwise velocity profiles continued to be flat at the next three stations (θ = 15°, 30° and 45°). At θ = 15° the negative slip velocity region was 0.24 of the hydraulic diameter and the maximum slip velocity was observed near the inner wall and it was equal to U_b x 0.32.

At θ = 30°, the thickness of the negative slip-velocity region was 0.15 of the hydraulic diameter. The maximum slip-velocity was further increased to 0.38 x U_b, occuring again near the inner wall.

At θ = 45°, no measurements were taken near the inner wall where a clearly identifiable particle-free region was observed. At this station the highest U_θ values were found to be near the outer wall region. Apart from that, the particle streamwise velocity profiles continued to be flat. The thickness of the negative slip velocity region was 0.15 of the hydraulic diameter.

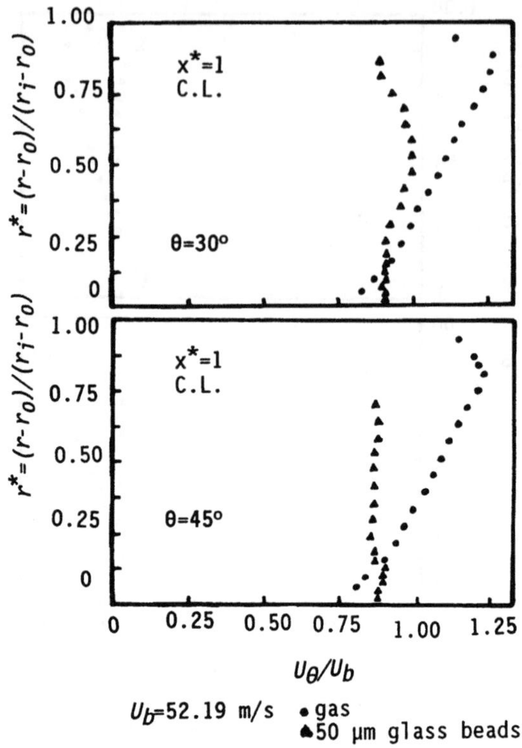

Fig.5. Mean streamwise velocity profiles at θ=30° and 45° stations (along the center line), U_b= 52.19 m/s.

Figures 6 and 7 show plots of U_θ/U_b against r* for the 50 μm and 100 μm particles (U_b = 33.09 m/s). The qualitative agreement between these results and the ones for the high speed flow is quite good. At θ = 0° the U_θ profiles for the 50 μm and 100 μm particles are the same for most of the center line. However, the 100 μm particle velocity is higher than the 50 μm velocity near the outer wall. Thus, a higher negative slip-velocity is observed for the large particles than the one for the 50 μm particles near the outer wall. This effect is present for all the stations examined in the present study. At θ = 15° and 30°, the 100 μm particle U_θ was slightly greater than the corresponding 50 μm particle velocity near the inner wall. This can be attributed to the flatter streamwise velocity profiles that appear in the large particle measurements.

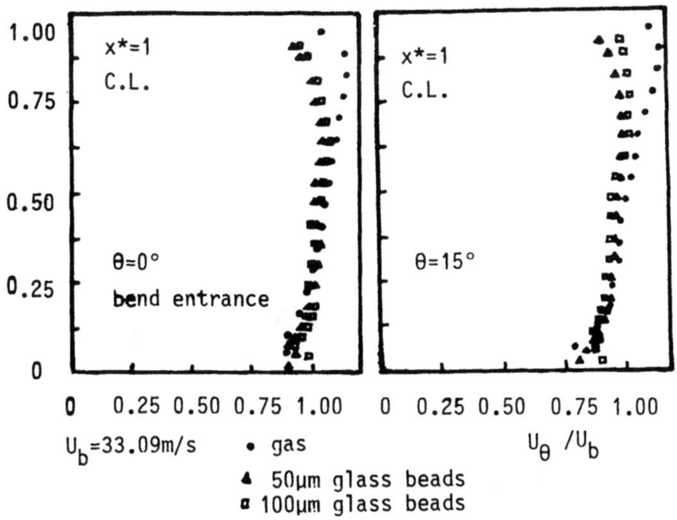

Fig.6. Mean streamwise velocity profiles at θ=0° and 15° stations (along the center line), U_b = 33.09 m/s.

Figures 8 and 9 present plots of the mean radial velocity against r^* for the low speed flow (33.09 m/s). At θ = 0°, the radial velocity of air is directed towards the inner wall over the entire cross-section. The same thing happens for the radial velocity of the 50 μm and 100 μm particles.

Thus, we can say that the large radial pressure gradient at the bend entrance has a significant effect on the particle radial velocities. However, its effect is less pronounced on the particle velocities than it is on gas velocities.

So U_r/U_b at the bend entrance (center line) lies between 3.1% and 11.2% for the gas phase, 1.8% and 6.2% for the 50 μm particles and 0.3% and 5% for the 100 μm particles. Thus, the large particles are the least affected by the pressure gradient.

At θ = 15°, the radial velocity of the particles (center line) is much higher than the radial velocity of the air and is directed towards the outer wall. So U_r/U_b lies between 0 and -3.1% for the gas, -8.2% and -16.5% for 50 μm particles and -16.5% and 21% for the 100 μm particles.

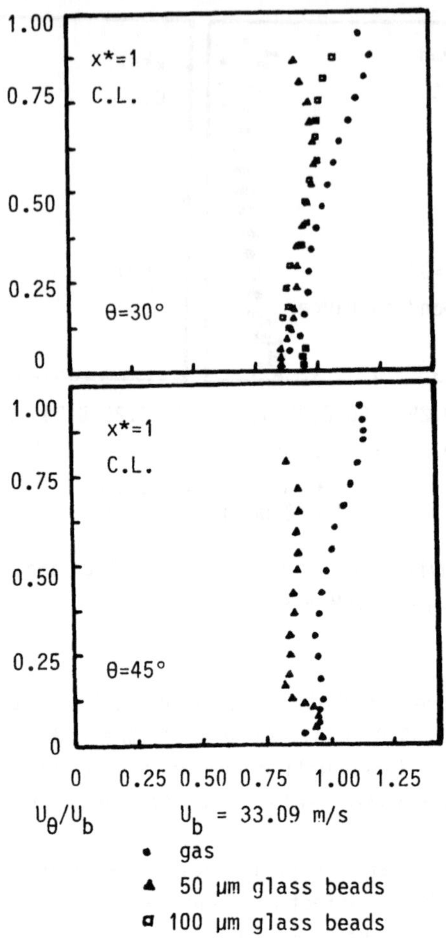

Fig.7. Mean streamwise velocity profiles at θ=30° and 45° stations (along the center line), U_b= 33.09 m/s.

At θ = 30° the level of the particle radial velocity is further increased. A sudden decrease of U_r is recorded in the region close to the outer wall. The thickness of this region is 0.12 x D. So, we can conclude that particle wall interaction is a controlling factor for the outer wall region of the flow. Particle-wall collisions are causing a clear change in the angular distribution. In the remainder of the center line at the θ = 30° station, U_r/U_b values range between -5.2% and -12.9% for the gas, -26% to -34% for the 50 μm particles and -35% to -48.5% for the 100 μm particles.

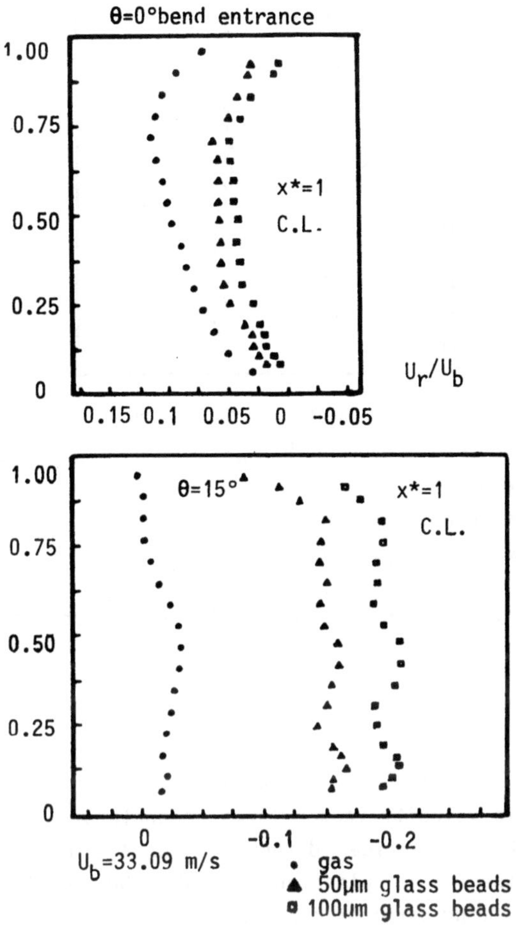

Fig.8. Mean radial velocity profiles at θ=0° and 15° stations (along the center line), U_b= 33.09 m/s.

At θ = 45°, a sudden decrease of U_r is observed close to the outer wall. The region where particle-wall ineraction is a controlling factor in the flow has a thickness of 0.2 x D. Very close to the outer wall, U_r is positive (towards the inner wall), which indicates a sudden change in the direction of particle movement. The change is due to the particle-wall collisions. The values of U_r/U_b go as high as -15.7% for the gas and -37.1% for the 50 μm particles.

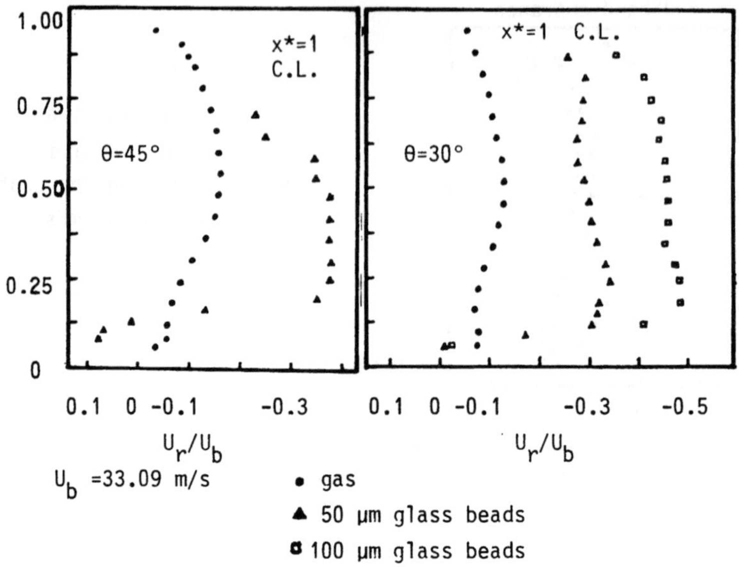

Fig.9. Mean radial velocity profiles at θ=30° and 45° stations (along the center line) U_b=33.09 m/s.

Since the U_θ and U_r particle profiles are quite flat, we can estimate an average direction for the velocity vector. For the 50 μm particles at θ = 15°, 30°, and 45°, the particle velocity vector forms an angle of 6.3°, 12°, and 22°, respectively with the vertical direction. For the 100 μm particles, the angle is 4° for the θ = 15° plane and 3.8° for the θ = 30° plane. We can conclude that the 100 μm particle trajectories are very close to being straight lines. Another observation drawn from all the results presented in this section is that only the outer wall is impinged by the particles.

Figures 10 and 11 present plots of U_r/U_b against $r*$ for the high speed flow (U_b = *52.19 m/s*). The qualitative agreement between the two sets of results was excellent. At θ = 0°, U_r/U_b ranged between 5 and 11.5% for the gas, 2.5 and 4.3% for the 50 μm particles. At the bend entrance, both the particle and gas radial velocity were directed towards the inner wall in a similar manner with the low speed flow.

At θ = 15°,30° and 45°, the 50 μm particle radial velocity ranged between -13.2 and -20%, -24 and -34% and -37 and -49%, respectively.

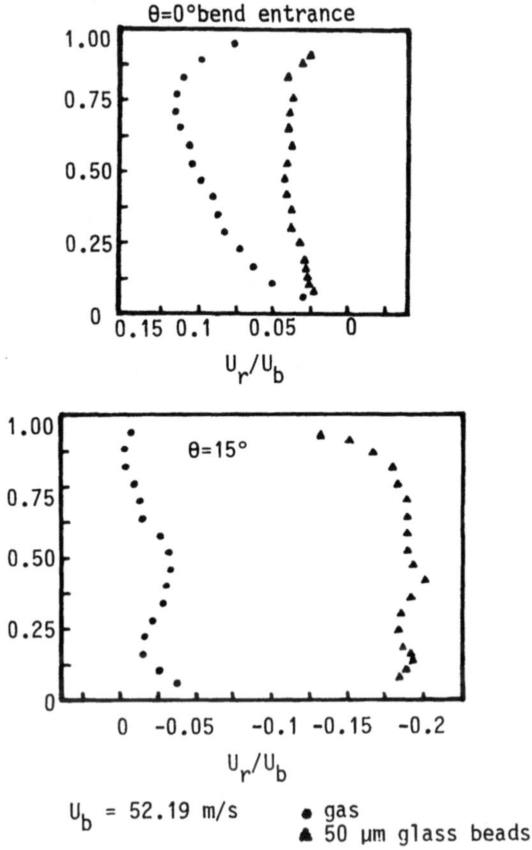

Fig.10. Mean radial velocity profiles at θ=0° and 15° stations (along the center line), U_b=52.19 m/s.

For the 50 μm particles at θ = 15°, 30° and 45°, the particle velocity vector forms an angle of 4.3°, 12.5° and 18.3°, respectively, with the vertical axis. So, there is greater tendency for the particles to follow straight line trajectories with increasing Reynolds numbers.

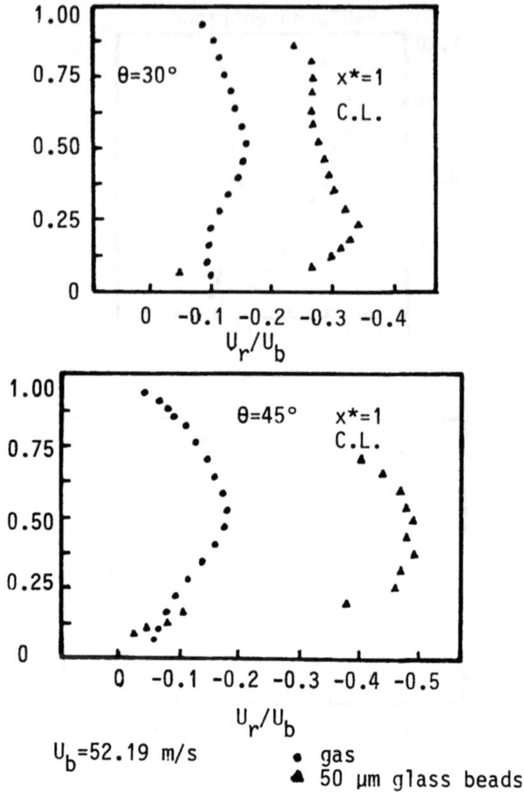

Fig.11. Mean radial velocity profiles at θ=30° and 45° stations (along the center line), U_b=52.19 m/s.

3.2 "Mass flux" measurements

Table 3 presents "mass flux" distributions, the "mass flux" being defined as the number of solid particles detected by the probe volume per second per unit area, along the center line at the following stations: θ=0°, 15°, 30°, 45°, 60° and 75°. Upstream of the bend entrance at x=-3.0 the "mass flux" was found to be very uniform. At this station the radial velocity, U_r, was very small (close to zero). The influence of the bend on the flow upstream was first present at $z^*=-0.4$. The secondary flow was directed towards the inner wall, over most of the above mentioned cross-section. At $\theta=0°$ a magnified version of the radial velocity profiles at $z^*=-0.4$ was observed. The detailed measurements of the gas phase appear in KLIAFAS [12] and KLIAFAS et al [13].

At $\theta=0^\circ$ particle distribution along the center line is quite uniform. However, slightly higher particle concentrations are recorded near the outer wall than near the inner wall. The radial pressure gradient at the bend entrance which causes U_r to be directed towards the inner wall is responsible for the above phenomenon. Thus, we can argue based on the results at $\theta=0^\circ$, that particle transport is partly controlled by the mean flow field. At $\theta=15^\circ$, much higher particle concentrations appear near the outer wall than those near the inner wall. The same thing happens at $\theta=30^\circ$, but in a more pronounced manner. At $\theta=45^\circ$ and 60° the particle concentration near the outer wall is further increased. It is between $\theta=30^\circ$ and 60° that most of the particle-wall collisions occur. At $\theta=45^\circ$ a particle-free region is identified close to the inner wall . The thickness of this particle-free region is $0.25 \times D$. At $\theta=60^\circ$ and 75° the thickness of the particle-free region is increased to $0.5 \times D$ and $0.68 \times D$ respectively. This is the reason why detailed measurements at these two stations are not presented. The particles were mainly concentrated in the outer wall region.

Table 3. Number of particles per second along the center line at different cross sections (50 μm).

r^*	θ					
	0°	15°	30°	45°	60°	75°
0.02 (2mm from O.W.)	0.9	27.0	32.0	33.0	43.0	21.0
0.05	1.8	8.3	32.0	21.0	52.0	20.0
0.08	1.9	6.8	23.0	21.0	14.2	14.0
0.16	1.5	4.5	3.0	8.1	13.0	17.0
0.24	1.3	3.9	3.1	2.4	15.0	17.0
0.32	1.5	4.6	2.9	3.3	13.0	0.6
0.40	1.8	5.0	1.7	4.5	1.0	0.5
0.48	2.0	5.1	1.8	4.0	0.6	0.5
0.56	2.2	4.7	1.3	3.7	0.5	0.4
0.64	3.0	4.7	1.5	1.5	0.4	0.4
0.72	2.8	6.3	1.4	0.6	0.4	0.4
0.80	4.0	7.0	1.5	0.4	0.3	0.4
0.88	5.2	5.1	1.0	0.3	0.3	0.3
0.94	6.2	2.9	0.3	0.3	0.2	0.3
0.96 (4 mm from I.W.)			0.2	0.2	0.2	0.2

4. Conclusions

The results presented in this paper are unique in the area of two-phase flows in curved ducts. However, now it is the time to further investigatethese flows by: a) measuring the air velocities in the presence of solids, b) examine the three dimensionality of the flow, c) measure the incident and rebound angles of the solid particles with the wall, d) include different sizes of particles. The results of this paper should help the understanding of the physical phenomena occuring in two-phase flows and also enable the material scientists and engineers to predict the erosion characteristics in a 90° bend.

From the experimental results presented in the paper we can pinpoint that:

(1) The velocity profiles for the gas and solid phase cross over near the outer wall, where the glass beads seem to maintain a finite velocity. The thickness of the negative slip-velocity region is 0.24 x D at the θ=0° and 15° stations, and 0.15 x D at the θ=30° and 45° stations. The solids velocity profiles are found to be quite flat. Higher negative slip velocities are observed for the 100 μm particles than those for the 50 μm particles.

(2) The large radial pressure gradient at the bend entrance has a significant effect on the particle radial velocities. At θ=0°, the radial velocity is directed towards the inner wall for both the 50 μm and 100 μm particles. The large particles are the least affected by the pressure gradient. U_r/U_b goes as high as -37% at the θ=45° station. At θ=30° and 45°, particle-wall collisions cause a clear change in the direction of U_r in the region close to the outer wall.

(3) The 100 μm particle trajectories are very close to being straight lines. In addition, there is a greater tendency for the particles to follow straight line trajectories as Re number increases.

(4) It is between θ=30° and 60° that most of the particle-wall collisions occur. Only the outer wall is impinged by particles. At θ=45°, 60°, and 75°, a particle free region at θ=45°,60° and 75° is 0.25, 0.50 and 0.68 of the hydraulic diameter, respectively.

The scope of the present study was to examine the fluid mechanics of the gas-particle mixture in a 90° bend. There are many factors that influence the behaviour of solid particles in two-phase flows, such as gravity forces, particle-particle and particle-wall interactions, and deposition of particles on the duct wall, just to name a few. Of those, particle-wall interaction proved to be a controlling factor in our study. Only the erosion effect comes into the picture. The inner wall was, in general, erosion free. No erosion was observed on the side walls, which were made of glass. This is to be expected according to MILLS et al [15], who reported that the glass (typical brittle material) suffers severe erosion under normal impact. The importance of the gravity force becomes apparent when we refer to the 100 μm particle measurements. The 100 μm particle trajectories come very close (but not quite) to being straight lines.

In the present study, a deposition free area (on all walls) extending (2.5m in length) was observed for both particle sizes used. This is due to the high air bulk velocities employed in the present study. Experimental results of BOWERS [16] and PATTERSON [17] taken for our conveying velocity range, air to solids ratio and bend radius agree with our findings and they predict a very long deposition free distance.

References

1. Y.Levy, F.C. Lockwood : Comb. and Flame, Vol.40, 333 (1981).
2. D. Modarress, H.Tan : Exp. In Fluids, Vol.1, 113 (1983).
3. W.M.Farmer : Appl. Opt., Vol.11, 2603 (1972).
4. F.Durst : ZAMP, Vol.24, 619 (1973).
5. W.D. Bachalo : Appl. Opt., Vol.19, 363 (1980).
6. Y. Kliafas, A.M.K.P. Taylor, J.H. Whitelaw : Exp. In Fluids, Vol.5, 159 (1987).
7. Y. Kliafas, A.M.K.P. Taylor, J.H.Whitelaw: A.S.M.E Meeting, FED Vol.35, 41 (1986).
8. W.D.Bachalo, M. Hauser : Opt. Eng., Vol.23,583 (1984).
9. W.S. Yeung : Wear, Vol.55,91 (1979).
10. F. Pourahmadi, J.A.C. Humphrey: Phys. Chem. Hydradyn., Vol.4, 191 (1983).
11. R.I. Crane, R.L. Evans: J. Aero. Sci., Vol.8, 161 (1977).
12. Y. Kliafas : Lawrence Berkeley Lab., LBL-17244 (1984).
13. Y. Kliafas, W. Bachalo, M. Holt: AIAA Conf., Paper No.85-4066 (1985).
14. D.K. McLaughlin, W.G. Tiederman : Phys. Fluids, Vol.16, 2082 (1973).
15. D. Mills, J.S. Mason : 1st Int. Conf. Protection of Pipes, G1-1 (1975).
16. H.M. Bowers, J.G. Wright : 4th Int. Conf. Pneumatic Transp. of Solids, A2-7 (1978).
17. R.C. Patterson : Trans. ASME, Vol.81, No43, 43 (1959).

Hyperbolic Solution Adaptive Mesh Generation

G.H. Klopfer

1. Introduction

In recent years there has been much activity in the development of solution adaptive mesh procedures as one method of increasing the efficiency and accuracy of numerical solutions of the Euler and Navier-Stokes equations. Most of the adaptive mesh procedures are based on a set of elliptic equations describing the location of the grid points subject to certain smoothness and orthogonality constraints. The computational effort to solve these equations is of the same order as that required to solve the flow equations. This additional cost of generating the adapted mesh negates much of the advantage of adapted meshes, i.e. computational efficiency. An alternative method of generating a solution adapted mesh is presented. The method is based on the hyperbolic grid generation procedure of [1] and is much faster than those based on elliptic grid generators.

The following sections discuss the basic theory of hyperbolic grid generation equations in two-dimensions and the additional terms required to obtain the clustering while maintaining grid smoothness and some level of orthogonality between grid lines. Some computational results are also presented for a viscous shock-on-shock interaction at hypersonic speeds about a two dimensional blunt body.

2. Hyperbolic Grid Generation

STEGER and CHAUSSEE [1] developed a method for hyperbolic grid generation in two dimensions by showing which two of many constraint equations for a mesh form a hyperbolic system of partial differential equations. There are several choices and the one combination that seems to work in many cases is that which puts constraints on the mesh orthogonality and mesh cell volume (area in 2-D)

Orthogonality $\quad \vec{R}_\varsigma \bullet \vec{R}_\eta = 0 \quad$ (1a)

Cell Volume $\quad V = (\vec{R}_\varsigma \times \vec{R}_\eta) \bullet \vec{e}_1 \quad$ (1b)

where $\quad \vec{R} = 0\, \vec{e}_1 + y\, \vec{e}_2 + z\, \vec{e}_3$

$\vec{e}_1, \vec{e}_2, \vec{e}_3$ are the Cartesian unit vectors and ς, η are the computational coordinates whose values are such that $\Delta\varsigma = \Delta\eta = 1$.

Expanded out, (1a) and (1b) become

$$z_\varsigma z_\eta + y_\varsigma y_\eta = 0 \tag{2a}$$

$$-y_\varsigma z_\eta + z_\varsigma y_\eta = V \tag{2b}$$

These equations can be integrated in either the ς or η direction. Let η be the direction normal to the body surface and ς be the direction along to the surface and j, k the corresponding indices. Steger and Chaussee solve (2) by linearizing and then using the Beam-Warming scheme to march in the η direction.

If (2) is solved directly for z_η, y_η without linearizing, we obtain with the addition of a second order smoothing term (called artificial dissipation) with the coefficient μ

$$\vec{R}_\eta - \vec{F} = \mu \frac{\partial^2}{\partial\varsigma^2} \vec{R} \tag{3}$$

where

$$\vec{F} = \left[\frac{-Vy_\varsigma}{z_\varsigma^2 + y_\varsigma^2} \; , \; \frac{Vx_\varsigma}{z_\varsigma^2 + y_\varsigma^2} \right]^T$$

This equation can be solved by the following algorithm

$$\vec{R}^{j+1} - \vec{R}^{j} = (1-\theta)\,\vec{F}^{j} + \theta\,\vec{F}^{j+1}$$
$$+ (1-\theta)\,\mu^{j}\,[\vec{R}_{k-1} - 2\vec{R}_{k} + \vec{R}_{k+1}]^{j} \tag{4}$$
$$+ \theta\,\mu^{j+1}\,[\vec{R}_{k-1} - 2\vec{R}_{k} + \vec{R}_{k+1}]^{j+1}$$

which is a generalized two-step implicit method.

The addition of second order spatial (ς) smoothing as opposed to fourth order smoothing is for intuitive reasons. For mesh generation procedures, accuracy in solving (2) is not the prime consideration; rather mesh smoothness and regularity are. The justification is given below.

A modified equation analysis of the present scheme, (4), produces the following:

$$\vec{R}_{\eta} - \vec{F} = (\theta - \tfrac{1}{2})\,\frac{\partial^2 \vec{R}}{\partial \eta^2} + \mu\,\frac{\partial^2 \vec{R}}{\partial \varsigma^2} + O(\Delta^3) \tag{5}$$

The modified (5) shows that the leading truncation error term is second order, dissipative, and called the numerical dissipation. In fact, the leading terms on the right side of (5) are similar to the Laplace operator used by STEGER and SORENSON [2] for the elliptic mesh generation, if the viscosity coefficients are scaled appropriately. If the scaling is

$$\theta - \tfrac{1}{2} \sim \frac{\nu_1}{z_{\eta}^2 + y_{\eta}^2} \tag{6}$$

and

$$\mu \sim \frac{\nu_2}{z_{\varsigma}^2 + y_{\varsigma}^2} \tag{7}$$

the modified equation becomes

$$\nu_1 (z_\zeta^2 + y_\zeta^2) \frac{\partial^2 \vec{R}}{\partial \eta^2} + \nu_2 (z_\eta^2 + y_\eta^2) \frac{\partial^2 \vec{R}}{\partial \zeta^2}$$

$$= (z_\zeta^2 + y_\zeta^2)(z_\eta^2 + y_\eta^2) \left\{ \frac{\partial \vec{R}}{\partial \eta} - \vec{F} \right\} + O(\Delta^3) \tag{8}$$

These equations are similar to those solved by Steger and Sorenson in their elliptic mesh generation procedure if the coordinate lines are orthogonal. The forcing function (right-hand side of (8)) is also similar in that it is a constraint on mesh orthogonality and mesh cell volume. Another way of looking at (8) is by rearranging to obtain

$$\frac{\partial \vec{R}}{\partial \eta} = \vec{F} + \frac{\nu_1}{(z_\eta^2 + y_\eta^2)} \left[\frac{\partial^2 \vec{R}}{\partial \eta^2} + \frac{\nu_2}{\nu_1} \frac{(z_\eta^2 + y_\eta^2)}{(z_\zeta^2 + y_\zeta^2)} \frac{\partial^2 \vec{R}}{\partial \zeta^2} \right] \tag{9}$$

Equation (9) shows that the forcing function for the hyperbolic method (right-hand side of (9)) is modified by the addition of the Laplace operator which is a result of the truncation error of the present procedure. By modulating ν_1 and ν_2, the relative importance of the original forcing function F and the Laplacian operator can be controlled.

By the appropriate choice of the artificial dissipation added to the scheme and the appropriate scaling of both the artificial and numerical viscosity coefficients, two things are forced to occur in the present procedure. Firstly, the leading truncation errors are always dissipative. Secondly, the leading truncation errors act by themselves as forcing functions with desirable mesh properties for the hyperbolic method.

Equation (4) is a system of scalar tridiagonal equations

$$\tilde{A} \vec{R}_{k+1}^{j+1} + \tilde{B} \vec{R}_k^{j+1} + \tilde{C} \vec{R}_{k-1}^{j+1} = \vec{D} \tag{10a}$$

where

$$\tilde{A} = \tilde{C} = 1$$

$$\tilde{B} = -2 - v_k^{j+1} w_k^j$$

$$\vec{D} = - v_k^{j+1} \left(\vec{F}_k^{j+1} + \vec{G}_k^j \right)$$

$$\vec{G}_k^j = w_k^j \ \vec{R}_k^j + \left(w_k^j - 1\right) \vec{F}_k^j$$

$$+ \ (w_k^j - 1) \ (\vec{R}_{k+1}^j \ - 2\vec{R}_k^j + \vec{R}_{k-1}^j) \ / \ v_k^j \tag{10b}$$

$$v_k^j = (z_\varsigma^2 + y_\varsigma^2) \ / \ \nu_1$$

and

$$w_k^j = 2 \ (z_\eta^2 + y_\eta^2) \ / \ (2\nu_2 + z_\eta^2 + y_\eta^2)$$

Since the coefficients of (10) depend on values at the (j+1)th level, it must be solved iteratively. Typically three to five iterations are sufficient. The volumes are defined to be

$$V^{j+1} = (dS)^{j+1} \ \sqrt{z_\varsigma^2 + y_\varsigma^2}^{\,j+1} \tag{11}$$

where $(dS)^{j+1}$ is a user defined increment in the marching direction between j+1 and j. The marching process is started by specifying a mesh distribution along the body surface. Equation (10) is then marched away from this surface. The viscosity coefficients can gradually be increased as the marching proceeds away from the surface so that the artificial and numerical viscosity are in balance and so that the Laplacian operator predominates over the forcing function F in the far field. Near the surface F should dominate, so that the mesh is orthogonal at the surface as required by many numercial schemes for accuracy. It was found useful to modulate both viscosity coefficients as a function of the ς coordinate. In particular, the viscosity had to be increased or decreased from its mean value at the jth level according to whether the constant (η) coordinate line had positive (concave) or negative (convex) curvature.

Since $(dS)^{j+1}$ is user specified, it is possible to generate a grid between fixed boundaries as well as to cluster the grid points in the marching direction. Two examples for unclustered meshes are shown in Fig. 1. The first example is a cavity inside a channel and the second is a thin fin in a channel.

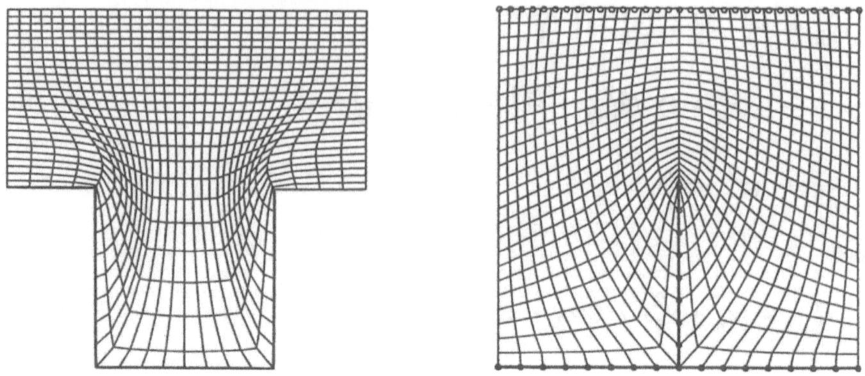

(a) cavity in channel (b) thin fin in channel.

Figure 1.- Meshes produced by the hyperbolic grid generator.

3. Hyperbolic Adaptive Mesh Generation

It is possible to cluster the mesh only in one direction by specifying (dS) with the present hyperbolic mesh generator procedure. For general two-dimensional flows fields, the one dimensional adaptivity is not sufficient. A method was developed to remove the one dimensional adaptation restrictions of the hyperbolic method. The method involves changing the basic differential equation, (3), to include extra forcing terms to allow the extra degree of freedom needed for control of the clustering in the ς direction. The extra forcing terms were derived from a procedure called mean value relaxation, developed by EISEMAN [3] and improved by CONNETT et al. [4]. The clustering is accomplished by a weighted averaging of the mesh points with the weights determined by the solution to be adapted.

For one-dimensional clustering mean value relaxation is similar to the tension spring adaptation procedure where the spring constants are determined by the solution, however, the basic idea is different. The tension spring procedure was used by NAKAHASHI and DEIWERT [5], among others. In the work of [5] a forcing term was included to prevent excessive deviation from orthogonality. This forcing term is based on the torsion spring analogy. Since the tension-torsion spring procedure is similar to the mean value relaxation procedure and easier to understand, it will be used in the following description of the modified hyperbolic grid generation.

The differential equation, (3) with the extra forcing terms is

$$\vec{R}_\eta - \vec{F} = \mu \frac{\partial^2 \vec{R}}{\partial \varsigma^2} + \frac{\partial}{\partial \varsigma} \omega \frac{\partial}{\partial \varsigma} \vec{R} - c\,(\vec{R} - \underset{\sim}{\vec{R}}) \tag{12a}$$

where ω is the tension spring constant and c is the torsion spring constant $\underset{\sim}{\vec{R}}$ are the location of the node points for a locally orthogonal mesh. These points can be determined from the $\vec{R}^j$ values and $(dS)^{j+1}$.

Equation (12a) can be solved in the same manner as (3) to arrive at the same system of algebraic equations as (10a). However, the coefficients are different and are as follows:

$$\begin{aligned}
\tilde{A} &= 1 + \omega_k^{j+1}\, v_k^{j+1} \\
\tilde{C} &= 1 - \omega_{k-1}^{j+1}\, v_k^{j+1} \\
\tilde{B} &= -2 - v_k^{j+1} \left\{ \omega_k^{j+1} - \omega_{k-1}^{j+1} + w_k^j\,(1 + c_k^j) \right\} \\
\vec{D} &= -v_k^{j+1}\,(F_k^{j+1} + \vec{G}_k^j\,) \\
\vec{G}_k^j &= w_k^j\, \vec{R}_k^j + w_k^j\, c_k^j\, \vec{R}_k^j + (w_k^j - 1)\, \vec{F}_k^j \\
&\quad + [(1 + \omega_k^j v_k^j)\, \vec{R}_{k+1}^j - \{2 + v_k^j\,(\omega_k^j - \omega_{k-1}^j)\}\, \vec{R}_k^j \\
&\quad + (1 - \omega_{k-1}^j v_k^j)\, \vec{R}_{k-1}^j]/v_k^j
\end{aligned} \tag{12b}$$

This procedure is not much more complicated than the original hyperbolic grid generator. To each node point, k, some weighting value (ω) of the solution is assigned. The weight function is determined by some rule discussed in the next section. The torsion spring constant c_k^j is solution dependent

$$c_k^j \propto \sqrt{(z_k^{j+1} - \tilde{z}_k^j)^2 + (y_k^{j+1} - \tilde{y}_k^j)^2}$$

and allows for some control over the grid orthogonality.

4. Weight Function

The link between the solution and the adaptive mesh is the weight function which determines the mesh clustering. Typically, ad hoc methods are used to construct the weight function from the solution.

Let s be the arc length along the ς-coordinate so that

$$s_{k+1} - s_k = [(y_{k+1} - y_k)^2 + (z_{k+1} - z_k)^2]^{1/2}$$

A typical weight function [5] is

$$\omega_k = 1 + A\ \bar{f}^B$$

where

$$\bar{f} = \frac{f_k - f_{min}}{f_{max} - f_{min}}$$

and

$$f_k = \frac{\partial q_k}{\partial s}$$

and q is the flow field variable chosen for the adaption procedure. A and B are functions of the maximum and minimum grid spacing,

$$A = \frac{\Delta s_{max}}{\Delta s_{min}} - 1$$

and B is obtained iteratively such that

$$\Delta s_k = \Delta s_{min} \text{ where } (\omega_{k+1} - \omega_k) \text{ is minimum}$$

A hypersonic shock-on-shock interaction solution obtained with the uniform mesh (Fig. 2a) is given in terms of Mach contours in Fig. 2b. A typical adapted mesh obtained for a hypersonic shock-on-shock interaction solution (Fig. 3) is shown in Fig. 4. Due to the complicated flow structure it is difficult, if not impossible, to obtain a "well-behaved" mesh with arbitrary amounts of clustering. The use of solution curvature or a linear combination of solution gradients and curvature for the weight functions does not help.

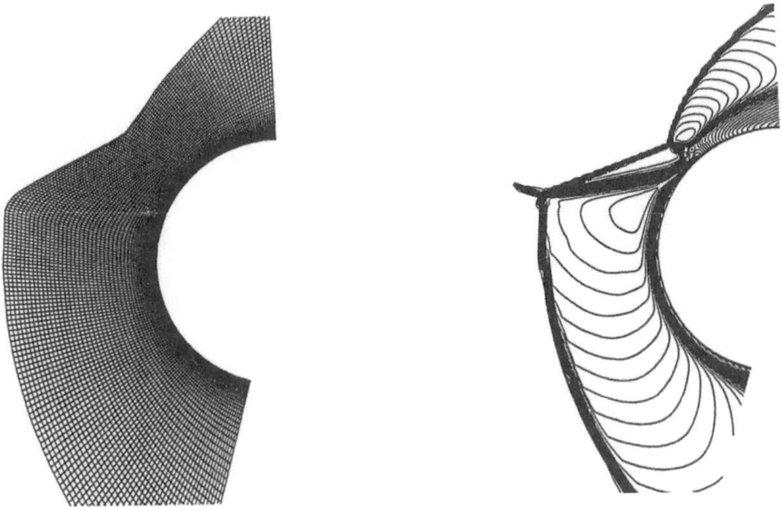

(a) (b)

Figure 2.- Computational mesh (a) and Mach contours (b) of a viscous hypersonic shock-on-shock interaction at $M_\infty = 8.03$ and $Re_D = 387{,}500$.

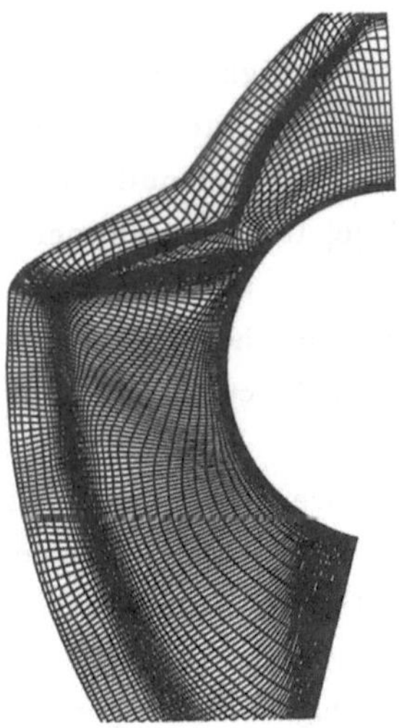

Figure 3.- Solution adapted mesh using the "classical" weight functions.

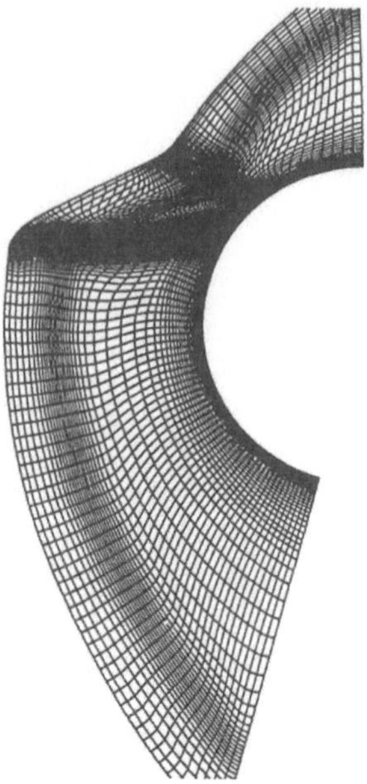

Figure 4.- Solution adapted mesh using the new weight functions.

The difficulty is that the weights are determined by the solution gradients, but the solution gradients are by themselves not a good measure of an optimally adapted grid. The ground rules for an optimally adapted mesh were set by VINOKUR [6] for a one-dimensional stretching function. The weights are redefined to follow the guide lines of [6]. The idea is that if the weights are properly chosen, then the adapted mesh clustering will be similar to the optimal stretching functions given by Vinokur.

In this paper the weights are chosen such that the solution gradients determine only the location of the extreme values of the weights and the distribution between the extremal values are determined by Vinokur's criteria. For the tension spring analogy the weight distribution is exponential. Furthermore, for high clustering ratios, it is important that the contour lines of maximum/minimum weights be smooth and continuous. Any abrupt discontinuities of the contour lines are avoided by interpolation or extrapolation to the outer boundaries of the mesh. While this is wasteful of mesh points it allows for arbitrary amounts of clustering and results in a "well-behaved" adapted mesh.

A solution adapted mesh obtained with the above weights is shown in Fig. 4. A solution in terms of Mach contours obtained with the adapted mesh is given in Fig. 5.

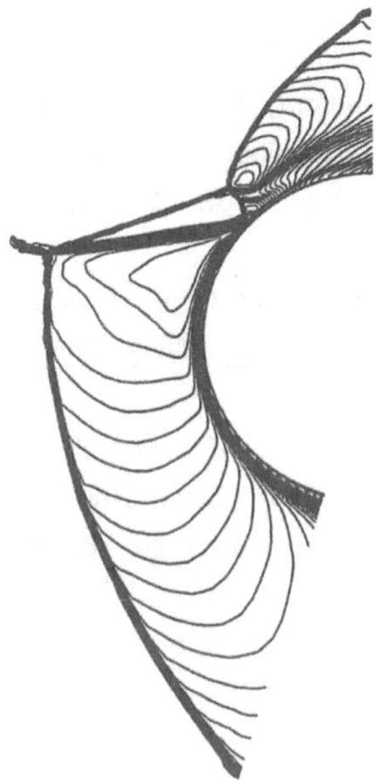

Figure 5. Mach contours obtained with the new solution adapted mesh. Same flow conditions as in Figure 2.

5. Concluding Remarks

A solution adaptive mesh procedure based on hyperbolic grid generation equation is presented. It is shown that the traditional methods of determining the weights from the solution are inadequate if arbitrary amounts of clustering are required. The solution should only be used to locate the contours of maximum/minimum clustering. The contour lines must be smooth and continuous. Unfortunately this requires interactive programming. To completely automate this process is difficult. The hyperbolic solution adapted mesh procedure is several times faster than those based on elliptic grid generation methods and is the only known method of obtaining arbitrary amounts of clustering on structured meshes.

6. References

1. Steger, J. L., and Chaussee, D. S., "Generation of Body Fitted Coordinates Using Hyperbolic Partial Differential Equations," SIAM J. Sci. Stat. Comput., Vol. 1, 1980, p. 431-437.

2. Steger, J. L., and Sorenson, R. L., "Automatic Mesh-Point Clustering Near a Boundary in Grid Generation with Elliptic Partial Differential Equations," J. Comp. Phys., Vol. 33, No. 3, 1979, pp. 405-410.

3. Eiseman, P.R., "Adaptive Grid Generation by Mean Value Relaxation," in Advances in Grid Generation (New York: ASME, 1983.)

4. Connett, W. C., Agarwal, R. K., and Schwartz, A. L., "An Adaptive Grid-Generation Scheme for Flow Field Calculations," AIAA 25th Aerospace Sciences Meeting, Paper No. 87-0199, January 1987.

5. Nakahashi, K., and Deiwert, G. S., "A Three-dimensional Adaptive Grid Method," AIAA 23rd Aerospace Sciences Meeting, Reno, Nevada, January 1985.

6. Vinokur, M., "On One-Dimensional Stretching Functions for Finite Difference Calculations," J. Comp. Phys., Vol. 50, pp. 215-234, 1983.

From the Kármán-Trefftz Profile to Cast-7 —75 Years of the Aerodynamisches Institut

Egon Krause

1. INTRODUCTION: A TRIBUTE TO PROFESSOR MAURICE HOLT

When the author first heard about the Second International Conference on Numerical Methods in Fluid Dynamics, to be organized by Professor Holt at the University of California at Berkeley in September 1970, he was surprised for several reasons. Until 1970 there was no conference dealing with the subject in question. The idea to initiate a series of conferences for this field on a worldwide basis was greeted with great enthusiasm: Finally, the "numerical community" as Gino Moretti once called the scientists working with numerical methods, was given the chance to meet and discuss their problems in the frame of international conferences. To be admired was also the courage the organizers must have had in order to overcome the difficulties, typical for such ventures. The associated problems were already signalized by the fact, that little or nothing was known about the first conference of the series proposed. Rumours had it, that it had taken place in Novosibirsk in 1969, at that time almost unbelievable in view of the relatively sparse scientific contacts between East and West.

Nevertheless, the second conference was a great success. Altogether 65 papers were presented by scientists from the USA, USSR, France, Germany, England, The Netherlands, Canada, and Australia. That then the third conference would follow two years later in Paris was no surprise to anybody anymore. The Berkeley conference was a sound scientific affair. The participants had the impression, that it was just one of a well-established series, but not one which was just initiated. It opened new dimensions, mainly due to Professor Holt and his colleagues who worked with him on the conference committee.

The spark carried over: the American Institute of Astronautics held the first "Computational Fluid Dynamics Conference" in Palm Springs in 1973. Twenty papers were presented, not even one third of the 75 papers presented at the last, the eighth AIAA conference, organized in Honolulu in 1987. The GAMM (Society for Applied Mathematics and Mechanics) followed suit and held its first "Conference on Numerical Methods and Fluid Dynamics" in Cologne in 1975 with 30 contributions, mainly from European countries. Although the GAMM-conference committee tried very hard not to expand the size of the conference, over 50 papers were presented at the seventh conference, held in Belgium in 1987. Other countries followed in the same spirit. One of the most recent developments was the foundation of the Japanese Society of Computational Fluid Dynamics by Professor Koichi Oshima, who, as a member of the International Conference committee, worked closely together with Professor Holt. The Japanese Society held its first "International Symposium on Computational Fluid Dynamics" in Tokyo in 1985 with worldwide participation.

Discussion of numerical problems in fluid dynamics now almost takes place everywhere in regular time intervals, with scientists participating from all over the world. The spirit of these conferences perhaps is caught in the picture shown in Fig. 1. Depicted is Professor Maurice Holt, listening attentively to the opening address of the eighth international conference in Aachen, 1982, together with his colleague, Professor V. V. Russanov, of the Keldysh Institute of Applied Mathematics in Moscow.

Fig. 1. Professors Maurice Holt and Victor Russanov attending the opening of the Eighth International Conference on Numerical Methods in Fluid Dynamics, June 1982 in Aachen.

This initiative reminded the author very much of the first conference, organized by von Kármán in Aachen 1929 on the occasion of the completion of an additional wing of the Aerodynamisches Institut. The participants of that conference are shown in the picture of Fig. 2. Many of them were already famous

1. von Kármán, Th.
2. Prandtl, L.
3. Tanakadate
4. Hopf, L.
5. Hahn, E.
6. Frl. von Kármán, J.
7. Misztal, F.
8. Bollenrath, F.
9. Mrs. Glauert, M.
10.
11. Barillon
12. Tollmien, W.
13. Hansen, M.
14. Nikuradse, J.
15. von Mises, R.
16.
17. Miss Swain
18.
19. Schuler, M.
20. Eisner, F.
21. Schiller, L.
22. Reißner, H.
23. Bertozzi Olmeda
24. Gilles, A.
25.
26. Lorenz, H.
27. Katzmayr, R.
28. Hartenberg, R. S.
29. Kempf G.
30. Hilbes.
31. Mathar
32. Finzi, L. A.
33. Wattendorf, F. L.
34. Betz, A.
35. Ackeret, J.
36. Treer, M. F.
37.
38.
39. Nemes
40. Kampé de Fériet, J.
41. Low, A. R.
42. Roy, M.
43. Peters, H.
44.
45. Glauert, H.
46. Kayser, E.
47. von Baumhauer, A.
48.
49. Douglas
50. Hübner, W.
51.
52. Loew, G.
53. Küssner, H. G.
54. Burgers, J. M.
55.
56. Miss Chitty, L.
57. Goldstein, S.
58. Thom, A.
59 Töpfer
60 van der Maas, H.J.
61. Muttray, H.
62. Blumenthal, O.
63. Maccoll, J. W.
64. Rosenhead, L.
65. Pohlhausen, E.
66. Brand
67. Baeumker, A.
68. Busemann, A.
69. Joshihara
70. Hoff, W.
71.
72. Helmbold, H. B.
73. Lerbs, H. E.
74. Blenk, H.
75. Hirota
76. Lennertz, J.
77. Epstein
78.
79. Weinel
80. Petersohn, E.
81. Spannhake, W.
82. von Mathes, P.
83 Pöschl, Th.
84. Trefftz, E.
85. Frl. Schiller, M.
86. Troller, Th.
87. Körner, K.
88. Friedrichs, K.
89.
90. van der Hegge Zijnen, B. G.

Fig. 2. Participants of the first international conference on Fluid Dynamics, organized by von Kármán in Aachen, 1929.

for their work done prior to 1929, or were to receive international recognition for the work they did in the years that followed. Courant, Friedrichs (Nr. 88 in the back), and Levy had just published the domain-of-dependence condition, and Prandtl (Nr. 2) and Busemann (Nr. 68) also published their graphical method of characteristics, in 1929.

Since the 70th birthday of Professor Maurice Holt seems to be an appropriate occasion, a brief description of work on aerodynamic problems carried out at the Aerodynamisches Institut in the 75 years of its existence will be given in the following. The discription is purposely restricted to works in aerodynamics because of lack of space.

2. EARLY WORK

Before von Kármán came to Aachen in 1913, Professor Hans Reissner occupied the chair for mechanics from 1906 to 1913, following Professor A. Ritter (until 1899) and Professor A. Sommerfeld from 1900 to 1906. Encouraged by Professor Junkers he was the first to set up aerodyanamic experiments in Aachen. He built the first Aachen wind-tunnel, depicted in the photograph of Fig. 3. It was an Eiffel-type tunnel, later to be changed into a Göttingen-type tunnel.

Fig. 3. First wind-tunnel in Aachen, built by Professor Hans Reissner.

Professor Reissner, who first was mainly interested in problems of stability and control of aeroplanes, became more involved in problems of design and construction of aeroplanes at later times [1]. In 1908 he built the first monoplane, with which successful flights over a distance of several hundred meters were

Fig. 4. Wing tests for the "Reissner-canard", an all-metal aeroplane built by Professor Hans Reissner in Aachen 1912.

Fig. 5. The "Reissner-canard", one of the first all-metal aeroplanes with selfsupporting wings, built by Professor Hans Reissner in Aachen 1912, completed several successful flights.

completed in a height of four to six meters. The first flights took place April 12, 1909 in the vicinity of Aachen. In 1912 he built another monoplane, which later became known as the "Reissner-canard". Also this aeroplane made several successful flights near Aachen. It is worthwhile to point out that the aeroplanes Professor Reissner built, were already all-metal aeroplanes, with selfsustaining wings made out of corrugated sheet metal. The picture in Fig. 4 shows one of the wings on a test rig in 1912, and Fig. 5 shows the "Reissner-canard" with wing and engine mounted in the rear of the pilot's seat.

Before Professor Reissner left Aachen in 1913, he together with Professor Junkers had persuaded the government, to establish an "Aerodynamisches Institut" at the Technical University Aachen. Von Kármán became its first director and professor for mechanics and aerodynamics on April, 1, 1913. It did not take long, until von Kármán's genius began to show impact on the work in the newly founded institute. He soon rebuilt the wind-tunnel to a Göttingen-type tunnel and began with theoretical investigations. When Junkers, who also had left Aachen, asked him to develop a mathematical method with which abritrary wing sections could be designed, von Kármán and Trefftz found a solution to the problem soon, which, indeed, could be applied to realistic wing sections. Due to the events of World War I, the paper could only be published in 1918 [2]. Nowadays it is referred to as the Kármán-Trefftz method.

The years that followed were to become one of the most fruitful scientific periods of the institute [3]. Since engine powered flight was prohibited by the Treaty of Versailles, gliding became the subject of study and research. The students of aeronautics in Aachen formed a society for flight sciences, which was strongly supported by von Kármán. In 1920 Dr. Klemperer, a member of this group, flew world record with a glider built in the institute. The actual profit of this work followed much later: Through the demand for the development of high-performance gliders, aero-dynamics, structures and meteorology were continuously supported in the course of time, and present-day gliders are aeroplanes with unmatched miraculous aerodynamic characteristics.

The "Abhandlungen", a series of collected papers, published by members of the Aerodynamisches Institut, bare witness of the work in those days. Karl Pohlhausen published his approximate integration of the differential equation of the boundary layer in 1922; W. Klemperer wrote an article on slotted wings at about the same time. He published his theory of gliding four years later in 1926. In the following year, J. Lennartz wrote a theoretical treatise on wing-fuselage interactions. K. O. Friederichs, at that time a member of the institute, coauthored von Kármán in an article on the computation of cantilever wings in 1929. A theory on propellers published by F. Misztal in 1932 was followed by numerous papers on wind-tunnel and measuring techniques.

Professor von Kármán left Aachen eventually in 1934. His successor became Professor Carl Wieselsberger, in December 1931. In addition to heading the institute he was offered the chair for fluid mechanics and applied mathematics.

3. THE TIME IN BETWEEN

Professor Wieselsberger worked with Prandtl from 1912 to 1922 and followed von Kármán to Japan, where he helped to build wind-tunnels for the Japanese industry. Experimental work related to aeronautics was his main interest when he came to Aachen. Here he installed a six-component balance, which he had developed in Japan. The balance is still in use. Although initially research was mainly concentrated on improvement of propellers, Wieselsberger soon turned to high-speed aerodynamics. Immeditely after taking over the institute, he built the first supersonic wind-tunnel in Aachen. The test section had a cross-section of $10x10cm^2$. The contour of the nozzle was determined with the graphical method of characteristics mentioned earlier. In order to guarantee a sufficiently smooth surface, the nozzle was coated with calcium sulphate. It is reported, that accurate contours could only be obtained, if the drying velocity was properly chosen. The tunnel is described in Ref. [4]. It was soon found out, that an x-shaped shock system caused by condensation appeared in the nozzle. As initially the reason for the appearance of the shock system was not known, it was referred to as "ghost". The condensation shocks were systematically investigated by Dr. R. Hermann, at that time also a member of the institute. After World War II he continued his studies in the United States.

In those days measuring techniques for pressures and aerodynamic forces in compressible flows were still in their infancy. Professor Wieselsberger made many contributions to this vitally important subject. He soon built a second supersonic wind-tunnel with a cross section of $20x20cm^2$, which in the years to come, was used to study aerodynamic characteristics of projectiles at supersonic speeds. For example, the effectiveness of the stabilizers of the V2 was investigated in this tunnel. It is also noteworthy, that the supersonic wind-tunnels, built in Peenemünde, were designed after the concept of the Aachen tunnel. The problem of wall interference, which did not find a satisfying solution until now, was already discussed by Wieselsberger in 1939, [5].

When World War II broke out in September 1939, most of the experimental facilities of the institute had to be dismantled within hours and were moved to Göttingen, where the most important ones were installed in the Aerodynamic Experimental Establishment (AVA). It was here, where Professor R. Sauer, also a member of the institute, gave a series of lectures on gasdynamics, which in 1943 was published by the Springer Verlag under the title "Introduction to Theoretical Gasdynamics". The book was later to become one of the classics in the field.

Professor Wieselsberger passed away unexpectedly in April 1941. Professor Sauer became acting director of the institute. He was allowed to move it back to Aachen in the same year.

One year later, Professor F. Seewald, a von Kármán student and former head of the German Experimental Establishment for Aeronautics (DVL), became director of the institute and professor for applied mathematics and fluid mechanics. Because of the events of the war, not much work could be done, and in the winter of 1943/44 the experimental facilities of the institute were moved to Sonthofen in the Alpes.

4. BEGIN AGAIN

The institute remained in Sonthofen until 1947, when it finally returned to Aachen. Aerodynamic research was not permitted in the following years. But in the early fifties, Professor Seewald could persuade the government agencies, that aeronautic research was too important to be neglected, and he was able to found the German Experimental Establishment for Aeronautics (DVL) anew, this time not in Berlin, but in Aachen. The first supersonic wind-tunnel with a cross-section of $6x8cm^2$ was completed in 1953, and a low-speed Göttingen-type wind-tunnel with a test section diameter of 120 cm in 1955. In the same year a DVL-Institute for Applied Gasdynamics was founded. It was headed by Professor A. Naumann, who also became the director of the Aerodynamisches Institut, when Professor Seewald retired in 1963.

Naumann threw the whole thrust of his work into the construction of new supersonic wind-tunnels, the first was a blow down tunnel, which was completed within three years. Together with his colleague Professor A. Heyser he equipped the tunnels with new measuring techniques with particular emphasis on investigations of unsteady flows. For example, in 1965, Naumann was able to demonstrate shock oscillations on wing sections in transonic flow. Amplitudes of the order of 20 to 30 percent chord and noticeable phase differences between lower and upper side of the profile were detected with Mach-Zehnder interferometry in conjunction with high-speed photography.

An explosion-like development followed. Plans were drawn up to build a large experimental research center near Cologne. In 1966 the DVL-Institute for Applied Gas-dynamics was moved to Cologne. It is now part of the DFVLR-research center. It has wind-tunnels covering the whole speed range, and the cryogenic European Transonic Wind-Tunnel is going to be built in this center in the near future.

In Aachen, the Aerodynamisches Institut, under Naumann's guidance and direction, embarked on research in fluid dynamics and thermodynamics, concentrating on unsteady flow phenomena, related to aerodynamics: shock-vortex interaction, generation of vortices through shocks, and locally unsteady transonic flow about wing sections. From 1970 these investigations were sponsored in a long term research program by the German Research Society.

Professor Naumann retired in 1973. In addition to building up modern wind-tunnel techniques in Germany after World War II, he was one of the leading scientists to establish research in bio-fluid dynamics in Aachen.

Today, the Aerodynamisches Institut of the RWTH Aachen is an active university research center in fluid dynamics. After completion of its new building in 1980, it is engaged in aerodynamic research and in research of other branches of fluid dynamics with experimental, theoretical and numerical techniques. With a transonic-supersonic wind-tunnel with a $40 \times 40 cm^2$ cross section, flows up to Mach numbers of about $Ma_\infty = 3.5$ can be investigated. A new enthusiastic generation is pursuing research on flows about delta wings, vortex breakdown, unsteady flows about wing sections, flow separation on wings, and complex two- and three-dimensional turbulent shear flows in the spirit of the generations that passed. In order to give an example of recent investigations, some results of aerodynamic studies are shown in the Figs. 6-8. Fig. 6 shows a Mach-Zehnder interferogram of the flow about a CAST-7 profile, taken in the transonic wind-tunnel of the Aerodynamisches Institut at a freestream Mach number of $Ma_\infty = 0.76$ by T. Franke. The formation of a shock, visible in the interferogram on the upper side of the wing is also reflected in a comparison calculation, the results of which are shown in Fig. 7. Therein lines of constant density as obtained from a numerical solution of the Navier-Stokes equations, recently deviced by D. Hänel and M. Breuer of the Aerodynamisches Institut are shown for almost the same flow conditions indicated in Fig. 6. Fig. 8 shows a result of one of the most recent investigations: computation of hypersonic viscous flows about space-planes. Shown are the oil-flow patterns computed with a numerical solution of Navier-Stokes equations for a two-ellipsoid-configuration at an angle of attack of $\alpha = 30°$, the Mach number is $Ma_\infty = 8.15$, and the nominal Reynolds number is $Re_\infty = 2 \cdot 10^6$. The results were provided by R. Schwane of the Aerodynamisches Institut.

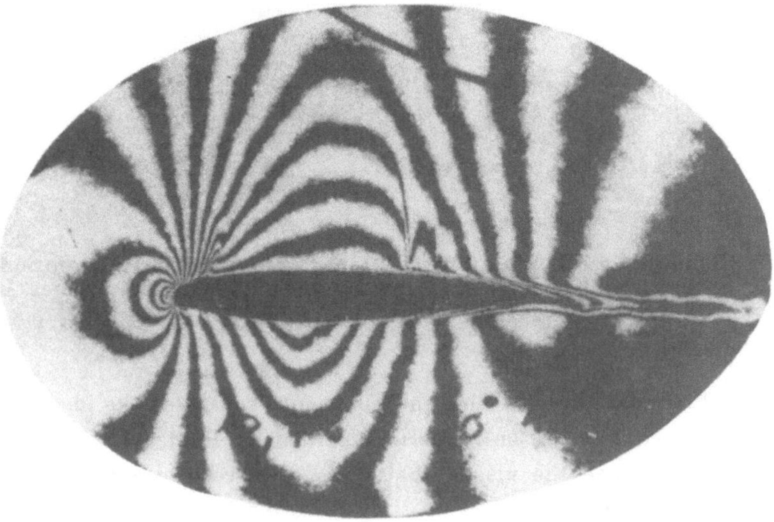

Fig. 6. Mach-Zehnder interferogram of transonic flow about the CAST-7 profile. $Ma_\infty = 0.76$; angle of attack $\alpha = 0°10'$ [6].

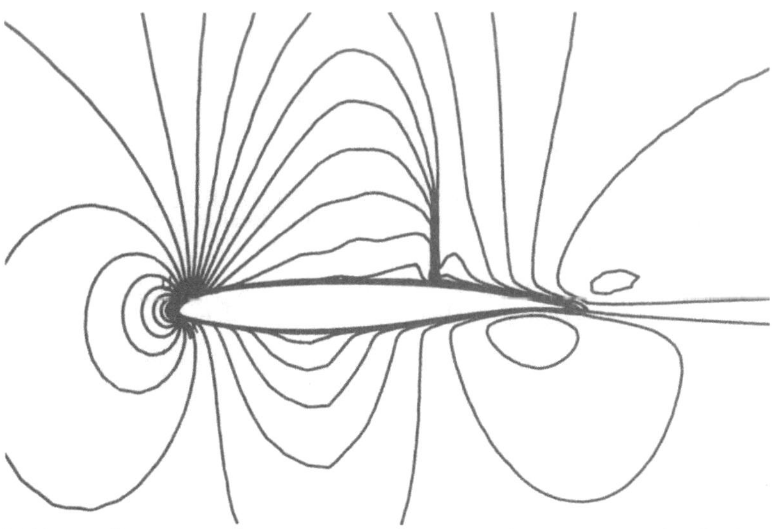

Fig. 7. Lines of constant density as computed with a numerical solution of the Navier-Stokes equations by D. Hänel and M. Breuer of the Aerodynamisches Institut for approximately the same flow conditions as given in Fig. 8.

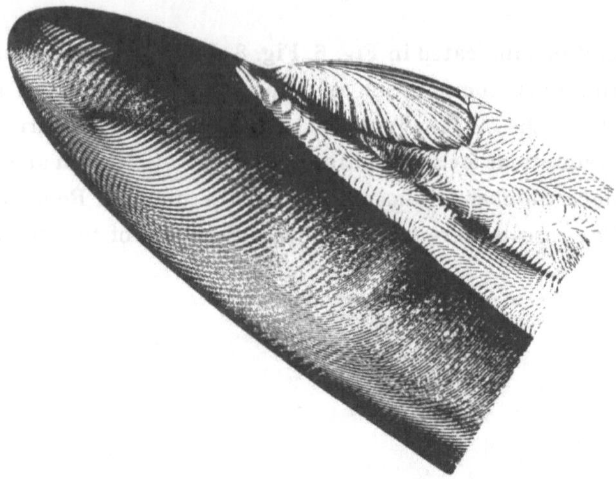

Fig. 8. Computed oil-flow patterns, obtained with a numerical solution of the Navier-Stokes equations by R. Schwane of the Aerodynamisches Institut. $Ma_\infty = 8.15$, $Re_\infty = 2 \cdot 10^6$, $\alpha = 30°$.

When Professor Maurice Holt and his friends conceived the idea of holding international conferences on numerical methods in fluid dynamics years ago, they must have had the vision that flows of extreme complexity some day eventually could be predicted.

At caelum certe patet. ibimus illac. (Ovid, Metam VIII, 186)

5. REFERENCES

1. A. Naumann: 50 Jahre Aerodynamisches Institut der Rheinisch-Westfälischen Technischen Hochschule Aachen. Abhandl. Aerodyn. Inst. RWTH Aachen, 1963.

2. Th. von Kármán und E. Trefftz: Potentialströmung um gegebene Tragflächenquerschnitte. Zeitschrift für Flugtechnik und Motorluftfahrt, Vol. 9, 1918.

3. E. Krause und W. Limberg: Theodore von Kármán - 16 Jahre Forschung am Aerodynamischen Institut. RWTH Themen, Vol. 2, pp. 12-15, 1981.

4. C. Wieselsberger: Die Überschallanlage des Aerodynamischen Instituts der Technischen Hochschule Aachen. Luftwissen 4, p. 301, 1931.

5. C. Wieselsberger: Über den Einfluß der Windkanalbegrenzung auf den Widerstand im Bereich der kompressiblen Strömung. FB 1172, 1939; LuFo 19, p. 124, 1942.

6. T. Franke: Experimentelle Untersuchung der Wandeinflüsse an schallnah angeströmten Tragflügelprofilen. Abhandl. Aerodyn. Inst. RWTH Aachen, Heft 28, pp. 12-18, 1988.

Progress and Future Directions in Computational Fluid Dynamics

Paul Kutler

1. ABSTRACT

Computational fluid dynamics (CFD) has made great strides in the detailed simulation of complex fluid flows, including the fluid physics of flows heretofore not understood. It is now being routinely applied to some rather complicated problems, and starting to impact the design cycle of aerospace flight vehicles and their components. In addition, it is being used to complement, and is being complemented by, experimental studies. In the present paper, some major elements of contemporary CFD research, such as code validation, turbulence physics, and hypersonic flows are discussed, along with a review of the principal pacing items that currently govern CFD. Several examples of pioneering CFD research are presented to illustrate the current state of the art. Finally, prospects for the future development and application of CFD are suggested.

2. INTRODUCTION

Fifteen years ago, the discipline of computational fluid dynamics (CFD) was in its infancy, and the foundation was being laid to establish it as an effective research, analysis and design tool of the future. The CFD has now moved into a phase in its life cycle where its potential is beginning to be realized. In the early stages the facets that encompass the discipline of CFD, such as algorithm development, grid generation, geometry definition, boundary and initial conditions, turbulence modeling, pre- and post-data processing, and computer technology were all ripe for technological advances. Progress in the discipline at its current stage are incremental, but still happening. Because of the demonstrated potential of CFD, the understood limitations of experimental and flight testing, and the rapid advances in computer hardware, the discipline of CFD has developed at an explosive pace.

The overarching goal of the discipline of CFD at NASA is to simulate in a reasonable amount of time the realistic flow about actual aerospace vehicles and their components at true flight conditions. A reasonable amount of time is, of course, dependent upon the need for that information but is generally accepted as less than an hour and ideally approximately 15 minutes on a state-of-the-art computer system. To attain this goal requires computer systems such as the Numerical Aerodynamics Simulation (NAS) Cray-2 with its large memory and fast clock time as well as accurate and robust CFD software. In the next section of this paper, current state-of-the art CFD software and demonstration examples accomplished over the past few years at NASA Ames will be described.

3. CFD DEMONSTRATION EXAMPLES

There are three major uses of CFD that have resulted in its widespread acceptance as an alternate aerodynamic prediction tool. They are to 1) predict fundamental fluid physics, 2) complement and supplement experimental testing, and 3) aid the aerospace vehicle design process. Examples are presented in this section that demonstrate the use of CFD in these three areas.

3.1 Predicting Fluid Phenomena

3.1.1 Vortex Breakdown

With the development of modern high-performance aircraft, including those with highly swept leading edges, has come additional flow complexities including the problem of control of the leading edge vortex. For highly swept wings this vortex plays a crucial role in the development and control of lift.

Extensive study of the vortical flow over a double delta wing has been carried out at the NASA Ames Research Center by FUJII [1] on the NAS Cray-2 by taking advantage of its large memory. The numerical simulation solves either the "thin-layer" Navier-Stokes or Euler equations. A typical result is shown in Fig. 1. In this figure, off-surface particle-path traces clearly demonstrate the breakdown of the primary vortex. The fine grid resolution provided by the large memory of the Cray-2 permitted an accurate description of this physical phenomenon. Note that the vortex from the strake region remains orderly and tightly coiled until about 80 to 85% of the chord, where an abrupt change of the flow is observed. The particle traces also indicate the existence of streamwise reverse flow along the core, a clear indication of vortex breakdown. This simulation capability will play an important part in the design and analysis of vortical flows for a wide range of aerospace applications.

Figure 1. Particle traces over a strake-delta wing demonstrating vortex breakdown.

3.1.2 Turbulent Boundary Layer

In recent years considerable progress has been made using the NAS system for the numerical simulation of the physics of turbulence using spectral methods to solve the complete, time-dependent Navier-Stokes equations without small-scale modeling. It is possible, from a high resolution workstation, to view the development and propagation of the numerically predicted turbulence quantities. Such a result, generated by ROBINSON [2] using 9.4 million grid points, is shown in Fig. 2 and represents the structure of turbulence in a simulated flat-plate boundary layer. The figure was originally developed in color, wherein each separate element of the boundary layer structure was shown in a different color. In that figure, the elongated white surfaces identified the low-pressure vortex cores, the red and blue areas identified where significant contributions to the Reynolds shear stress were occurring, red indicated regions of low-speed fluid being ejected outwards, and blue denoted regions of high-speed fluid being swept towards the wall. From the present figure, one can obtain a general idea of the computed results. These kinds of presentations help the scientist to more clearly understand the role that each of the physical variables play in the onset of turbulence, and can significantly aid in the development of statistical models to represent such turbulent flows in Reynolds averaged, Navier-Stokes codes.

3.1.3 Jet-Induced Lift Loss on Delta-Wing

With the advent of modern V/STOL aircraft, such as the British Harrier, a great deal of interest has been generated in the fluid dynamics of powered lift. To study

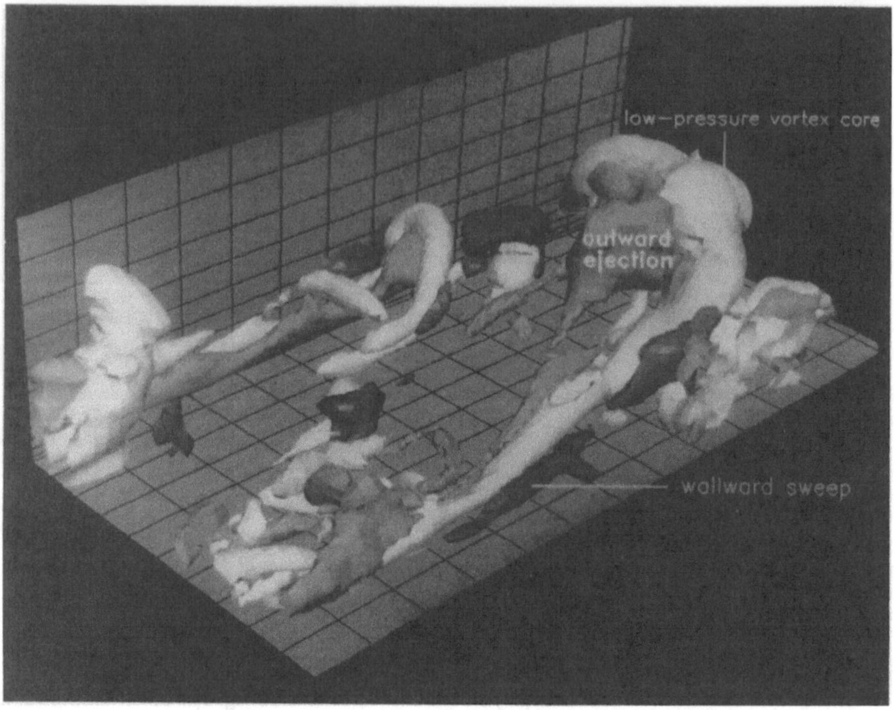

Figure 2. Numerical simulation of a turbulent boundary layer on a flat plate.

the fundamental mechanisms of such flows, RAO and VAN DALSEM [3] who solved the three-dimensional, unsteady, Reynolds-averaged, Navier-Stokes equations, simulated the effects of two jets emanating from a 60° delta wing in a cross-flow. The results are shown in Fig. 3 for a selected flow condition where the jet velocity is 15 times that of the free stream and the height of the wing above the ground is 0.25 times the span. Using a high-resolution workstation, it is possible to visualize the otherwise complex flow field in great detail. The vortex generated by the interaction of the jet with the cross flow and the ground plane results in a locally decreased pressure, causing a loss of lift on the wing. In effect the delta wing sees a negative angle of attack and is "sucked down" to the ground.

3.2 Complementing and Supplementing Experimentation

3.2.1 Ogive Cylinder at Large Incidence

Modern aircraft, with their characteristically long, slender nose shapes, and operation at high angles of attack, develop unusual flow patterns and forces which frequently affect controllability. As a result, an experimental research effort by

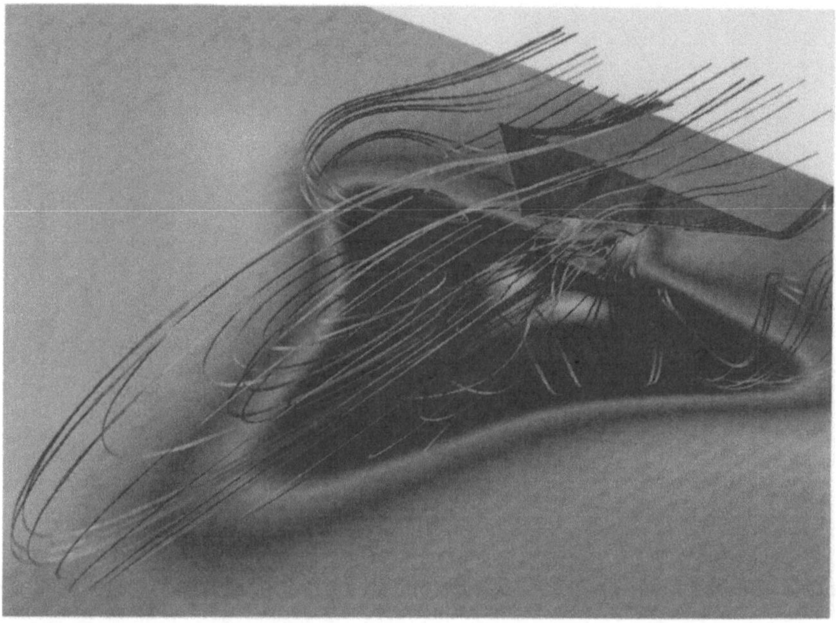

Figure 3. Particle traces generated by jets emanating from a delta wing in a cross flow.

DEGANI and ZILLIAC [4] was undertaken to study the unsteadiness of the flow around an ogive cylinder (L/D =12) at 40° incidence. This study was prompted by computations using a three-dimensional, time-dependent thin-layer Navier-Stokes code in which an unsteady behavior of the flow on the leeside was discovered that produced time-varying normal forces. Closer examination of the solution showed what appeared to be shear-layer vortices being shed into the leeside vortex flow field at a frequency of 1200 hz. Subsequent experiments in a low-turbulence wind tunnel confirmed the flow-field unsteadiness. Figure 4 shows a smoke-laser sheet visualization of the leeside flow field. High-speed movies of the ripples in the free shear-layer show that they are shed at frequencies near the frequency of the computation. Questions about the origin and structure of the unsteady flow field are presently being investigated in further detail experimentally using hotwire probes and surface-mounted pressure transducers and computationally using several Navier-Stokes codes.

3.2.2 Transonic Wing-Body Configuration

An example in which CFD played a major role in many facets of an experimental research program is the transonic wing-body configuration experiment of Olsen in the Ames Research Center high-Reynolds-number-II facility. As part of the experimental research program to provide data to validate the Ames Research

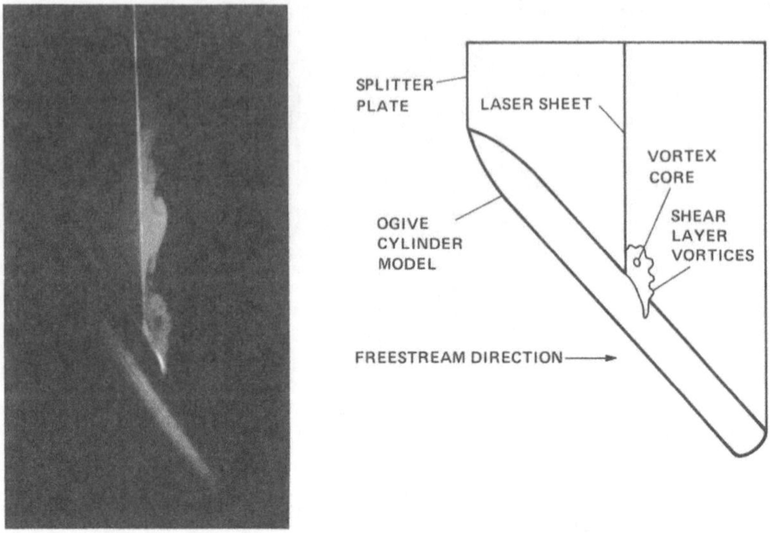

Figure 4. Smoke laser sheet on leeside of ogive cylinder at large angle of attack.

Center TNS code (a three-dimensional, zonal, Reynolds-averaged, Navier-Stokes program) and investigate the flow about a modern fighter-like, wing-body combination, three separate CFD codes were used. First, a panel method code was used to evaluate the potential effects of wind-tunnel wall interference, predict the onset of supercritical flow, and evaluate subsonic model loads. Second, an axisymmetric transonic full-potential code was used to redesign the nose of the body to eliminate body-tunnel shock interference and to specify the location of pressure taps to resolve the body flow field. Third, a transonic full-potential wing code was employed to ensure the presence of supercritical flow on the wing for the Mach numbers to be investigated, to guide the placement of pressure taps to resolve the wing flow field, and to predict the transonic model loads.

The model is shown in Fig. 5, and the experimental results, and comparisons with numerical simulations, for the wing alone are presented in Fig. 6. Future experimental work will include testing the complete wing-body configuration; future computational work will include grid-refinement studies to determine solution independence.

3.2.3 Turbulence Modeling

Current computer power limitations preclude solution of the complete Navier-Stokes equations for flow about realistic configurations; thus, it is necessary to solve a reduced equation set. The Reynolds-averaged, Navier-Stokes equations,

MODEL INSTALLATION

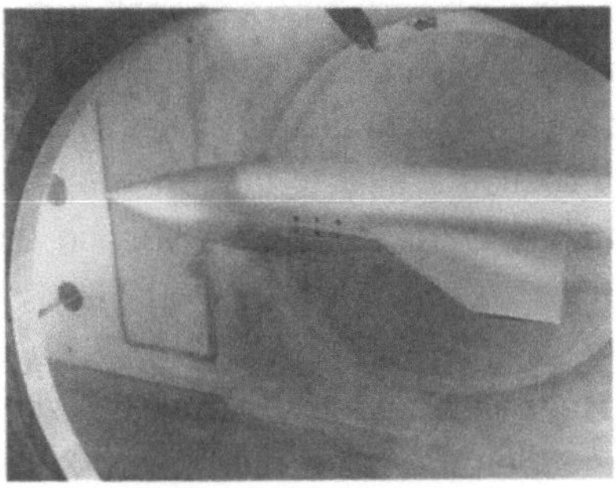

MODEL
BODY: OGIVE CYLINDER
WING: NACA 64A008
L.E. SWEEP ANGLE = 37°
Æ = 3

TEST CONDITIONS
SOLID TUNNEL WALLS
M_{∞} = 0.5 TO 0.82
$Re_c = 1 \times 10^6$ TO 10×10^6
α = 0 TO 15°

MEASUREMENTS
MODEL: p, OIL FLOW
TUNNEL WALLS: p, OIL FLOW, BOUNDARY LAYER THICKNESS (MOUNTING WALL)
FLOW FIELD: u, v USING LDA

Figure 5. Transonic wing-body combination mounted in the Ames Research Center high Reynolds number facility.

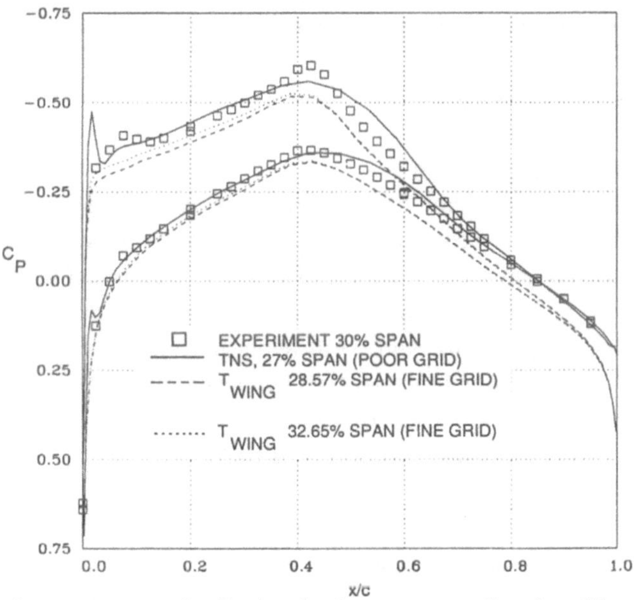

Figure 6. Surface pressure distribution for low-aspect ratio wing (M = 0.8, α = 1.9, Re = 9 million).

however, require a turbulence model for closure. The turbulence model emulates all scales of turbulence generated in the thin viscous region near the surface of the configuration. Thus turbulence modeling is an important element in the numerical simulation of many flow problems. The development of turbulence models requires a strong interaction between computationalists and experimentalists.

An example of a building-block experiment performed by BACHALO and JOHNSON [5] for extracting turbulence data, and how it was used in conjunction with computational fluid dynamics to guide the development of a "zonal" turbulence model for transonic shock-wave/boundary-layer interactions is shown in Fig. 7. The test model consisted of a circular body fitted with a circular-arc section. Transonic flow developed over the circular arc section similar to that on an airfoil, and shock-wave interactions of varying strengths were studied by varying the free-stream Mach number.

Computations of the flow field using a Reynolds-averaged, Navier-Stokes code revealed deficiencies in the turbulence model. It used a model developed primarily for attached boundary layers, and the shock wave location was predicted incorrectly. As a result, the pressure recovery was seriously overpredicted. The experimental mean and turbulence profile data were used to explain the differences and guide the development of an improved turbulence model. Using new modeling concepts in conjunction with the experimentally obtained turbulence data, a significant improvement in the results was observed.

3.3 Aiding the Design Process

3.3.1 National Aerospace Plane

The United States has as one of its aeronautical goals to pursue the research and development of the technology necessary to permit an aerospace vehicle to routinely cruise into and out of the atmosphere with takeoff and landing from conventional runways. This effort is called the National Aerospace Plane program, and will require a hypersonic vehicle to attain the speed of Mach 25 to reach Earth orbital velocity. Computational fluid dynamics plays a key role in this program because of the lack of ground-based experimental facilities to simulate the flow about the vehicle along some parts of its trajectory.

Hypersonic flows, although incorporating the same advantages as supersonic flows (i.e, a limited domain of dependence) contain additional complications that are yet to be fully evaluated. These include more complicated boundary-layer transition and turbulence physics, real-gas effects, merged shock and boundary layers, stronger shock waves, and radiation effects.

Computer software is currently under development to simulate the flow about the aerospace plane. The governing equations are the three-dimensional

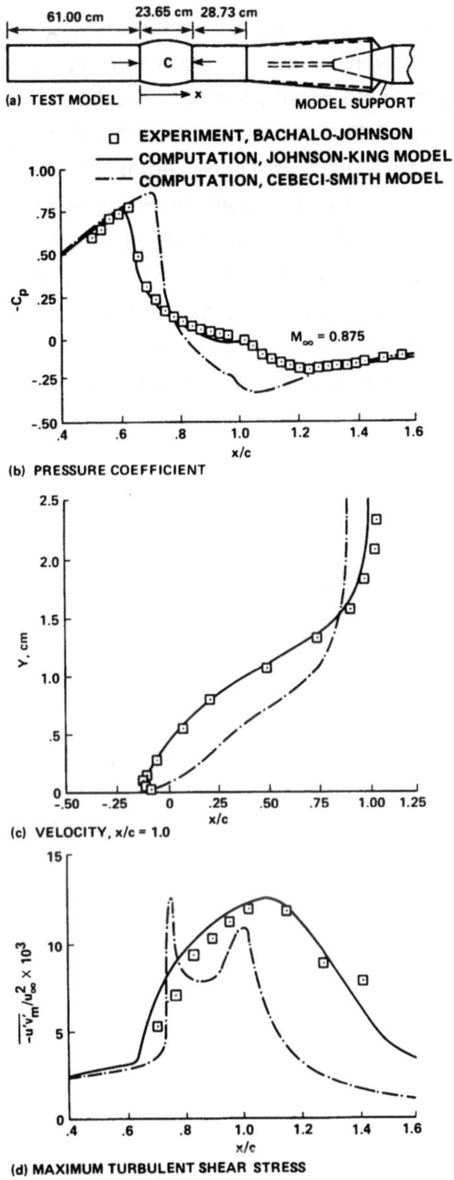

Figure 7. Shock-wave, boundary-layer interaction.

Reynolds-averaged, Navier-Stokes. In one approach the unsteady version of the equations are solved; in another the parabolized equations (PNS) are solved .

A typical solution about a generic hypersonic configuration using the PNS approach developed by LAWRENCE [6] is shown in Fig. 8. The flow conditions consist of Mach 12.4 and angle of attack of 0°. The figure was originally developed in color, wherein a gradation of pressure was depicted by a color scale. The red color corresponded to a high-pressure region and the blue color corresponded to a low-pressure region. The black lines superimposed on the surface of the vehicle represent streamlines. The shock structure at the aft end of the vehicle was shown by the color band in that vicinity. A high-pressure region on the concave undersurface exists in the vicinity of the inlet. This externally compressed flow enters the engine modules where it is mixed with fuel and burned. Comparisons of these numerically generated solutions with experimental data show excellent agreement.

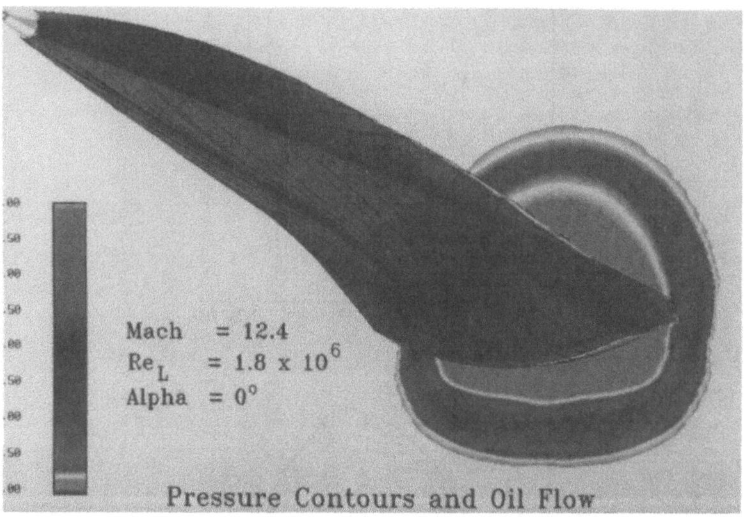

Figure 8. Surface pressure contours and particle streamlines for generic hypersonic vehicle.

3.3.2 Space Shuttle Main Engine

Operating the Space Shuttle Main Engine (SSME) on the test stand at power settings greater than 104% results in flow problems with components of the liquid hydrogen hot gas manifold. The flow problems encountered within the SSME, and the subsequent computational analysis and redesign have served as an indication of the practicality of modern CFD technology to solve "real-world" problems in an effective manner. Early experience with the SSME indicated that there were problems of reliability and thrust performance, and a large computational simulation research program was undertaken at Ames Research Center by Ames scientists and Rocketdyne scientists (the manufacturer of the engine).

A computer code was developed that could treat the complex internal geometry of the SSME powerhead. KWAK et al. [7] developed a three-dimensional incompressible Navier-Stokes code (INS3D) for this application that provided an extensive low-speed simulation capability. Working with engineers from Rocketdyne Corp., this code was applied to simulate the flow within the hot-gas manifold, transfer ducts, and the main injector. Results from this code indicated that the flow within the present three-duct design was inefficient. Large, separated flow regions were exposed by the calculations within some of the components, and it was found that the center transfer duct transmitted only 9% of the total flow.

Calculations and cold flow experiments on a proposed two-duct design showed significantly improved flow characteristics. Figure 9 shows the surface-pressure map for the redesigned powerhead arrangement. As a result of this computational demonstration, Rocketdyne has determined that CFD simulation will be utilized to develop and analyze all future SSME designs.

3.3.3 Space Shuttle Fast Separation

The computation performed by BUNING et al. [8] for the time-dependent flow over the integrated Space Shuttle launch configuration consisting of the Space Shuttle, the external tank, and the solid rocket boosters, is a good example of a contemporary CFD simulation problem. In this case, the numerical simulation is being used to analyze flight performance, and the feasibility of Space Shuttle separation and crew escape during launch. A solution was obtained by solving the unsteady three-dimensional Reynolds-averaged, Navier-Stokes equations using a composite grid and an algebraic turbulence model. A comparison of computed and experimental results are shown in Fig. 10. Although the connecting support structure between the orbiter and the external tank was not modeled, the computed results compare well with the experimental values. Future computational work will extend the simulation to include staging, where the orbiter separates from the solid rocket booster and external tank assembly.

4. ELEMENTS PACING CFD

There are two primary elements that are pacing the discipline of CFD: physical modeling and computer systems [9].

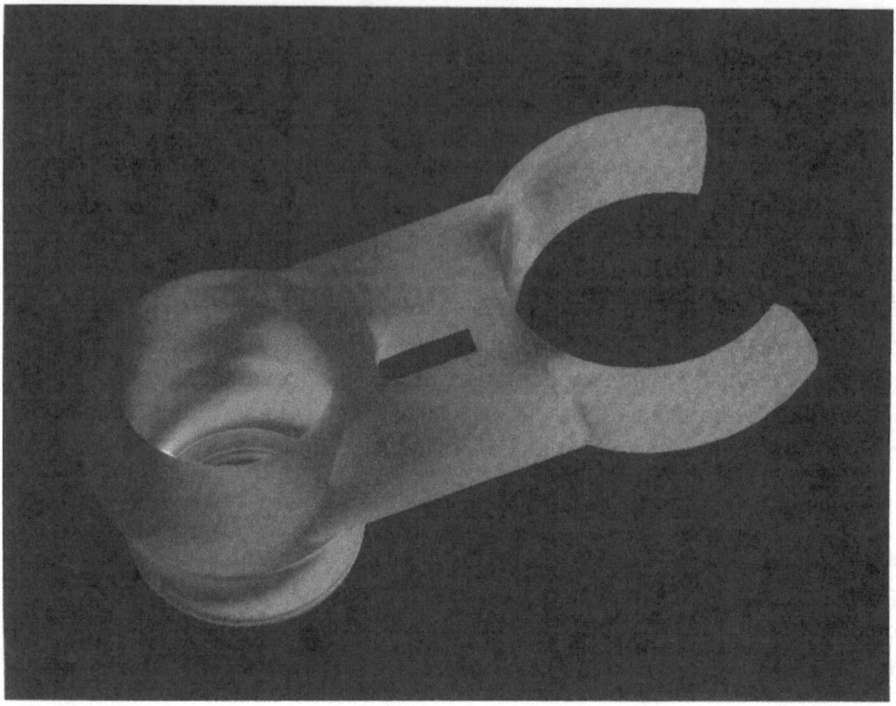

Figure 9. Pressure map of two-transfer duct design of Space Shuttle-Main-Engine powerhead.

With the interest in hypersonic flight, new challenges are facing the CFD scientist [10]. These include developing flow-field simulation codes that account for real-gas behavior. Associated with that is the need for turbulence models for these hypersonic flows. In addition, because hypersonic vehicles cruise at high altitudes, it is important to understand boundary-layer transition and be able to correctly account for it in the simulation tools.

A necessary step, of course, in the development process of any new computer code is validation. Therefore, either experiments or flight tests that can providedata in the real-gas flight regime must be performed to validate or calibrate newly developed hypersonic flow analysis computer codes.

For hypersonic simulations, it is necessary to know the correct rate constants used in the finite-rate chemistry equations that are coupled with the fluid flow equations. In the past, these constants were determined experimentally. However, with the advent of large computer systems and advances in chemical

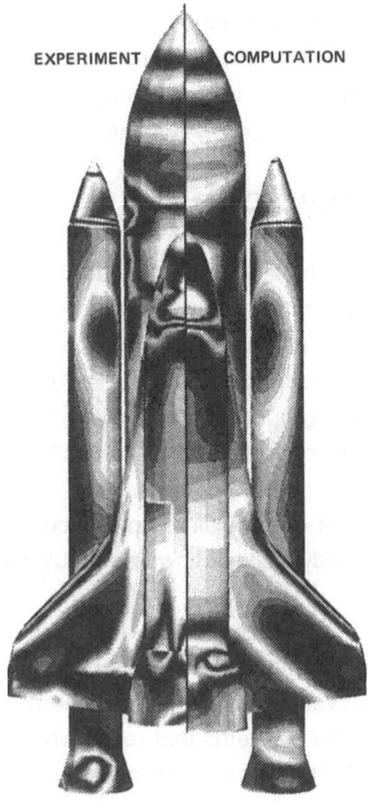

Figure 10. Computed and experimental surface pressure data for Space Shuttle, external tank, and solid rocket boosters.

reaction modeling tools, the discipline of computational chemistry is providing accurate rate constants for use in CFD simulation codes.

For the other speed regimes, turbulence modeling remains the pacing element in CFD flow simulations. Until computer systems are large enough to perform direct or at least large-eddy simulations on realistic configurations at flight Reynolds numbers, accurate turbulence models will be required. There currently exists no universal turbulence model; therefore, nearly each peculiar flow requires its own model. As such, this places an enormous burden on the experimentalist to obtain the necessary data to develop the appropriate turbulence model for a particular flow.

The CFD scientist as well as many other simulation scientists are limited by what they can do numerically on existing computer systems. They are limited by the storage and speed available on those systems. Other associated computer limitations include network bandwidths for transmission of data; high-speed, mass-storage devices for archiving the large data bases generated; and post-processing software and hardware for analyzing the massive data bases generated from unsteady, three-dimensional solutions.

5. FUTURE DIRECTIONS FOR CFD

A considerable number of challenging technical problems exist for which CFD will be beneficial and sometimes mandatory [11]. In the future, CFD will be applied to several interesting and challenging problems that will push the state of the art in the discipline. Typical problems include flows about 1) powered lift aircraft operating in and out of ground effect; 2) hypersonic vehicles including inlet, engine, and exhaust; 3) rotorcraft in hover, transition, and forward flight; and 4) turbomachinery including pumps, compressors, and turbines. To meet these future challenges, two areas of particular importance must be addressed. These are the simulation of unsteady viscous flows and multidisciplinary simulation.

The simulation of unsteady viscous flows about realistic aircraft configurations is now possible using computational tools. Several computer codes for simulating these flows have been developed by various researchers. It is now possible to begin studying unsteady flow problems that result when high performance aircraft fly at high angles of attack. It is possible to model complex phenomenon such as asymmetric vortex shedding and vortex breakdown or bursting. It is also possible to model flows about STOVL ARC during dynamic landings, with the goal of predicting A/C behavior. The goal is to develop these codes to enable, for example, the prediction of aircraft performance near their performance boundaries.

Equally challenging is the ability to integrate several physical models into a single-computer simulation program. For example, a typical hypersonic application would include the coupling of the finite-rate chemistry equations with the Navier-Stokes equations. In the past, simple couplings have occurred such as linking a structural response code to a fluid dynamics code to study aeroelastic problems or coupling a flow code with an optimization routine for wing design. The future holds promise for the birth of a new and more computationally intensive discipline called Computational Aeroscience. With more complete coupling, this new discipline will allow the simulation of entire aerospace vehicle systems involving the complex integration of fluid dynamics, propulsion, structural mechanics, thermo-gas dynamics, electromagnetics, and controls.

A major pacing item in meeting these challenges is computer system speed and the ability to use this speed. The achievement of practical unsteady, fully viscous simulations and multidisciplinary aeroscience simulations will require approximately 1000 times more computing power than is now available. Furthermore, future simulations and computer systems are becoming so complex that new more sophisticated software tools must be made available if the full utility of these powerful systems are to be realized. Two promising technologies that can be applied to meet these challenges are massively parallel processors and artificial intelligence (AI) systems.

Most of the CFD development to date has taken place on single processor computers. With advances in clock speed and memory size, brought about largely by more efficient chips, the overall processing speed and power has increased by orders of magnitude during the past 10 yr. The physical limits of such development are now being approached, caused by inherent electron speeds and problems of cooling [12]. Although new opportunities may be provided by the development of practical means of achieving near-room temperature superconductivity, the direction most promising in the future for satisfying the insatiable thirst of scientists for CPU cycles is parallel processing.

Most of the current generation of supercomputers have more than one processor, up to eight in the case of the recently announced Cray YMP. Conventional supercomputer technology is likely to increase in parallelism at a moderate rate with steady factors of 3 or 4 improvement in performance. The conventional designs are being challenged, however, by massively parallel architectures offering thousands of processors and the potential for orders-of-magnitude improvements in the near term. The impediment to the rapid application of massively parallel computers to CFD problems is the limited amount of system software currently available. How to formulate the equations of CFD into algorithms suitable for solution on a multi-thousand processor machine, as well as how to manage the data internally among the processors are first order research problems at the present time. Progress is being made in these areas.

At Ames Research Center, the Navier-Stokes equations are being solved on Thinking Machine's 16,000 processor Connection Machine computer (It is designed for 64,000 processors) installed in the NAS system. Great strides remain to be made in this area toward the goal of a new generation of supercomputers able to handle problems that are now not amenable to practical solution.

A fully productive scientific-user interface, needed to meet the dual complexities of massively parallel systems and complex simulations, does not exist and must be developed. What is needed is a computational-fluid-dynamicist's workbench containing graphical, numerical and symbolic computing capabilities that can be integrated into the NAS-like environments. The techniques and potential capabilities of AI offer a promising direction for the development of the workbench and CFD applications. Expert systems are already being explored to help the CFD scientist divide a complex flow region into several zones, each of which can then be discretized using relatively simple grids. At the other end of the numerical simulation process, visualizing the computed results, AI will be used to investigate and analyze the resulting flow patterns, and identify and display those deemed to be of most interest. Such tools will provide the aerodynamicist with the means to quickly identify problem areas, speed the solution process, and possibly even suggest alternative configurations. Expert system programs will also be developed to aid non-CFD scientists learn to use complicated flow simulation programs.

6. SUMMARY

These previous examples have illustrated the power of CFD to advance the state-of-the-art in predicting fundamental fluid physics, complementing and supplementing experimental testing, and aiding the aerospace vehicle design process. Computers are assuming an increasingly important role in aerospace research and development. Computational fluid dynamics is rapidly becoming an extremely powerful tool in the design process as well as the understanding of complex fluid physics. Substantial payoffs in both arenas have been demonstrated. The validation of CFD codes has been shown to now be a particularly important element of the CFD process, to build the confidence necessary to further extend its practicality. The accurate and rapid simulation of turbulence is the key pacing item in the ultimate application of CFD to problems of aerospace vehicle design and performance prediction. Hypersonic flows, with their necessary coupling of real-gas and flow simulations, is now a major element of the CFD research and development program, and will play a critical role in the design of all future hypersonic flight vehicles. Finally, limitations inherent in serial processing machines will soon be reached, and it will be necessary to explore massively parallel solution methodologies to continue the rapid growth of CFD technology.

7. REFERENCES

1. K. Fujii, W. R. Van Dalsem, L. B. Schiff, and J. L. Steger: Numerical Simulation over a Strake-Delta Wing, AIAA Paper 87-1229 (Honolulu, Hawaii, 1987).

2. S. K. Robinson, and S. J. Kline: Spatial Character and Time-Evolution of Coherent Structures in a Numerically Simulated Boundary Layer, First National Fluid Dynamics Congress, Cincinnati, Ohio, July, 1988.

3. K. V. Rao, and W. R. Van Dalsem: Downwash Effects of Multiple Jets on a Delta Wing in Ground Effect, in preparation.

4. D. Degani, and G. G. Zilliac: Experimental Study of Unsteadiness of the Flow Around an Ogive Cylinder at Incidence, abstract submitted to AIAA Applied Aerodynamics meeting.

5. W. D. Bachalo, and D. A. Johnson: An Investigation of Transonic Turbulent Boundary Layer Separation Generated on an Axisymmetric Flow Model, AIAA Paper 79-1479 (Williamsburg, VA, 1979).

6. S. L. Lawrence, and R. U. Jettmar: Hypersonic Flow Field Prediction using Space Marching and Time Dependent Upwind Codes, in preparation.

7. D. Kwak, D. L. C. Chang, S. P. Shanks, and S. Chakavarthy: A Three-Dimensional Incompressible Navier-Stokes Flow Solver Using Primitive Variables, AIAA Journal, 24, 3, Mar. 1986, pp. 390-396.

8. P. Buning, I. Chiu, S. Obayashi, Y. Rizk, and J. L. Steger: Numerical Simulation of the Integrated Space Shuttle Vehicle in Ascent, AIAA Atmospheric Flight Mechanics Meeting (Minneapolis, Minnesota, August, 1988).

9. P. Kutler: A Perspective of Theoretical and Applied Computational Fluid Dynamics, AIAA Journal, 23, 3, March, 1985.

10. D. L. Dwoyer, P. Kutler, and L. A. Povinelli: Retooling CFD for Hypersonic Aircraft, Aerospace America, 10, October, 1987.

11. P. Kutler, J. L. Steger, and F. R. Bailey: Status of Computational Fluid Dynamics in the United States, AIAA-87-1135-CP, June, 1987.

12. S. Karin, and N. P. Smith: The Supercomputer Era, Harcourt Brace Jovanovich, Publishers, New York, 1987.

Computer Models Needed to Analyze the Hazard of Direct Containment Heating (DCH)

Tim M. Lee

Abstract

Accidents at Three Mile Island and at Chernobyl heightened the public awareness and at the same time changed the industry's perception of nuclear safety; core melt accidents are no longer considered incredible for nuclear reactor plants, although the probability for such accidents is still believed to be very low. One of the safety issues that is currently under extensive investigation is direct heating and pressurization of the reactor containment atmosphere by molten core materials ejected from the reactor vessel under pressure. Direct containment heating (DCH) is of concern because of its potential for early failure to the reactor containment that provides the last barrier against the release of radionuclides to the environment. DCH involves many complicated processes: an unsteady, compressible, and turbulent flow of multi-phase and multi-component fluids; chemical reactions; heat transfer; jet disintegration; debris- structure interactions; etc. Several research activities are ongoing in an attempt to quantify the effects of DCH. Extrapolation of test results is difficult because of the lack of a scaling law. A more suitable approach is the development of analytical models of the associated phenomena and the computation of the extent of DCH using an integrated containment response code. The models needed for such analyses, and their relative importance are discussed.

1. Introduction

The discovery of nuclear fission in 1939 opened for mankind a new source of energy. Less than half a century after the discovery, there already are some 400 nuclear power units in operation that generated 16% of the world's electric power output last year [1]. France had the highest proportion with 70% nuclear compared with 18% for the United States. Although relatively abundant and inexpensive, nuclear power carries with itself a unique problem; splitting of uranium and plutonium atoms produces multitudes of radionuclides, some of which have halflives of many years, that could pose a serious health hazard to the public if released.

A typical nuclear power unit consists of a nuclear reactor encased in a containment building (Figure 1), and a power conversion system that generates electricity. It is common to have multiple units at one plant site. In light water reactors, the reactor core that contains nuclear fuel (typically UO_2 pellets) in a large number of

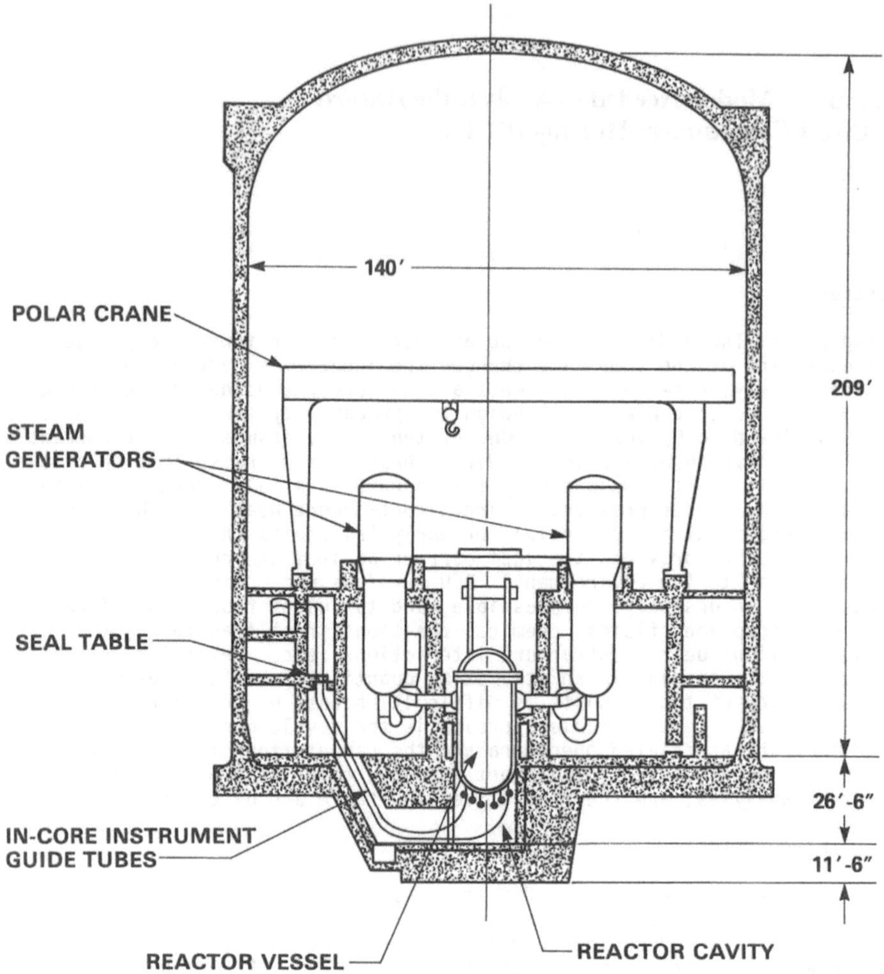

Figure 1
SCHEMATIC ILLUSTRATION OF ZION REACTOR CONTAINMENT BUILDING

zirconium alloy tubes is moderated and cooled by circulating ordinary (light) water at high pressure. Heat removed form the core is then carried by the coolant to the power conversion system. An important consideration in a nuclear plant is that the reactor core will continue to generate heat, albeit at a lower and decaying scale, after it is shut down so that cooling must be maintained until long after the plant shutdown to avoid melt-down of the reactor core. The fuel rod tubing, the reactor vessel and associated piping, and the containment building provide successive barriers to contain radioactive materials. Light water reactors (LWR) are subdivided into two types, boiling water reactors (BWRs) and pressurized water reactors (PWRs), depending on whether or not the coolant (water) is allowed to boil inside the reactor system during normal operation. BWR containments are equipped with pressure suppression pool and generally have smaller free volumes. BWRs have a depressurization system that can be used to depressurize the reactor system, if necessary, during an accident.

To minimize the potential for the release of highly toxic radioactive materials, nuclear power plants are afforded protection by engineered safety systems. The concept of maximum credible accidents, or design basis accidents, was introduced to establish requirements upon which these safety devices were designed. The safety systems so designed are expected to prevent the reactor core from melting and to contain radioactive materials within the plant. Core melt accidents, were considered incredible at the time the design-basis accidents were developed. Accidents at Three Mile Island and at Chernobyl changed this perception; core melt accidents are no longer considered incredible, although the probability for such accidents is still believed to be very low. Major research on phenomena associated with core melt accidents came to being following the accident at Three Mile Island. One of such phenomena that is currently under extensive investigation is direct heating of the reactor containment atmosphere associated with high pressure ejection of the core melt from the reactor vessel [2].

In certain accident scenarios, core melting can take place while the reactor system remains pressurized. When the bottom head of the reactor vessel is breached by molten core materials that slumped and collected there, the core melt will be ejected under pressure into the cavity below the reactor vessel. Subsequent blowdown of a mixture of steam and gases from the vessel is likely to entrain and disperse the ejected core debris out of the reactor cavity into surrounding containment volumes as fine particles, quickly transferring thermal energy to the containment atmosphere. In addition, metallic contents of the dispersed debris particles can react with oxygen and steam in the atmosphere to generate a large amount of chemical energy and hydrogen, heating and pressurizing the containment still further. This complicated physical and chemical process is referred to as direct containment heating, or simply DCH.

The containment pressure rise due to DCH could be rapid and very substantial. Simple heat balance calculations show that the integrity of containment could be challenged if more than 20-30% of the reactor core materials are involved in DCH [3]. The peak pressure would be attained within seconds following the melt ejection [4]. A large number of aerosols, containing radioactive fission products, would be generated in the process of pressurized melt ejection [5]. The airborne radionuclides would barely have time to settle or decay by the time the peak pressure is reached so that, should the containment fail, radiological consequences could be very large.

2. The DCH Process

2.1 Initial Conditions

Although the sequence leading up to DCH starts with core melting inside the reactor, events preceding the ejection of core melt are being addressed by another Nuclear Regulatory Commission (NRC) program, Core Melt Progression [2] and therefore, are outside the scope of the DCH programs. For this review, initial conditions of the melt ejection will be assumed to be given. Such conditions include the pressure in the reactor system; the amount, the temperature, and composition of the melt; and the size of the vessel break. The sequence of high pressure melt ejection and DCH events is presented schematically in Figures 2A through 2D and discussed below to the best of the present understanding.

The state of the reactor and the reactor cavity shortly before the vessel breach is depicted in Figure 2A. The core melt accident would

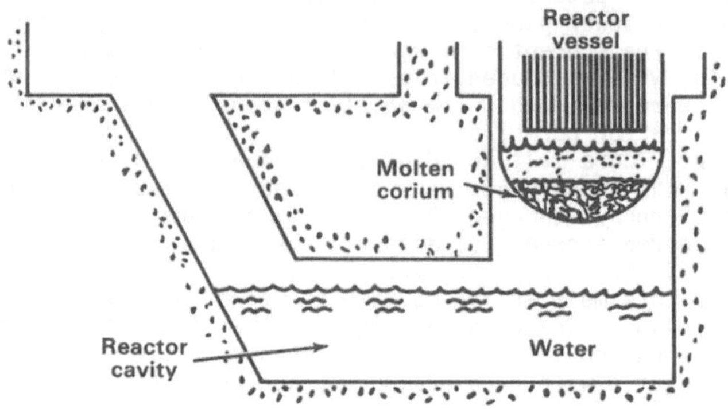

Figure 2A
PRIOR TO CORIUM EJECTION

have progressed for some time and a certain amount of molten core materials would have been collected at the bottom of the reactor. Depending on the accident scenario, the reactor system could remain pressurized at a level somewhere between the containment pressure that is near atmospheric and the normal operating pressure of the reactor that is about 16.5 MPa (2400 psig) for PWRs. A higher reactor pressure would result in more severe DCH because the core melt would be ejected and dispersed more forcefully, and because there would be larger inventories of mass and energy stored in the reactor system that are available to pressurize the containment when discharged.

Uranium oxide has a melting point of about 3100K but the temperature of the core melt would likely be lower because eutectics formed with zirconium, steel, and control rod materials melt at much lower temperatures. Core melt at a higher temperature should contain more thermal energy; it would tend to be fragmented into finer particles because of a lower surface tension. The smaller the particle sizes, the larger the contact area would be fore chemical reactions and heat transfer. A higher melt temperature, therefore, would lead to more severe DCH. As the temperature rises above 1300K, zirconium begins to react with steam generating hydrogen. The reaction is strongly exothermic so that core melting could be accelerated by the reaction. Iron also oxidizes exothermically in a steam environment at high temperature [3]. These reactions could be limited by steam starvation and/or insufficient time available for the reactions. Significant amounts of metallic Zr and Fe, therefore, could be present in the core melt. Metallic contents as well as the amount and the temperature of the core melt depend on how the accident was initiated and how it has progressed.

As the temperature of the core melt is higher than the melting point of steel that the reactor vessel is made of, molten core materials collected at the bottom of the reactor vessel would attack and fail the vessel. Figure 2B shows the phase of the melt ejection immediately

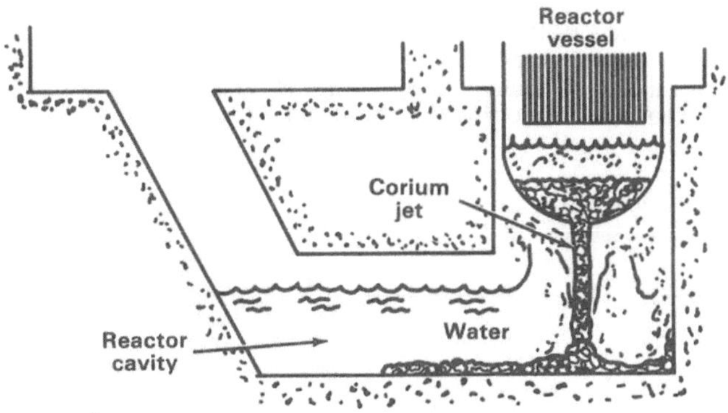

Figure 2B
ONSET OF CORIUM EJECTION

following the vessel rupture when the core melt would be ejected vigorously. Several phenomena in this phase of the event must be noted because of their significant impact on DCH. These are hole ablation [6], jet disintegration [6], and melt-water interactions.

It is commonly postulated that the vessel would fail at one of many instrument tube penetrations although the possibility of other modes of failure cannot be discounted. Whatever the initial size, the hole will be ablated and enlarged as the flow of high temperature melt gushes through. Tests at Sandia National Laboratories [4] showed that a 2.54 cm diameter hole was enlarged to an irregular hole of 6-7 cm diameter by the flow of a high temperature melt that lasted 0.2 seconds. A larger break area would cause more rapid containment heating and produce a higher peak containment pressure because both core melt and blowdown gases would be discharged to the containment at higher rates.

2.2 High Pressure Melt Ejection

When the core melt is ejected from the reactor, hydrogen dissolved in the melt will effervesce at a sudden drop in pressure. Gas effervescence is the main cause for breakup of the core melt jet before it impinges on the cavity floor. There is not sufficient data available at this time to model this effect. The process is complicated further when, toward the late stage of the melt ejection, the overlying gases blow through the draining melt [6]. Blowthrough enhances atomization of the ejected melt.

In some cases, water may be present in the reactor cavity when the melt ejection takes place. What would happen when a jet of high temperature core melt comes in contact with water is uncertain at this time; speculations range from benign termination of the accident to energetic steam explosions. Some people believe that the core melt will be quenched and cooled to reach a stable and manageable state, but there is evidence that the effect may not be so quiescent. In the two tests [7] conducted at Sandia with water filled cavities, the test cavities were blown apart in halves by pressure spikes when the melt was ejected. Although the phenomenon has been observed for years, the process of steam explosions is not yet well understood.

Regardless of whether a steam explosion takes place, ejection of high temperature core melt into a water-filled reactor cavity would be accompanied by a rapid steam generation that would expel the remaining water from the cavity probably preventing it from quenching the melt. This stage of the melt ejection is shown on Figure 2C.

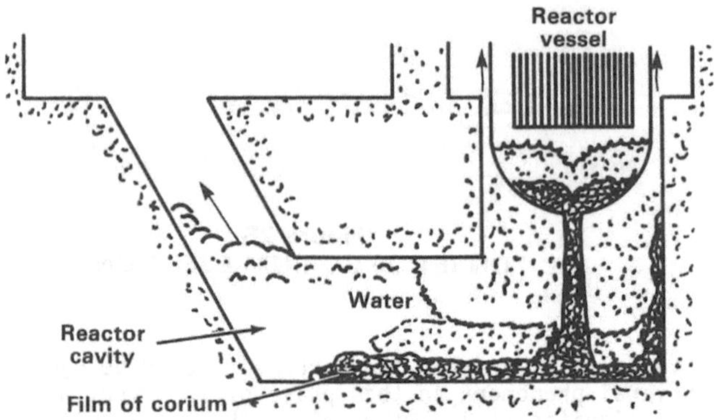

Figure 2C
WATER DISPERSAL FROM CAVITY

The dominant mechanism for dispersal of core debris out of the reactor cavity is believed to be entrainment of the ejected melt in a high speed flow of gases discharged from the reactor vessel following the melt ejection. The actual process of core debris dispersal, based on experimental observation, seems to be a combination of particle levitation, film sweepout, splashout, and film entrainment (Figure 2D).

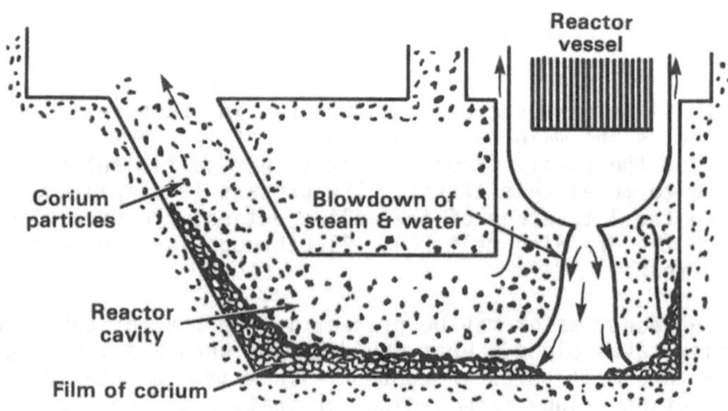

Figure 2D
VESSEL BLOWDOWN/CORIUM SEEPOUT

Two phenomena that take place in and around the reactor cavity have major impact on core debris transport and containment heating: they are ex-vessel metal-steam reactions and debris-structure interactions, which are discussed below.

2.3 Ex-vessel Metal-Steam Reactions

Air in the reactor cavity would quickly be purged by steam when the blowdown starts. As the melt particles are levitated and carried forward, the metallic contents of the core melt, mainly Zr and Fe, will be oxidized exothermically to generate H_2 [4]. These reactions are likely to be very rapid because of the high temperature and of the large contact area for the reactions. Scoping calculations conducted by Brookhaven National Laboratory indeed showed that the reactions would be more than half complete by the time melt particles exited from the reactor cavity.

The ex-vessel metal-steam reactions are important because of two considerations. First, they liberate part of the chemical energy even in the absence of oxygen. But more importantly, hydrogen generated in such reactions can migrate to other parts of the containment much easier than the dispersing core debris, combining with oxygen available there to release the remaining chemical energy for containment heating. Substantial containment pressurization is, therefore, still possible even if most of the ejected core melt were trapped by structures in and around the reactor cavity.

2.4 Debris-Structure Interactions

When a debris particle impacts on the surface of a structure it could bounce off, skid, splash, or be trapped on the surface. If the local flow velocity is sufficiently high, the deposited melt could be subsequently re-entrained in the flow. The process is further complicated by the possibilities that molten core materials could freeze up on cold structural surfaces and that concrete would decompose and release CO_2 and water vapor when exposed to a high temperature. Outgassing would not only prevent the core debris from sticking on the concrete surface, but may also contribute to fragmentation of the melt particles. The particle sizes and the pattern of debris transport are controlled by the behavior of the core debris in interacting with the structures as it is carried forward by the gas flow.

Once out of the cavity region, the gas flow decelerates very quickly because of sudden increase in the flow area while the debris particles continue on their trajectories until being intercepted by equipment or structures in their way. The local flow velocity in this region may no longer be high enough to re-entrain the melt deposited on the structure. The trapped melt may drip down by gravity falling to the floor. The residence time, or the time period during which core debris remains airborne, controls the rate at which heat is imparted to the containment atmosphere because heat transfer and chemical reactions are most active when debris particles are moving through the atmosphere. The residence time as well as the rates of thermal and chemical interactions can only be determined by detailed debris transport analyses that may prove difficult in view of the complexity of the containment configuration.

2.5 Fission Product Release

When the containment fails from over-pressurization, most of the radionuclide release will come from materials that are airborne at the time of the failure. These include noble gases and aerosols. Copious amounts of aerosols are generated in the process of pressurized melt ejection [5]. Such aerosols are likely to contain refractory as well as volatile fission products because the aerosols are generated by mechanical atomization in addition to condensation of vaporized materials. Some of the non-volatile fission products are known to have very high biological consequences [9]. Contributions to the release from gaseous fission products are fairly well understood, but mechanical aerosol generation phenomena are not. Aerosol generation is complicated and is closely coupled with such processes as jet disintegration, debris-structure interactions, chemical reactions, etc.

3. Computer Models

3.1 Scaling Laws

Two approaches are available to quantify a new engineering problem: conduct tests and extrapolate the test results using a scaling law, or develop analytical models of phenomena to be incorporated in a computer code for integrated analyses of the problem. For the problem at hand, neither approach is an easy task because of the complexity of the problem. Brookhaven has attempted the former approach in its study of hydrodynamics of core debris dispersal from the reactor cavity and identified eleven dimensionless parameters that must be kept constant to preserve physical similarities [10]. Brookhaven's study does not, however, include effects of heat transfer and chemical reactions. The number of similarity parameters is likely to be increased when these effects are included. It is clearly impossible to design a test that matches so many parameters with the reactor accident under investigation. Direct application of the test data, therefore, has limitations. The large number of similarity parameters also makes it difficult to attempt correlations among these parameters that are needed for extrapolation of the test data to accident conditions because the number of tests needed to generate the data base for such a purpose would be large. However, scaling studies may be helpful to certain codes that use parametric input in establishing guidelines for the choice of these physical quantities that are otherwise left to the user's discretion.

3.2 Lumped-Parameter Codes

The NRC and its contractors are also working on the latter approach. Currently, there are several computer codes available for the analysis of DCH: they include CONTAIN-IDHM [4] by SNL, DHCVIM [11] by BNL, HMC/DIRHET [14] by the University of Wisconsin and PARSEC [12]/HARDCORE [13] by Argonne National Laboratory. These codes are all of the lumped-parameter variety. They all calculate mass and heat balances, and include rate equations for chemical reactions and heat transfer. Rates of mass transfer and convective heat transfer are generally determined using correlations for Nusselt, Reynolds, Prandtl, Sherwood, and Schmidt numbers.

Despite many similarities, there are significant differences in details among these codes, especially in the treatment of core debris transport that probably is the most important aspect of the DCH analysis. DHCVIM, HMC, and PARSEC/HARDCORE all attempted a mechanistic approach for the debris transport. A Kutateladze number [8] of 10 is commonly adopted as the threshold above which film entrainment is assumed to take place. Median particle sizes are calculated by the Weber criterion considering hydrodynamic break-up of liquid droplets in a high speed gas stream. A Weber number of 6 is commonly used. DHCVIM and PARSEC/HARDCORE have a provision to treat multiple particle sizes. Particle velocities are calculated by a momentum equation that includes interfacial drag and gravity forces. Slip velocities between the particles and the gas, particle trajectories, and the residence time of the particles can all be calculated in this approach. In CONTAIN, particle sizes, slip velocities and the residence time are all user input. Homogeneous mixtures are assumed in CONTAIN calculations of inter-compartment flows.

The gas flow velocity in the reactor cavity is calculated by one-dimensional approximation; secondary flows and turbulence are not considered. PARSEC/HARDCORE uses a two-phase compressible flow model which predicts that the flow of the mixture would be near critical at the cavity exit. DHCVIM uses single phase compressible flow model while both CONTAIN and HMC determine the gas velocity by dividing the volume flow by the cross-sectional area of the cavity. The volume flow is corrected for the expansion due to heating by ejected core melt, but the acceleration of compressible fluid by the conversion of internal energy to kinetic energy is not accounted for in the CONTAIN and HMC treatment. The containment volume is treated as one Eulerian cell and debris particles are traced through Lagrangian coordinates in DHCVIM and PARSEC. All codes make certain assumptions on debris interactions with the containment walls. CONTAIN and PARSEC included the cloud effect in their modeling of radation heat transfer. HMC omits radiation heat transfer, while PARSEC neglected convective heat loss, to the containment walls. PARSEC has a model to calculate debris retention in the cavity by freeze-up.

These codes were recently used in pre-test calculations of DCH-2, the second DCH test conducted in the Surtsey facility at Sandia. All codes substantially over-predicted "containment" pressurization for this test [18]. The discrepancies observed are to be expected in view of the rather simplistic modeling in these codes for very complicated phenomena. It is obvious that a lot more sophistication in the analytical tool is needed for the analysis of the DCH effects.

4. Examination of Computer Models

Computer models needed to address phenomena discussed above are examined below with a brief assessment of the prospect for their availability. It appears that many of these models will need a containment code which is more detailed than the lumped parameter variety.

4.1 Discharge Phase

In addition to time rates of core melt discharge and gas blowdown that includes a choked flow model, phenomena need to be modeled in this phase of the process include (a) hole ablation, (b) jet-disintegration, and (c) gas blowthrough. Sandia has proposed an analytical model for hole ablation [6] and an empirical correlation for gas blowthrough [6] that appear to agree well with their own experiments. Incorporating these models in a computer code should be straightforward. Jet disintegration, however, is difficult to model because data are scarce regarding solubility of hydrogen in the core melt and the effervescence of dissolved hydrogen at a lower pressure. More tests are needed in this area. Sandia has also proposed correlations for pneumatic atomization when gas blowthrough takes place. Particle sizes predicted by the pneumatic atomization model may need to be modified to account for the effects from jet disintegration due to effervescence of the dissolved gas and from debris interactions with structures.

4.2 Fluid Flow

Most of containment codes, including CONTAIN and HMC, uses a simple incompressible fluid flow equation that is satisfactory for containment analyses under design basis accidents; the gas mixture is assumed homogeneous and the gas flow from one cell to the other is calculated from the pressure difference at the "junction" corrected for the friction and the form losses. This approach, unfortunately, is not adequate for the analysis of the DCH process that includes a near sonic diabatic flow in a reactor cavity that is entraining melt particles and breaking them up while carrying them forward. Flow velocities calculated from the incompressible equation could be an order of magnitude lower even if a correction were made of the increased specific volume due to heat addition. As the entrainment rate and the interfacial drag on the debris particles are both proportional to the square of the gas velocity, the use of an incompressible flow equation could grossly under-predict debris dispersal.

High speed movies taken by Brookhaven of their debris dispersal tests showed that the fluid flow inside the model cavity is circulative and highly turbulent. The local gas velocity that is relevant to the entrainment of melt particles, therefore, could be several times higher than that calculated by one-dimensional approach. For the analysis of interactions inside the reactor cavity it is highly desirable that a three-dimensional, two-phase flow model that includes energy equations be used. Several computer codes with such capability are available [16], [17], but application to the present problem will be extremely challenging.

4.3 Effect of Water

The amount of heat expended in converting water to steam would seem to be a simple heat balance calculation except that the core melt is not likely to mix uniformly with water in a high pressure melt ejection. The extent to which the ejected core melt can be quenched by water in the reactor cavity, therefore, is not an easy problem even disregarding the possibility for steam explosions. This process is not currently modeled mechanistically, and such modeling appears to be difficult. Currently, a parametric approach is used for the fraction quenched, with benchmarking against the available experimental data, which are sparse at this time. Any consideration to include a steam explosion model in the DCH analysis appears premature at this writing.

4.4 Debris Transport

The most difficult aspect of the DCH process is core debris transport that include debris-structure interactions. The main reason why DCH-2 pretest calculations were not more successful was the inability of participating codes to analyze the interactions between dispersing melt particles and the side wall of the Surtsey vessel that simulated the containment volume. There are several detailed codes that can trace the motion of debris particles, but they will need models for debris-structure interactions. Sandia attempts to work with the KIVA code [17] for the core debris transport analysis. There was an earlier attempt to use COBRA-NC [16]. Debris-structure interaction models are likely to be elaborate and will require a large data base to develop. A full three-dimensional, finite difference, two-phase, multi- component code that includes elaborate debris-structure interaction models may eventually prove to be too expensive and complicated. However, the ex-vessel metal-steam reactions that generate hydrogen may make detailed debris transport analyses unnecessary, at least in part.

4.5 Heat Transfer and Chemical Reactions

The magnitude of the peak containment pressure rise is governed by competing effects of heat addition to, and heat loss form, the containment atmosphere. The rates of chemical reactions and convective heat transfer are derived with the assumption of a single droplet suspended in an infinite air or steam atmosphere. Effects from the neighboring particles are not included. It is not clear at this time how much the cloud effect influences the extent of containment heating. As in water quenching, this effect is not likely to be amenable to a mechanistic modeling. A correction factor based on experimental data may be the only approach possible. The concern here is scaling of such a correction factor. The cloud effect is equally important to radiation heat transfer; models used by CONTAIN and PARSEC should be a reasonable starting point.

Ex-vessel hydrogen generation is an important phenomenon that could circumvent the difficulty of modeling debris transport because transport of hydrogen that carries forward a large fraction of chemical energy available for containment heating is much easier to model. There are, in fact, several detailed codes developed at Los Alamos National Laboratory (KIVA [17] and HMS-BURN [15]) that are immediately available for such application. Hydrogen burning models in HMS-BURN, however, should be modified for combustion at higher temperatures. KIVA has yet to incorporate a H_2 burning model.

4.6 Aerosols

To calculate the radiological consequences in case of a containment failure, concentrations of radionuclides suspended in the containment atmosphere at the time of failure need to be known. An important variable in such calculations is aerosol generation. The mode of aerosol generation due to condensation of vaporized materials can be modeled fairly satisfactorily if partial pressures of these materials are available. The other mode of aerosol generation by mechanical disintegration, however, is difficult to model because it combines the effects of atomization, gas effervescence, and debris shattering. Transport of aerosols can be accommodated if the DCH code has a detailed core debris transport model, but additional models are needed to address aerosol generation and the decay of fission product isotopes.

5. Conclusions

DCH is a challenging engineering problem. It is a relatively new issue that has a significant implications for nuclear plant safety. The most practical approach to quantity the effects of DCH appears to be through computer modeling. The DCH process is extremely complicated, especially with regard to core debris transport. It is uncertain that all relevant phenomena could ever be fully modeled. However, ex-vessel metal-steam reactions and hydrogen generation could alleviate the need for a detailed core debris transport model if it can be established that thermal energy associated with the ejected core melt that is trapped by structures in the vicinity of the reactor cavity makes only a small contribution to DCH. Mitigative effects of water and containment structures are yet to be demonstrated. The USNRC is also investigating feasibility and desirability of depressurizing the reactor system during a core melt accident to preclude high pressure melt ejection.

References

1. U.S. Council for Energy Awareness: Energy Daily 16., No. 90 (1988)
2. T. M. Lee: NUREG-1265, Chapter 4, (USNRC, 1987).
3. K. D. Bergeron, D. C. Williams; Nuc. Eng. & Dsn. 90 (1985)
4. D. C. Williams et. al.: Proc. Am. Chem. Soc. Severe Accident Chem. Symposium (1986).
5. W. W. Tarbell et. al.: NUREG/CR-4512 (Sandia Nat'l Lab. 1986)
6. M. Pilch, W. W. Tarbell: NUREG/CR-4383 (SNL 1985)
7. W. W. Tarbell et. al.: NUREG/CR-3916 (SNL 1988)
8. A. E. Bergles et. al.: In Two-phase Flow & Heat Transfer in the Power & Process Industries 33 (Hemisphere Pub. Corp. 1981)
9. R. Wilson et. al.: Rev. of Mod. Phy. 57, No. 3, Part II, S18 (1985)
10 N. Tutu, T. Ginsberg: NUREG/CR-5146 (Brookhaven Nat'l Lab. 1988)
11. T. Ginsberg, N. Tutu: AICHE Symposium Series 83, No. 257 (1987)
12. J. J. Sienicki: Internal Memo. to B. W. Spencer on DCH Tests (Argonne Nat'l Lab. 1986)
13. J. J. Sienicki, B. W. Spencer: Internat'l Symposium on Multiphase Transport & Particulate Phenomena, Miami, FL (1986).
14. I. Huhtiniemi et. al.: UWRSR 50 (1986)
15. J. R. Travis: NUREG/CP-0071, P19-9 (USNRC 1985)
16. M. J. Thurgood: NUREG/CR-3262 (Pacific NW Lab. 1986)
17. A. A. Amsden et. al.: LA-10245-MS (Los Alamos Nat'l Lab. 1985)
18. T. M. Lee: NUREG/CP-0091 4, 209 (USNRC 1988)

Numerical Simulation of Shock Focussing

D. Leutloff, K.G. Roesner, and R.C. Srivastava

1 Abstract

The two-dimensional time-dependent flow of a gas without viscosity and heat-conductivity is simulated numerically by using the strong conservative form of the hyperbolic system of partial differential equations which describe the basic equations of gasdynamics. The initial conditions of the developing flow field are chosen such that a rarefaction wave and a converging shock wave should appear. The motion is initiated by a bursting diaphragm which leads to a shock formation with focussing effect. Different initial conditions are investigated to show the dependence of the intensity of the shock wave in the focus from geometrical and physical parameters. Diagrams of pressure distributions for different pressure ratios show the complicated wave pattern created by interfering wave systems. Calculations are done for a plane shock wave which is reflected by a parabolic mirror-plate.

The mathematical method used is "shock capturing" on the basis of the finite difference approximation on an equally spaced Cartesian grid according to an invariant scheme due to RUSANOV and described by SHOKIN |1|. The artificial viscosity which is added to avoid the oscillating character of the solution near the shock wave is very small, so that the influence on the travelling waves is kept under control.

2. Introduction

The analysis of shock wave propagation is interesting in itself because it leads to a bench mark problem which can be used to estimate the quality of a shock tracking solver. Thus the numerical simulation of such kind of flow is an excellent tool for the applied mathematicians to look for the inherent features of a numerical scheme approximating the

Eulerian equations. On the other side a lot of applications are connected with the problem of shock wave formation and shock wave propagation in homogeneous media. As examples the phenomena of sonic boom reflection from the topography of the earth and focussing effect of sonic booms created by special maneuvering of aircrafts can be mentioned. Typical for all these phenomena is the fact that the shock front gets deformed because of the reflection from curved rigid walls.

In the numerical experiments the strength of the shock wave is chosen such that no instabilities should occur. The gas in which the shock front is created is assumed to be at rest and homogeneous that the only reason for a deformation of the shock front is due to the complicated form of the geometry of the reflecting rigid walls or the geometrical shape of the bursting membrane at t=0.

Two different situations are analysed on the basis of a numerical simulation of the shock formation: In the first case it is assumed that a parabolically shaped diaphragm bursts at time t=0 and sets into motion part of the high pressure gas which builds up a rarefaction wave and a shock front which travels into the low pressure region of the two-dimensional space. From the initial value distribution of density, pressure, velocity, and internal energy the formation of a converging shock wave is expected. In the focus of this shock wave extremely high pressure values are encountered.

In the second case two plane shock fronts are created which travel into opposite directions along the x-axis. One shock wave meets a rigid wall which has the form of a hyperbola. At this "mirror" the shock front is reflected and forms after some time a converging shock wave pattern which shows a focus. An interaction with the second reflected shock wave from a plane wall takes place after a while and the merging of two shock waves can be observed.

The mathematical approach of this physical situation is possible on the basis of a difference scheme which has an invariant character especially with respect to rotation around an axis perpendicular to the (x,y)-plane. Earlier calculations |2| have shown that the artificial viscosity which has to be added is very small and does therefore not affect the propagation time of the shock front.

2. Shock focussing and cusp formation

Using the approximation of an acoustical wave in a two-dimensional region |3| one can realize that a curved sound wave which is travelling into a homogeneous gas is forming a typical focus followed by a caustic which can be interpreted as an envelope of the rays of the wave system. Using the nonlinear approach on the basis of the Eulerian equations the same phenomenon is to be expected. An important question is the reliable simulation of the flow charactestics in the focal region where a caustic leads to a very high pressure of the gas. The flow field under consideration is time-dependent and covers an interval of time until a cusp of the converging shock waves is formed.

3. Basic equations in strong conservative form

As the problem of shock focussing is treated in rectangular coordinates we use the Eulerian equations in the strong conservative form introducing the three column vectors W,F, and G:

$$\frac{\partial W}{\partial t} + \frac{\partial F(W)}{\partial x} + \frac{\partial G(W)}{\partial y} = 0 \tag{1}$$

$$W = \begin{bmatrix} \rho \\ m \\ n \\ E \end{bmatrix}, \qquad F = \begin{bmatrix} m \\ (3-\gamma)\frac{m^2}{2\rho} + (\gamma-1)\ (E - \frac{n^2}{2\rho}) \\ \frac{m\,n}{\rho} \\ (\frac{\gamma m E}{\rho}) + \frac{(1-\gamma)(m^2+n^2)m}{2\rho^2} \end{bmatrix},$$

$$G = \begin{bmatrix} n \\ \frac{m\,n}{\rho} \\ (3-\gamma)\frac{n^2}{2\rho} + (\gamma-1)\ (\ E - \frac{m^2}{2\rho}\ ) \\ \frac{\gamma n E}{\rho} + \frac{(1-\gamma)\ (m^2+n^2)n}{2\rho^2} \end{bmatrix}.$$

The independent variables are denoted by x, y, and t. The density ρ and the velocity components u and v in x- and y-direction are the dependent variables. $E = \frac{1}{2}\,\rho\,(u^2 + v^2) + \frac{p}{\gamma-1}$ is the energy density of the gas, and the components of the mass flux vector are given by $m=\rho u$, and $n=\rho v$. γ is the ratio of the specific heats. The pressure is related to the other flow variables by the relationship $p = (\gamma-1)(E - \frac{1}{2\rho}(m^2+n^2))$. The equation of state is assumed to be $p = (\gamma-1)\cdot\rho\cdot\varepsilon$, where ε is the internal energy per unit mass. According to |2| the system of equations (1) is rewritten in dimensionless form leading to the following explicit finite difference approximation:

$$\frac{W^{n+1}_{j,k} - W^n_{j,k}}{\tau} + \frac{F^n_{j+1,k} - F^n_{j-1,k}}{2\,h_1} + \frac{G^n_{j,k+1} - G^n_{j,k-1}}{2\,h_2} = \tag{2}$$

$$= \frac{K_1}{2}\Big[(\Omega_{11})^n_{j+\frac{1}{2},k} \frac{W^n_{j+1,k} - W^n_{j,k}}{h_1} - (\Omega_{11})^n_{j-\frac{1}{2},k} \frac{W^n_{j,k} - W^n_{j-1,k}}{h_1}$$

$$+ (\Omega_{12})^n_{j+\frac{1}{2},k} \frac{W^n_{j+\frac{1}{2},k+\frac{1}{2}} - W^n_{j+\frac{1}{2},k-\frac{1}{2}}}{h_2} - (\Omega_{12})^n_{j-\frac{1}{2},k} \frac{W^n_{j-\frac{1}{2},k+\frac{1}{2}} - W^n_{j-\frac{1}{2},k-\frac{1}{2}}}{h_2}\Big]$$

$$+ \frac{K_2}{2}\Big[(\Omega_{22})^n_{j,k+\frac{1}{2}} \frac{W^n_{j,k+1} - W^n_{j,k}}{h_2} - (\Omega_{22})^n_{j,k-\frac{1}{2}} \frac{W^n_{j,k} - W^n_{j,k-1}}{h_2}$$

$$+ (\Omega_{21})^n_{j,k+\frac{1}{2}} \frac{W^n_{j+\frac{1}{2},k+\frac{1}{2}} - W^n_{j-\frac{1}{2},k+\frac{1}{2}}}{h_1} - (\Omega_{21})^n_{j,k-\frac{1}{2}} \frac{W^n_{j+\frac{1}{2},k-\frac{1}{2}} - W^n_{j-\frac{1}{2},k-\frac{1}{2}}}{h_1}\Big]$$

where τ is the time step, $h_1=h_2=h$ is the step size in x- and y-direction.

$K_i = \frac{\tau}{h_i}$, $i = 1,2$, and

$$(\Omega_{ml})^n_{j\pm\frac{1}{2},k} = \Omega_{ml}(\,W_{j\pm\frac{1}{2},k}\,) = \frac{(\Omega_{ml})^n_{j\pm1,k} + (\Omega_{ml})^n_{j,k}}{2}.$$

The matrices Ω_{ml} are defined as:

$$\Omega_{ml} = A_m\,A_l + C_{ml}\,,\ (m,l = 1,2) \qquad C_{12} = C_{21} = 0\,, \qquad C_{11} = C_{22} = \mu\cdot I$$

where we choose $\mu = 2\,\mu_0\,\frac{h^2}{\tau^2}$, μ_0 is constant and is chosen to be 0.04, and I is the unit matrix. $A_1(W)$ and $A_2(W)$ are matrices of the following form:

$$A_1(W)=\begin{bmatrix} 0 & 1 & 0 & 0 \\ \frac{(\gamma-1)n^2-(3-\gamma)m^2}{2\rho^2} & \frac{(3-\gamma)m}{\rho} & \frac{(1-\gamma)n}{\rho} & (\gamma-1) \\ -\frac{mn}{\rho^2} & \frac{n}{\rho} & \frac{m}{\rho} & 0 \\ -\frac{\gamma mE}{\rho^2}-\frac{(1-\gamma)(m^2+n^2)m}{\rho^3} & \frac{\gamma E}{\rho}+\frac{(1-\gamma)(3m^2+n^2)}{2\rho^2} & \frac{(1-\gamma)mn}{\rho^2} & \frac{\gamma m}{\rho} \end{bmatrix}$$

$$A_2(W)=\begin{bmatrix} 0 & 0 & 1 & 0 \\ -\frac{mn}{\rho^2} & \frac{n}{\rho} & \frac{m}{\rho} & 0 \\ \frac{(\gamma-1)m^2-(3-\gamma)n^2}{2\rho^2} & \frac{(1-\gamma)m}{\rho} & \frac{(3-\gamma)n}{\rho} & (\gamma-1) \\ -\frac{\gamma mE}{\rho^2}-\frac{(1-\gamma)(m^2+n^2)n}{\rho^3} & \frac{(1-\gamma)mn}{\rho^2} & \frac{\gamma E}{\rho}+\frac{(1-\gamma)(m^2+3n^2)}{2\rho^2} & \frac{\gamma n}{\rho} \end{bmatrix}$$

4. Numerical procedure and boundary conditions

Using the stability criterion derived by Rusanov |2| the time step has to be restricted according to the estimation:

$$\tau \le \frac{\alpha\, h}{\sqrt{2}\,\sigma}\,; \qquad \alpha < 1 \tag{3}$$

σ is the maximum of the spectral radii of the matrices A_1 and A_2. The physical domain is a square region with a grid of 42x42 lattice points (Fig.1).

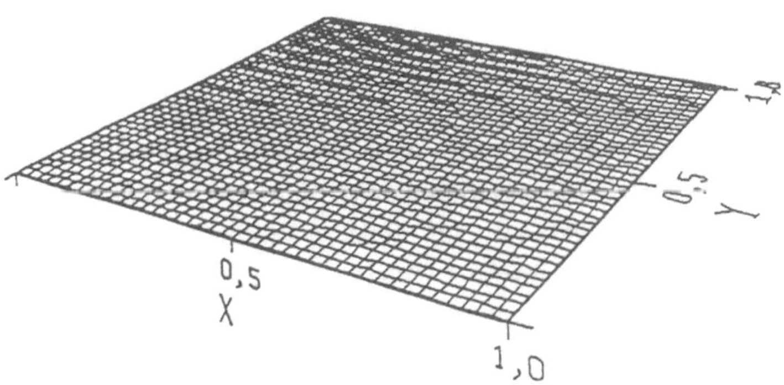

Fig.1 Computational grid (44x44 grid points) and physical region (42x42 grid points)

To simulate at the boundaries zero gradient conditions four additional grid lines are added and the flow field quantities are transferred to the boarder lines from the last inner grid line after calculation. The geometrical shape of the diagphragm which divides the flow field into a low pressure and a high pressure region is defined analytically by:

$$(y-0.5)^2 = 30\ x - (1-\xi^2)\ x^2. \tag{4}$$

The parameter ξ was given two values: $\xi=1$ corresponds to a parabola, and $\xi=1.3$ represents a hyperbola with its apex at $x=0$. Fig.2 shows a perspective view of the initial pressure distribution in the (x,y)-plane.

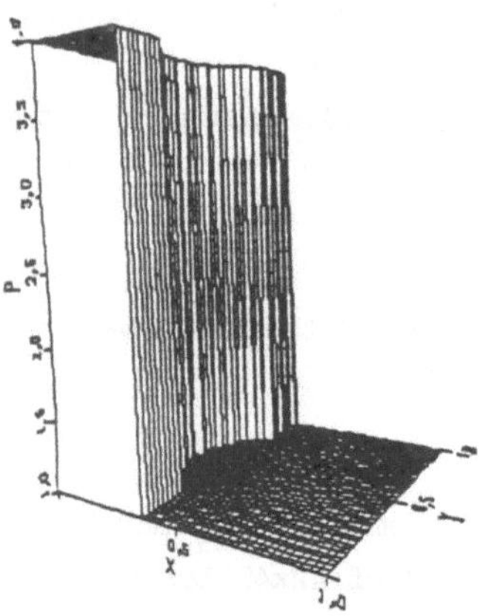

Fig.2 Initial distribution of the pressure in the (x,y)-plane for the pressure ratio of 4

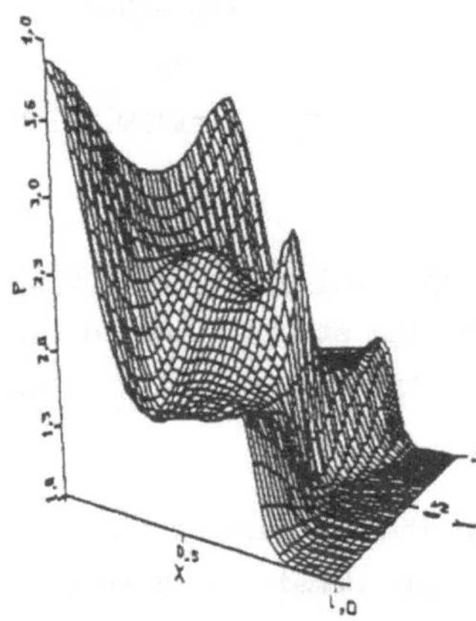

Fig.3 Calculated pressure distribution in the computational plane at time t=0.4

5. Numerical results

The first numerical experiment simulates the time-dependent flow of a gas which is divided by a parabolic membrane into two regions of different pressure. The pressure ratio in the example is chosen to be 4. When the fictitious diaphragm is released a rarefaction wave and a shock front are moving into the gas at rest. After a dimensionless time of t=0.4 the wave pattern forms a caustic with a focal region where the pressure is raised to higher values than elsewhere in the flow field. In the geometrical focus of the parabola of the initial distribution a high peak of the pressure can be observed. Fig.3 gives a three-dimensional plot of the pressure distribution in the computational plane for t=0.4. A plot of isobars is given in Fig.4. For the pressure ratio 4 the isobars are marking the focus and the caustic for the time t=0.4. Another numerical experiment was carried out to study the behavior of a plane shock wave which is reflected by a hyperbolic "mirror". Two pressure distributions are chosen. In Fig.5 the special kind of initial distribution is shown. The pressure ratios were chosen to be 4 and 8.

To calculate for one time step the flow quantities in all grid points a CPU-time of 8 seconds was necessary. For two different pressure ratios the calculated pressure fields are shown in Fig.6-Fig.9. In the perspective view of Fig.6 the pressure distribution is represented for the pressure ratio 4 at the time t=0.4. In front of the rigid wall the pressure raises to about twice the value of the given initial distribution. The maximum of the pressure is found in the geometrical focus of the initial distribution.

The isobars for this pressure ratio are given in Fig.7. The straight lines in the right hand side of Fig.7 show the reflected plane shock wave from the rigid wall at x=1.

The corresponding numerical data for a higher pressure ratio of 8 are shown in the Fig.8 and Fig.9, respectively. The geometrical shape of the reflecting mirror is defined by the constant ξ=1.3. This leads to a hyperbolic shape of the plate.

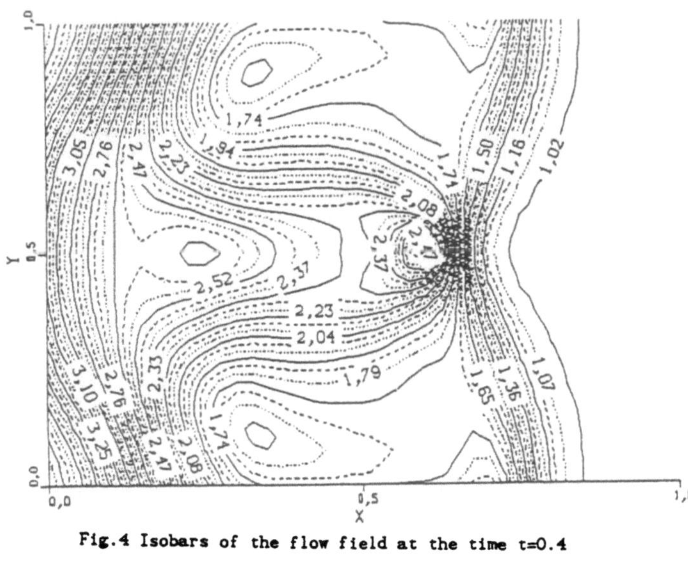

Fig.4 Isobars of the flow field at the time t=0.4

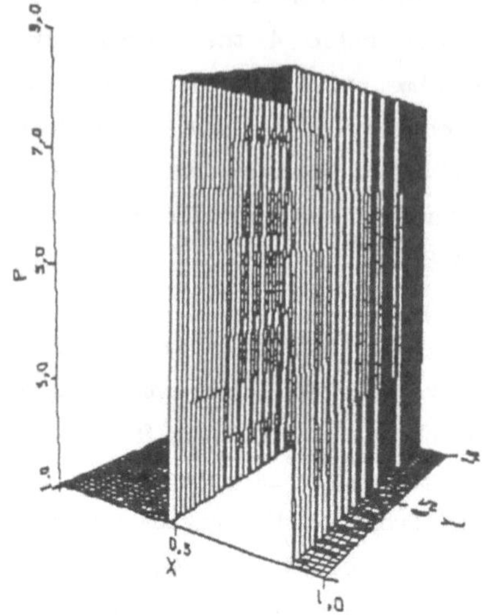

Fig.5 Initial condition for the shock wave reflection at a hyperbolic mirror-plate

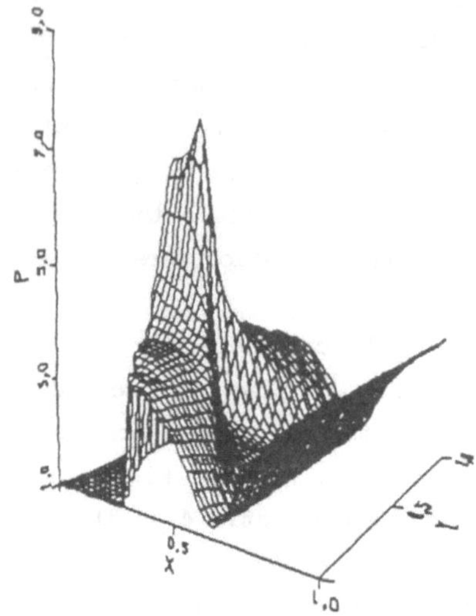

Fig.6 Pressure distribution for an initial pressure ratio of 4 at time t=0.4

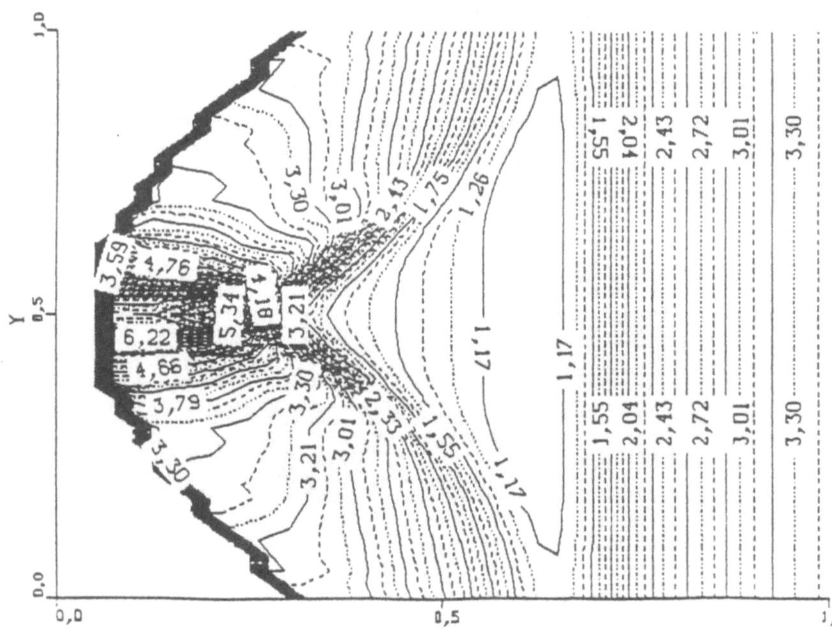

Fig.7 Plot of isobars for a pressure ratio 4 at time t=0.4 after reflection at the mirror-plate

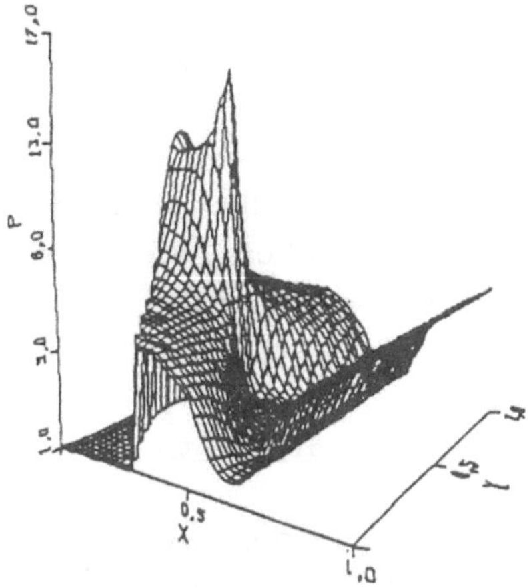

Fig.8 Perspective view of the pressure distribution for the initial pressure ratio 8 at time t=0.4

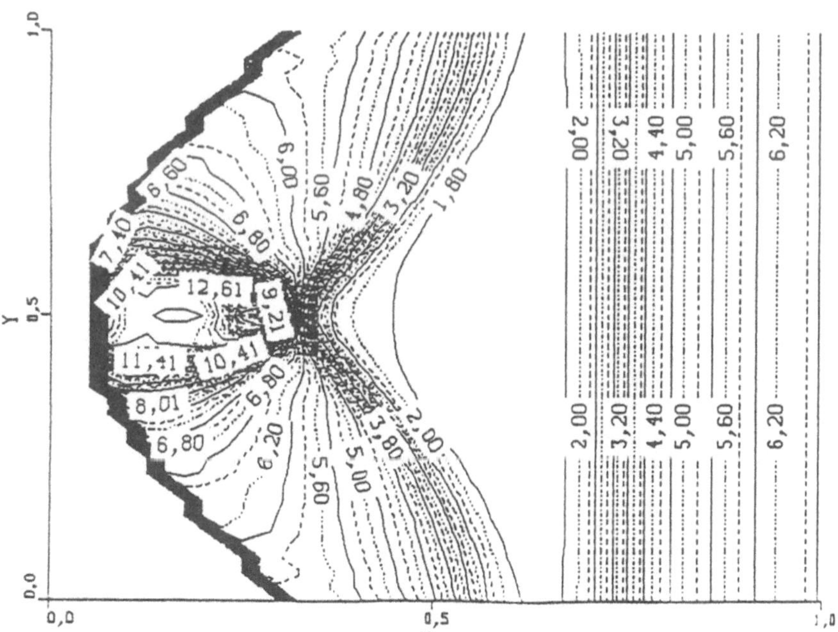

Fig.9 Isobars of the time-dependent flow field at time t=0.4 for the initial pressure ratio 8

6. Conclusion

An invariant explicit scheme which was used for the shock capturing procedure is a good tool to simulate even on a very coarse grid the shock-shock interaction in the focus of special pressure distributions. The CPU-time for 200 time steps is about 28 minutes on a VAX-8530 system. Only a small amount of the artificial viscosity is added to avoid the oscillating character of the shock front.

References

1. Yu. I. Shokin. The method of differential approximation. Springer Verlag, (1983)
2. D. Leutloff, K.G. Roesner, and R.C. Srivastava. Comp. & Fluids Vol. 16, No. 2, 175-182, (1988)
3. B. Sturtevant, and V.A. Kulkarny. J. Fluid Mech. 73, 651-671, (1976)

The Interaction Between a Pair of Counter-Rotating Potential Vortices in Vertical Ascent and a Free Surface

Daniel L. Marcus and Stanley A. Berger

1. Introduction

The vortical structures shed by a lifting surface have long been the object of study as an area of both fundamental interest and practical application. In particular, the development, trajectories, large scale instabilities, and eventual decay of a pair of counter-rotating line vortices has been the subject of many investigations.

Recently, attention has focused on another area of concern, the signature left on the free surface of a liquid due to interaction with trailing vortices generated by a submerged lifting surface and rising to the free surface. Experiments show that the signature on the free surface manifests itself in two distinct modes: "striations" and "scars" (SARPKAYA and HENDERSON [4]). The "scars" are depressions in the free surface aligned with the axes of the vortices; "striations" are ridges perpendicular to the axes. Distinct regimes in the time evolution of the signatures correspond to different segments of the vortex trajectories and their lifetimes. Striations first appear when the vortex pair is at a distance of roughly one initial separation from the surface; scars appear when the vortices approach to within half an initial separation. These features, according to the experiments, are independent of the shape and aspect ratio of the generating lifting surface, and depend primarily on the initial depth, d, and Froude number, defined by

$$F^2 = \frac{\Gamma^2}{ga^3} \qquad (1)$$

where Γ is the vortex strength, $\underline{a}$ the initial pair separation, and g the acceleration due to gravity.

The signatures occur at a relatively late stage in the lifetime of the vortices. The onset of demise-related phenomena such as vortex breakdown and Crow instability may further complicate the flow and contribute to it ultimately becoming three dimensional.

In this paper, the motion of a pair of counter-rotating line vortices approaching a free surface is analyzed assuming the flow to be two dimensional and inviscid. Consistent with the former assumption, only the scarring phenomenon is considered.

2. Governing Equations and the Linearized Problem

The flow is assumed to be irrotational, inviscid, and incompressible, and to initially fill the semi-infinite half space: $y < 0, \ -\infty < x < \infty$. Initially the entire fluid is assumed to be at rest, and at all later times the disturbances are assumed to vanish at infinity. We first consider the case of a single vortex following an arbitrary path. The flow can be completely described by a potential function $\Phi(x,y,t)$ satisfying the following equations and boundary and initial conditions:

$$\Delta\Phi(x,y,t) = 0, \quad -\infty < x < \infty, \quad \eta(x,t) \geq y > -\infty, \quad t \geq 0 \tag{2a}$$

$$\left[\Phi_t = \frac{1}{2}\nabla\Phi\cdot\nabla\Phi - g\eta\right]_{y=\eta} \tag{2b}$$

$$\left[\frac{\partial\eta}{\partial t} + u\frac{\partial\eta}{\partial x} = v\right]_{y=\eta}, \quad \underline{u} = (u(x,y,t),\ v(x,y,t)) = \nabla\Phi \tag{2c}$$

$$\lim_{y\to-\infty} \nabla\Phi = 0 \tag{2d}$$

$$\Phi(x,y,0) = \Phi_t(x,y,0) = 0 \tag{2e}$$

$$\left.\begin{aligned}\Phi(x,y,t) = \Phi_0(x,y,t) + \frac{\Gamma}{2\pi}\,\mathrm{Re}\,\{-i\log(z-c(t))\}, \quad z = x+iy \\ c(t) = a(t) + ib(t), \quad b(t) < 0\end{aligned}\right\} \tag{2f}$$

where $a(t)$ and $b(t)$ are the x and y coordinates of the vortex trajectory.

The problem defined by (2a)-(2f) is nonlinear and can be solved exactly only by numerical techniques. Such an approach is described later. It is instructive, however, and useful for comparison to this subsequent numerical solution, to first consider the analytically tractable linearized problem. This results from assuming that the free surface displacements and the velocities everywhere in the flow are small.

The boundary conditions may then be linearized, and all quantities evaluated at the plane of the undisturbed free surface. Boundary conditions (2b) and (2c) can then be combined to yield the single equation

$$\Phi_{tt}(x,0,t) + g\Phi_y(x,0,t) = 0. \qquad (3)$$

An exact solution to this linearized problem can be found as described below.

As suggested by the analysis of WEHAUSEN and LAITONE [7] describing the motion of a three-dimensional source in the vicinity of a free surface, an image vortex is introduced, and one seeks a solution in the form

$$\Phi(x,y,t) = -\frac{\Gamma}{2\pi}\,\mathrm{Re}\,\{i\log(z-c(t)) + i\log(z-\bar{c}(t))\} + \Phi_1(x,y,t), \qquad (4)$$

where Φ_1 is harmonic in $y < 0$, satisfies (2a), (2d), and (2e), and

$$\Phi_{1tt}(x,0,t) + g\Phi_{1y}(x,0,t) = \frac{\Gamma}{2\pi}\left(\frac{\partial^2}{\partial t^2} + g\frac{\partial}{\partial y}\right)\mathrm{Re}\,\{i\log(z-c(t)) + i\log(z-\bar{c}(t))\}_{y=0}\,. \qquad (5)$$

Since the $\frac{\partial^2}{\partial t^2}$ terms on the right hand side vanish at the surface, this reduces to

$$\Phi_{1tt}(x,0,t) + g\Phi_{1y}(x,0,t) = -\frac{\Gamma g}{\pi}\left[\frac{x-a(t)}{(x-a(t))^2 + b^2(t)}\right]. \qquad (6)$$

Now consider the function

$$\Phi_{1tt}(x,y,t) + g\Phi_{1y}(x,y,t) + \frac{\Gamma g}{\pi}\left[\frac{x-a(t)}{(x-a(t))^2 + (y+b(t))^2}\right],$$

which is harmonic in $y < 0$, vanishes at the surface, and has vanishing gradient at infinity. By Liouville's theorem, the function is equal to zero in the entire lower half-plane, resulting in the equation

$$\left(\frac{\partial^2}{\partial t^2} + g\frac{\partial}{\partial y}\right)\Phi_1(x,y,t) = -\frac{\Gamma g}{\pi}\left[\frac{x-a(t)}{(x-a(t))^2 + (y+b(t))^2}\right]. \qquad (7)$$

The solution of (7) can be written in the form

$$\Phi_1(x,y,t) = -\frac{\Gamma g}{\pi}\int_0^\infty\int_0^t \frac{\sin(\sqrt{gk}\,(t-\tau))}{\sqrt{gk}}\, e^{k(y+b(\tau))}\sin(k(x-a(\tau)))\,dk\,d\tau \quad (8)$$

which is equivalent to Eq. 13.54 in WEHAUSEN and LAITONE [7].

One can then determine the position of the free surface from the expression

$$\eta(x,t) = -\frac{1}{g}\,\Phi_t(x,0,t) \quad . \qquad (9)$$

Upon substitution of (4) and (8) this becomes

$$\eta(x,t) = \frac{\Gamma}{\pi}\int_0^\infty\int_0^t \cos(\sqrt{gk}(t-\tau))e^{kb(\tau)}\sin k(x-a(\tau))\,dk\,d\tau \quad . \qquad (10)$$

For comparison purposes it is useful to begin an investigation of the above exact, linearized, solution by determining the free surface displacement that would result if the vortex pair followed the trajectory they would if approaching a solid plane wall. First we must determine what this trajectory would be. Moving each vortex in the velocity field of its neighbor and the image system above the plane results in the set of coupled ordinary differential equations

$$\dot{a} = -\frac{\Gamma}{2\pi}\frac{a^2}{b(a^2+b^2)} \ , \qquad (11a)$$

$$\dot{b} = \frac{\Gamma}{2\pi}\frac{b^2}{a(a^2+b^2)} \ , \qquad (11b)$$

the exact solutions of which are hyperbolae (LAMB [2]).

$$\frac{a^2-2}{\sqrt{a^2-1}} = C_0 + \frac{\Gamma}{4\pi}t \ , \qquad (12a)$$

$$\frac{b^2-2}{\sqrt{b^2-1}} = -C_0 - \frac{\Gamma}{4\pi}t \ , \quad \text{where } C_0 = \frac{a_0^2-2}{\sqrt{a_0^{2}-1}} \qquad (12b)$$

where a_0 is the initial pair separation. The vortex trajectories, instantaneous vortex positions, and surface displacements at two different times, t = 1 and 3, are shown in Figs. 1 and 2.

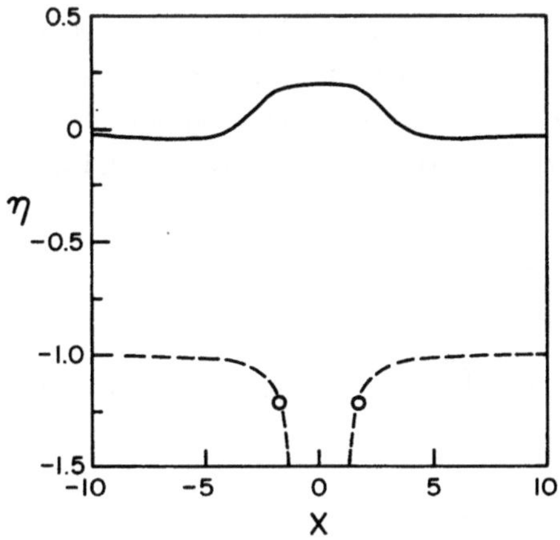

Figure 1. Surface displacement for vortices on plane boundary trajectories: t = 1.0. Dotted lines are the solution represented by (12a,b); circles indicate the instantaneous positions of the vortices.

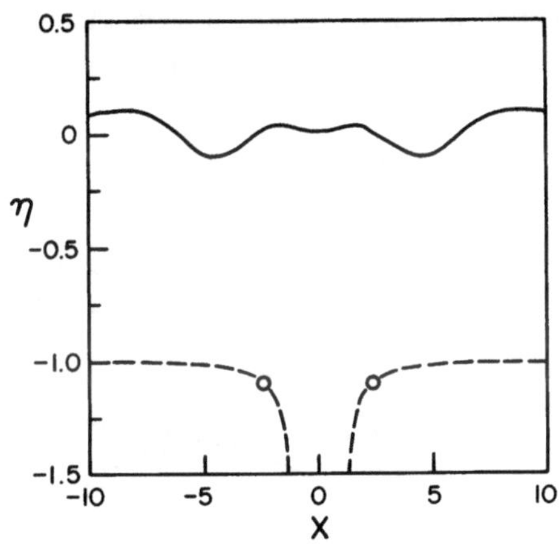

Figure 2. Same as Fig. 1, at t = 3.0.

This particular example does not account for the interaction between the vortex trajectories and the disturbance to the free surface, and therefore, at best, provides only a reasonable first approximation. To capture the effect of this interaction, each vortex must be moved in the velocity field of its neighbor, the image system above the plane, and that due to the disturbance potential associated with the instantaneous displacement of the surface. The following set of coupled integro-differential equations result from such an analysis:

$$\dot{a} = -\frac{\Gamma}{2\pi}\left(\frac{a^2}{b(a^2+b^2)}\right) + 2\int_0^t \int_0^\infty \sqrt{gk}\,\sin\sqrt{gk}\,(t-\tau)\,\sin^2(ka(\tau))e^{2kb(\tau)}dk\,d\tau\,. \tag{13a}$$

$$\dot{b} = \frac{\Gamma}{2\pi}\,\frac{2a^2+b^2}{a(a^2+b^2)} - 2\int_0^t \int_0^\infty \sqrt{gk}\,\sin\sqrt{gk}\,(t-\tau)\,\sin(2ka(\tau))e^{2kb(\tau)}dk\,d\tau\,, \tag{13b}$$

These were solved by a fourth-order Runge-Kutta scheme, Romberg quadrature being used for the integral term. A comparison of resulting trajectories with that for the plane boundary case is shown in Figure 3.

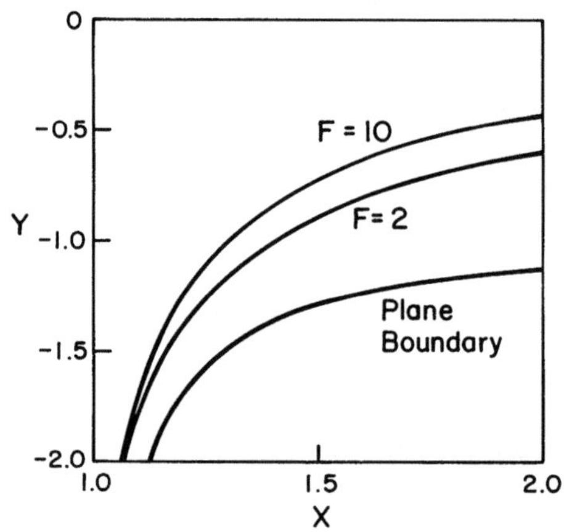

Figure 3. Comparison of vortex trajectories for a plane boundary with those for deformable boundaries.

3. <u>The Nonlinear Case</u>

Consider now the nonlinear problem. We wish to determine the unsteady potential field in the semi-infinite lower half plane whose upper boundary deforms in time subject to the full nonlinear boundary conditions.

The problem is complex because the trajectories of the vortices are in part determined by the instantaneous gradient of the disturbance potential, and their location in turn influences the evolution of the boundary and therefore the boundary conditions for the potential. Numerical methods are employed to obtain the exact solution to the problem.

We proceed somewhat differently than in the linear case. We dispense with the image system, and express the complete velocity potential as the sum of the potential associated with a pair of line vortices, and a disturbance potential due to the deformed surface. Length, time, velocity, and the velocity potential are nondimensionalized by a, $\sqrt{a/g}$, $\sqrt{ag}$, and $\sqrt{a^3g}$, respectively, where a is the initial pair separation. The problem can then be stated in the following equations:

$$\Delta\phi(x,y,t) = 0 \tag{14a}$$

$$\phi_x(0,y,t) = \phi_x(\infty,y,t) = \phi_y(x,-\infty,t) = \phi(x,y,0) = \phi_t(x,y,0) = 0 \tag{14b}$$

$$\left[\phi_t = -\frac{1}{2}\nabla\phi\cdot\nabla\phi - \eta(x,t)\right]_{y=\eta} \tag{14c}$$

$$\left[\eta_t + u\eta_x = v\right]_{y=\eta} \tag{14d}$$

$$\underline{u} = \nabla\left(\phi + \frac{F}{2\pi}\left(\tan^{-1}\left(\frac{x-a(t)}{y-b(t)}\right) - \tan^{-1}\left(\frac{x-a(t)}{y-b(t)}\right)\right)\right) \tag{14e}$$

$$\underline{u} = \frac{d\underline{x}}{dt}\ ,\ \underline{x} = (a(t),\ b(t)) \tag{14f}$$

where F is the Froude number. With this nondimensionalization, the Froude number appears in the velocity decomposition as a direct measure of vortex strength.

Equations (24) are solved by finite differences. Laplace's equation is discretized on a five point star in the traditional way:

$$a_{i,j}\phi_{i+1,j} + b_{i,j}\phi_{i-1,j} + c_{i,j}\phi_{i,j+1} + d_{i,j}\phi_{i,j-1} + e_{i,j}\phi_{i,j} = 0. \tag{15}$$

The kinematic and dynamic boundary conditions are expressed in finite difference form, with a Lax-type method, (16), used for the latter, and a predictor-corrector algorithm based on the MacCormack method, (17a-d), for the former:

$$\phi^{n+1}_{i,js(i)} = \frac{1}{2}\left(\phi^n_{i-1,js(i-1)} + \phi^n_{i+1,js(i+1)}\right) - \Delta t\left(\nabla\phi^n_{i,js(i)} \cdot \nabla\phi^n_{i,js(i)} - \eta^n_i\right) \quad (16)$$

$$\bar{\eta}^{n+1}_i = \eta^n_i - u^n_{i,js(i)} \frac{\Delta t}{\Delta x}\left(\eta^n_{i+1} - \eta^n_i\right) \quad (17a)$$

$$\eta^{n+1} = \frac{1}{2}\left[\eta^n_i + \bar{\eta}^{n+1}_i - u^n_{i,js(i)} \frac{\Delta t}{\Delta x}\left(\bar{\eta}^{n+1}_i - \bar{\eta}^{n+1}_{i-1}\right)\right] - v^n_{i,js(i)}\Delta t\,, \quad \text{if } u_{i,js(i)} > 0 \quad (17b)$$

$$\bar{\eta}^{n+1}_i = \eta^n_i - u^n_{i,js(i)} \frac{\Delta t}{\Delta x}\left(\eta^n_i - \eta^n_{i-1}\right) \quad (17c)$$

$$\eta^{n+1} = \frac{1}{2}\left[\eta^n_i + \bar{\eta}^{n+1}_i - u^n_{i,js(i)} \frac{\Delta t}{\Delta x}\left(\bar{\eta}^{n+1}_{i+1} - \bar{\eta}^{n+1}_i\right)\right] - v^n_{i,js(i)}\Delta t\,, \quad \text{if } u_{i,js(i)} < 0 \quad (17d)$$

where the subscript js(i) indicates a node at the free surface. It should be noted that when the convection velocity is positive, upwind differencing is used in the predictor equation and downwind differencing in the corrector; when the convection velocity is negative, this is reversed.

These difference equations are solved iteratively until the difference between two successive iterations of both the surface displacement and the boundary condition for the velocity potential fall within preassigned convergence limits.

At each time step the calculation proceeds in the following order: i) the dynamic boundary condition is advanced in time; ii) Laplace's equation is solved for the disturbance potential using the MSI procedure (SCHNEIDER and ZEDAN [6]), an iterative scheme, the backbone of which is an LU decomposition; iii) the velocity field is constructed as a superposition of the vortex-induced velocities and the gradient of the disturbance potential; iv) the new position of the free surface is calculated by advancing the kinematic boundary condition in time; v) steps i) through iv) are repeated until the convergence criterial are satisfied; vi) the finite difference cells within which the vortices lie are determined, and the velocity is then interpolated from the velocities at the four nearest grid points, similar to the procedure employed in the MAC method (HARLOW et al., [1]) for locating marker particles; vii) the position of the vortices is advanced by solving the trajectory equations (14f) using a 4th-order Runge-Kutta scheme.

Interior nodes, points surrounded on all sides by regular grid points, are treated by standard second-order differencing. For nodes adjacent to the free surface values of the disturbance potential are calculated by means of irregular finite-difference templates. To calculate derivatives at the surface Taylor series expansions about adjacent nodes are used. With the exception of the above, a uniform grid was employed in all the calculations. Because of the assumed symmetry of the flow, calculations need be performed in the right half-plane only.

The position of the free surface is found to be subject to numerical instabilities. A linear filtering scheme is adopted (SHAPIRO [5]) to deal with this. This procedure, employed successfully in surface wave problems (LONGUET-HIGGENS and COKELET [3]), when applied to both the kinematic and dynamic boundary conditions was effective in smoothing the instability.

4. Results and Discussion

Results are presented for two cases, F = 1.0 and 0.1. The accuracy and self-consistency of the numerical calculations were validated by grid and time-step refinement studies.

Case 1: F = 1.0

The time step and mesh width were 0.001 and 0.25, respectively. The evolution of the surface at time intervals of 1.0 is shown in Figure 4. The velocity fields and surface displacements at t = 2 and 6 are displayed in Figures 5 and 6.

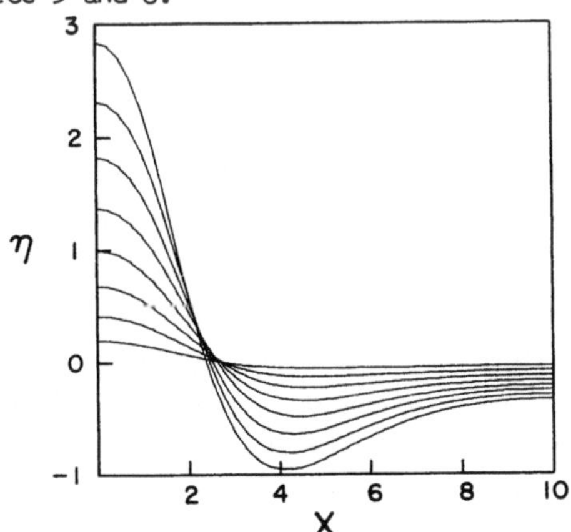

Figure 4. Surface displacement at successive times: non-linear boundary conditions, F = 1.0, t = 1-8 for time increments of 1 unit.

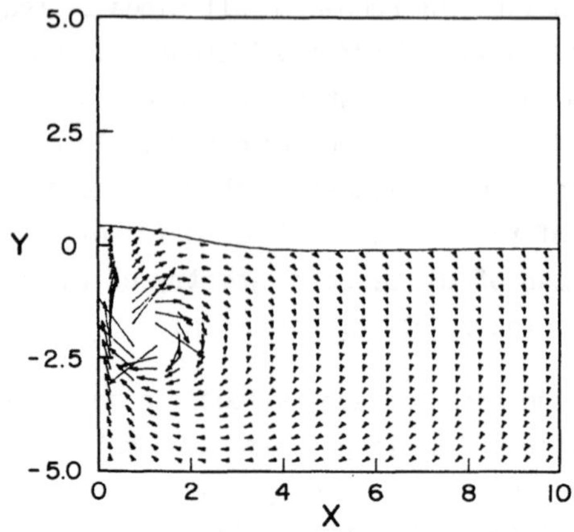

Figure 5. Velocity field and surface displacement: F = 1.0, t = 2.

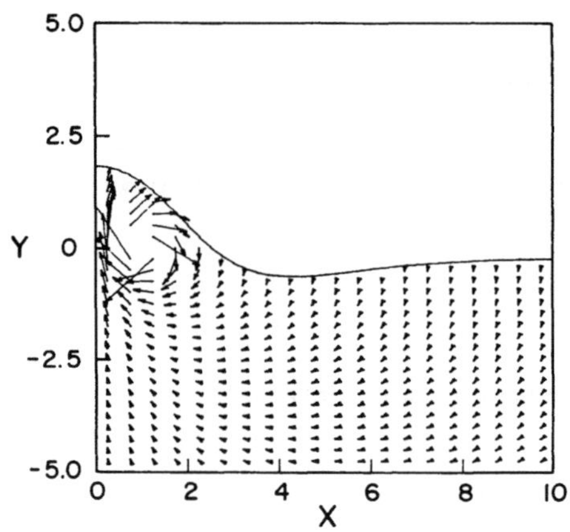

Figure 6. Same as Fig. 5, at t = 6.

We note that the vortices appear to rise primarily under their mutual influence, their trajectories not much affected by the presence of the free surface. The surface deforms into a mounded shape, the vortices rising above the level of the undisturbed surface. At t = 6 surrounding fluid has been entrained into the mound, the sides of which are becoming quite steep, with breaking appearing imminent. Failure of the numerical solution shortly thereafter was inevitable because the numerical scheme employed fails when the surface height becomes multi-valued at a given x-location.

We particularly note that a trough forms to the right of the vortex, and deepens as the vortex rises. This has the characteristics of the "scarring" observed in the experiments of Sarpkaya and Henderson.

Case 2: $F = 0.1$

This case is one of lesser vortex strength. The same time step and mesh width were used as for the previous case. Filtering was required at every time step from the initial start.

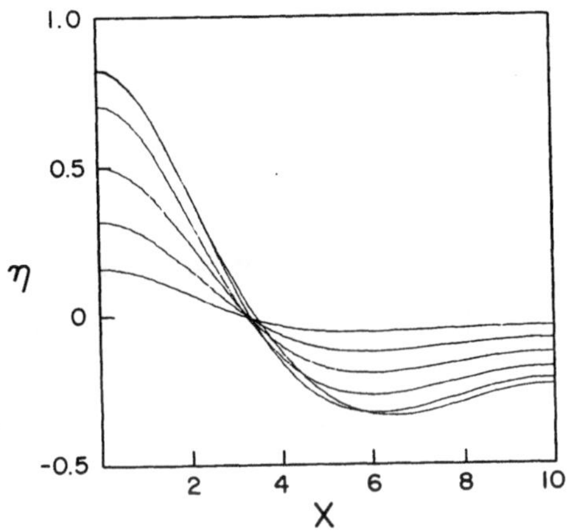

Figure 7. Surface displacement at successive times: non-linear boundary conditions, $F = 0.1$, $t = 1$-6 for time increments of 1 unit.

The evolution of the free surface at six time intervals is shown in Figure 7. The sides of the central upwelling have a gentler slope than for $F = 1$. The trajectory of the vortex is deflected away from the vertical and towards the trajectory assumed by a vortex pair approaching a solid plane boundary. As in the case of greater vortex strength, a scar forms as the vortex rises. It is broader and shallower, however, than in the previous case.

5. Conclusion

We have analyzed the inviscid interaction between a pair of trailing vortices and a free surface. A solution to the linearized problem has been presented. A numerical method of solution for the nonlinear problem has been described, and calculations carried out for a range of Froude numbers.

Strong vortices are found to be little affected by the presence of the surface, rising to form a bubble-like disturbance on the surface and bursting through. Weaker vortices are deflected away from the surface and towards the trajectory assumed by vortices approaching a plane solid boundary (Fig. 8)

The scarring phenomenon described by Sarpkaya and Henderson was observed in the calculations.

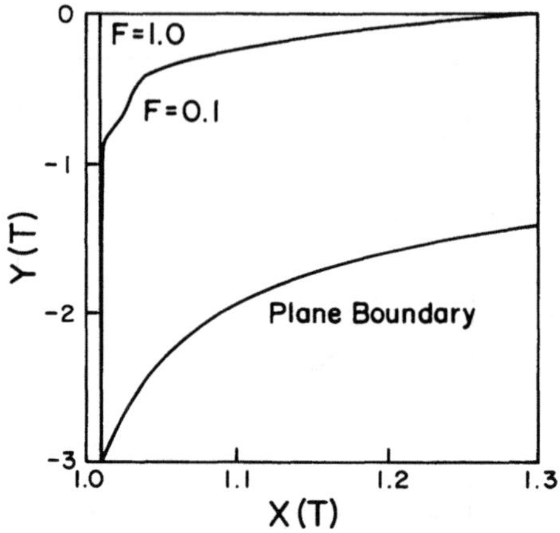

Figure 8. Vortex trajectories: F = 1.0, 0.1 and plane solid boundary cases.

6. Acknowledgements

The authors wish to acknowledge very helpful discussions with J.V. Wehausen and T. Sarpkaya. This work was supported by the Office of Naval Research under Contract No. N00014-85-K-0509.

References

1. F.W. Harlow, J.E. Welch, J.P. Shannon, B.J. Daly: The MAC Method: A Computing Technique for Solving Viscous, Incompressible, Transient Fluid Flow Problems Involving Free Surfaces, Los Alamos Scientific Laboratory Report LA-3425 (1966).

2. H. Lamb: Hydrodynamics. 6th Ed., Cambridge University Press (1932).

3. M.S. Longuet-Higgens, E.D. Cokelet: Proc. R. Soc. Lond. A 350, 1-26 (1976).

4. T. Sarpkaya, D.O. Henderson, Jr.: "Free Surface Scars and Striations Due to Trailing Vortices Generated by Submerged Lifting Surfaces", Paper 85-0445, presented at AIAA 23rd Aerospace Sciences Meeting, Jan. 14-17, 1985, Reno, NV.

5. R. Shapiro: Math. Comp. 29, 1094-1097 (1975).

6. G.E. Schneider, M. Zedan: J. Num. Heat Transfer 4, 1-19 (1981).

7. J.V. Wehausen, E.V. Laitone: Surface Waves, Handbuch der Physik IX, Springer-Verlag (1960).

Microbubble Formulation and Splitting in a Turbulent Boundary Layer for Turbulence Reduction

James C.S. Meng and James S. Uhlman, Jr.

1. Introduction

Bubble formation has traditionally been addressed by chemical engineers for bubbles generated in a stationary liquid. With the advent of microbubble drag reduction, the emphasis has been placed upon obtaining quantitative relationships of bubble formation in a high speed turbulent boundary layer (TBL) and on identifying if bubble splitting is a possible mechanism for absorbing turbulence energy and therefore turbulence reduction. Little is known about the bubble formation mechanisms, and as a result, no quantitative relationship exists among the gas ejection rate (Q_{pore}), pore diameter (d_{pore}), and liquid flow speed (U_o) for practical applications. We have conducted a study to address both issues of: what mechanisms dominate during a bubble formation in a TBL, how to establish quantitatively the relationship among Q_{pore}, d_{pore} and U_o and the resulting turbulence reduction.

The purpose of this approach is to offer some insight into the key mechanisms. This approach, although it is theoretical in nature, adopts as much empirical information as possible to aid in establishing the bubble size regimes, bubble size spectrum, bubble splitting and bubble transport in a TBL. Based on this result, a concept of the possible mechanism for microbubble drag reduction based on energetics, is also presented. We made an estimation of the total amount of turbulent kinetic energy needed to split bubbles in a turbulent boundary layer and compared that with the total available energy in a TBL. From this estimation, we can identify whether the bubble splitting is a plausible basic mechanism for reducing the turbulence.

2. Theoretical Aspects of Bubble Formation Dynamics

2.1 Hydrodynamics of Bubble Generation

2.1.1 Hydrodynamic Forces at Bubble Formation

The physics governing the formation or "breakaway" of a microbubble at a pore in a wall beneath a moving liquid are quite complex. In order to model the physics properly one would have to take into consideration effects of low Reynolds number viscous and inertial forces, wall effects, buoyancy, surface tension and free-surface phenomena including surface chemistry; all varying with time. To model such a problem analytically would require an immense effort. Our goals here are much more modest. By identifying various hydrodynamic forces and performing approximate force balances, both parallel and perpendicular to the wall, we can obtain equations which yield order-of-magnitude estimates of the bubble "break-away" size under various assumptions. Therefore, the first step is to identify and quantify the various forces, both tangential and normal to the wall. We will start with the tangential forces.

If one considers a bubble being exuded from a pore in a wall into a liquid flow parallel to the wall. the flow exerts a drag force on the bubble. Due to this drag force the bubble's shape will be altered, becoming skewed. This skewed bubble shape will introduce a component of surface tension force parallel to the wall which balances the drag force. The bubble will detach from the wall when some critical skew angle is reached or when either a lift force or the buoyancy overcomes the vertical component of surface tension, whichever occurs first.

For the tangential force balance case the balance is between the drag on the bubble and the horizontal component of surface tension force. The latter force was obtained by AL-HAYES and WINTERTON [3, 4]. They calculated the horizontal component of surface tension force from the following equation

$$F_{st} = - \{ \frac{58}{\theta_o+5} + 0.14\} \times \frac{1}{2} \pi R\sigma \sin\theta_o$$
$$\times (\cos\theta_r - \cos\theta_a)$$

where σ is the surface tension between air and water, θ_o is the equilibrium contact angle between the bubble and the wall, i.e. $\sin \theta_o = \frac{r}{R}$ and θ_r and θ_a are the receding and advancing contact angles (Fig. 2.1), respectively, when the bubble is in a cross flow. Values for the various angles involved were determined experimentally and the quantity in brackets is a correction factor that Al-Hayes and Winterton applied to their equation to improve the comparison between theory and experiment. Following Al-Hayes and Winterton one may assume that the angles are

$\theta_o \sim 40°$, $\theta_r \sim 30°$, and $\theta_a \sim 50°$, then $F_{st} = - 0.16 \pi R\sigma \sin\theta_o$. (2.1)

If one now assumes that the drag force which balances the surface tension is

$$F_D = \frac{1}{2} \rho C_D U^2 \pi R^2 \quad ,$$

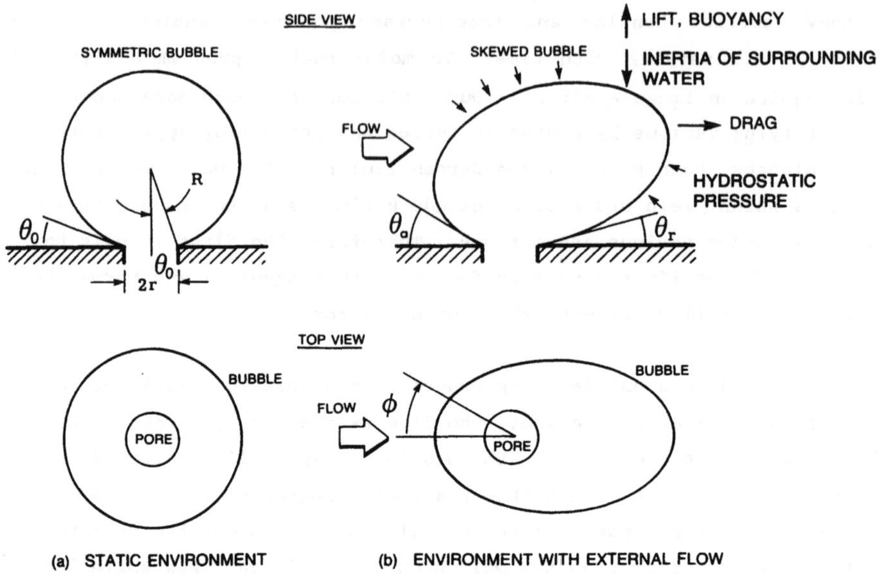

Figure 2.1. Bubble formation geometry and nomenclature in static and flowing environments.

substituting $C_D = 1.22$ and θ_o, θ_a, θ_r values, one obtains the following drag-surface tension force balance equation

$$\rho \frac{U^2R}{\sigma} = 0.17 \qquad (2.2)$$

Another formulation of the balance between the drag force on the bubble and the horizontal component of surface tension force can be derived by assuming that the contact angle of the bubble has a simple sinusoidal dependence. With this assumption

$$\theta = \theta_o + \delta\theta \cos\varphi$$

where φ is the angle about the base of the bubble clockwise from upstream (Fig. 2.1).

Hence

$$\theta_a = \theta_o + \delta\theta.$$

and

$$\theta_r = \theta_o - \delta\theta$$

The horizontal component of the surface tension force is then

$$F_{st} = r\sigma \int_0^{2\pi} \cos[\Theta_o + \delta\Theta \cos\varphi] \cos\varphi \, d\varphi$$

$$= -2\pi\, r\sigma \sin\Theta_o J_1 (\delta\Theta)$$

where J_1 is the Bessel function of order one (GRADSHTEYN and RYZHIK, [20] However,

$$J_1(\delta\Theta) . \frac{1}{2}\delta\Theta \qquad \text{for } \delta\Theta<<1,$$

thus we find

$$F_{st} \simeq - \pi r\sigma \sin\Theta_o \, \delta\Theta. \qquad (2.3)$$

According to the data of AL-HAYES and WINTERTON [3, 4] the quantities $|\Theta_o - \Theta_a|$ and $|\Theta_o - \Theta_r|$ appear to be fairly constant with a value of approximately 10°. thus $\delta\Theta$. 0.17. Thus (2.3) differs from (2.1) by $\frac{r}{R}$, which presumably is on the order one.

If the drag force is now assumed to follow Hadamard's law because the bubble Reynolds number is very low, then the drag force to surface tension force balance becomes

$$4\pi\mu RU \simeq \pi\ \sigma \frac{r}{R}^2 \quad \delta\Theta \qquad . \qquad (2.4)$$

Proceeding to the normal force balance case, the normal component of surface tension force can be calculated similarly to the tangential case, and we have

$$F_{sn} = 2\pi\ r\ \sigma \sin\Theta_o J_o (\delta\Theta) \simeq 2\pi\sigma \frac{r^2}{R} \ .$$

Another normal force to take into consideration is the low Reynolds number "Saffman" lift force (see SAFFMAN [40, 41]). its effect is to lift the bubble away from the wall. Strictly speaking, the Saffman lift result applies only to a particle in an unbounded fluid. However, it is applied here to a bubble near a wall for the purpose of engineering order-of-magnitude estimate.

VOLOSHKO et al [49] identified other normal forces to be;

the added mass term $\varepsilon_m \dfrac{\pi\rho r^4 v_o^2}{12R^2}$, the gravity/ buoyancy term $\frac{4}{3}\pi R^3(\rho - \rho')g$ and the gas injection momentum flux term $\rho' v_o^2 \pi r^2$, where ϵ_m is the added mass coefficient, ρ' is the gas density and v_o is the velocity of the gas as it exits from a pore. If in addition the Saffman lift term,

$C_L \pi\rho R^2 U(\nu k)^{1/2}$,

and surface tension terms are also included, following VOLOSHKO et al [49] a modified Voloshko equation governing bubble normal force balance becomes

$$\epsilon_m \frac{\pi\rho r^4 v_o^2}{12 R^2} + 2\pi \frac{r^2\sigma}{R}$$

$$= \frac{4}{3}\pi R^3 (\rho - \rho')g$$

$$+ C_L\pi\rho R^2 U(\nu k)^{1/2} + \rho' v_o^2 \pi r^2, \qquad (2.5)$$

where U is the flow past the wall at the bubbles' centerline, ν is the kinematic viscosity of the liquid and k is the effective shear in the liquid at the bubble centerline.

2.1.2 Bubble Size Regimes

Continuing the bubble breakaway analysis we have extended the basic bubble hydrodynamics to include analysis of the bubble geometries other than a semi-sphere. More important is the fact that we have developed a mathematical model to analyze the bubble size spectrum which may be a natural consequence of generating bubbles in a turbulent boundary layer.

In contrast to the commonly referred to bubble geometry, as depicted in Fig. 2.1, as much as four bubble regimes can be identified, as are indicated in Fig. 2.2. The sheared spherical cap case was attributed to MACINTYRE [28], while the other three cases were identified by SILBERMAN [46]. These four regimes can be delimited by comparison of the ratio of the gas pore exit velocity with the liquid velocity.

For the sheared spherical cap case Fig. 2.2(a), the height $h = R - (R^2 - r_{pore}^2)^{1/2}$, where R is the radius of curvature of the cap, and the cap volume $= \frac{\pi}{6} h(h^2 + 3r_{pore}^2)$ is assumed to be the bubble volume after it issheared off the pore. Since the spherical cap cannot be sheared off the pore faster than the fluid flow over the pore, we find that

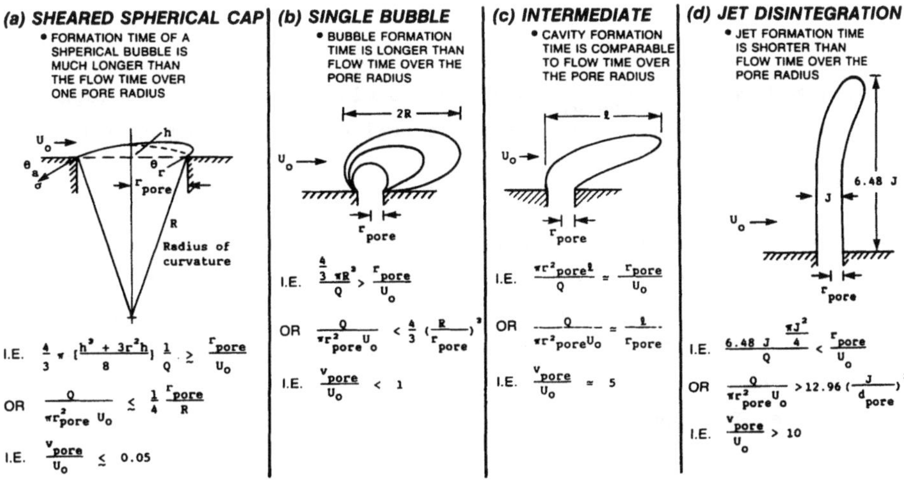

Figure 2.2. Bubble formation time analysis.

$$\frac{\pi\, h\, (h^2 + 3r_{pore}^2)}{6\, Q} \gtrsim \frac{r_{pore}}{U_o} ,$$

where Q is the volumetric gas flow rate.

Replacing h in terms of R and r_{pore} and taking the limit of small $\frac{r_{pore}}{R}$, the above becomes,

$$\frac{\pi r_{pore}^3}{4Q} \left(1 + \frac{1}{12} \left(\frac{r_{pore}}{R}\right)^2\right) \gtrsim \frac{R}{U_o} ,$$

or

$$\frac{Q}{\pi r_{pore}^2\, U_o} \lesssim \frac{1}{4} \frac{r_{pore}}{R} .$$

Judging from the Fig., $\frac{r_{pore}}{R} < 0.2$, and replacing Q by $\pi r_{pore}^2\, v_{pore}$, we obtain $\frac{v_{pore}}{U_o} \lesssim 0.05$.

In the single bubble regime, Fig. 2.2(b), the bubble formation time is longer than the time that it takes for the fluid to flow over the pore. Since the bubble in this case has the time to grow to a nearly spherical shape, the bubble volume is $\frac{4}{3} \pi R^3$. The time scale for the bubble growth is then $\frac{4}{3} \frac{\pi R^3}{Q}$ which is greater than the flow time, i.e.,

$$\frac{4}{3} \frac{\pi R^3}{Q} > \frac{r_{pore}}{U_o}$$

or

$$\frac{Q}{\pi r_{pore}^2 U_o} < \frac{4}{3} \left(\frac{R}{r_{pore}}\right)^3 .$$

Replacing Q by $\pi r_{pore}^2 v_{pore}$, and using the minimum value for the right hand side we find for the single bubble regime the following should apply,

$$\frac{v_{pore}}{U_o} \lesssim 1.$$

In the jet disintegration regime, Fig. 2.2(d), the jet formation time is shorter than the flow time as a result of high gas injection rate, so that it is the instability of the gas jet that will determine the final bubble size. RAYLEIGH [37] studied this problem and found that the wavelength of the maximum amplification of instability is 6.48 times the mean jet diameter J. Therefore the

jet formation time is $\dfrac{6.48\ J\ \frac{\pi J^2}{4}}{Q}$ and should be shorter than the flow time $\dfrac{r_{pore}}{U_o}$. In other words,

$$\frac{Q}{\pi r_{pore}^2 U_o} > 12.96 \left(\frac{J}{d_{pore}}\right)^3 .$$

Replacing Q by $\pi r_{pore}^2 v_{pore}$ and using the maximum value for the right

hand side, we find $\dfrac{v_{pore}}{U_o} \gtrsim 10$ for the jet regime.

The intermediate regime, Fig. 2.2(c), corresponds to a bent jet, and the bubble breakaway mechanism is due more to the liquid dynamics than that of the jet itself.

In this regime, the formation time analysis gives $\dfrac{v_{pore}}{U_o} \cong 5$. We have thus completed a description of different bubble regimes.

Naturally, different hydrodynamic mechanisms control the bubble formation in different regimes and different dependences of the bubble size on v_{pore}, U_o, and the surface tension will result. However, it is important to note that one single parameter, i.e., $\dfrac{v_{pore}}{U_o}$, emerges from this analysis as the key parameter demarcating the different regimes.

Summarized in Fig. 2.3 are the approximate values of bubble diameter criteria. The ordinate is the ratio d_{bubble}/d_{pore}, the abscissa is the "injection coefficient", or more precisely the ratio v_{pore}/U_o that was identifed to be the key parameter in the bubble formation time analysis. As is noted in Fig. 2.3, two distinct regions are identified;

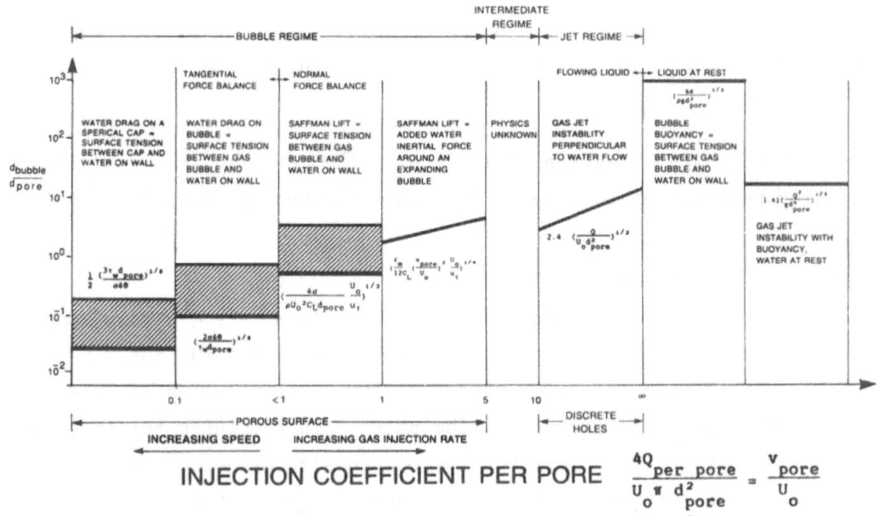

Figure 2.3. Approximate realms of bubble diameter criteria.

one, the liquid at rest which corresponds to the $\frac{v_{pore}}{U_o} \to \infty$ limit, and two, the flowing liquid case.

In the liquid at rest case, the bubble sizes are determined by two possible normal force balances (assuming the wall to be normal to the gravity force vector). Notice that in this case, the tangential force balance is not operative. First, at low gas injection rate, the bubble buoyancy is in equilibrium with the surface tension between the gas bubble and the wall, i.e.,

$$\frac{4}{3}\pi\left(\frac{d_{bubble}}{2}\right)^3 \rho_{liquid} g = \pi d_{pore}\sigma$$

or

$$\frac{d_{bubble}}{d_{pore}} = \left(\frac{6\sigma}{\rho g d_{pore}^2}\right)^{1/3} . \qquad (2.6)$$

Second, at higher gas injection rate, SILBERMAN [46] applied three assumptions to obtain the following relationship,

$$\frac{d_{bubble}}{d_{pore}} = 1.41\left(\frac{Q^2}{g d_{pore}^5}\right)^{1/5} . \qquad (2.7)$$

The three assumptions are that, the unstable wavelength obtained by Rayleigh in a stationary environment is valid in a shearing flow, the Bernoulli's equation can relate the jet velocity and jet height against gravity, and the diameter of the bubble is related to the liquid velocity U_o the same way as the gas jet velocity. The last assumption is the weakest.

In the flowing liquid case, the physics governing bubble size may be divided into at least three regimes which, in turn, can be further divided into smaller

groups. At the high injection limit, i.e., $\frac{v_{pore}}{U_o} > 10$, the gas jet instability perpendicular to water flow was obtained by SILBERMAN [46],

$$\frac{d_{bubble}}{d_{pore}} = 2.4 \left(\frac{Q}{U_o d_{pore}^2}\right)^{1/2} \quad . \tag{2.8}$$

Notice the dependence of bubble diameter on the flow rate Q. This result has been verified by Silberman and recently by REISCHMAN and HOLZMANN [38]. In practice, this high injection rate limit typically applies to an ejection using discrete drilled holes. The physics in the intermediate regime is unknown, an analytic formula for the bubble size cannot be easily obtained. As the injection coefficient reduces, we move into the single bubble regime, i.e., during the bubble formation the gas cavity appears to resemble a bubble geometry. In this regime, two distinct force balances exist; the normal and tangential force balances. It is intuitively convincing that in a rapidly flowing liquid, if the gas exit velocity equals approximately that of the liquid tangential velocity the mechanism of bubble breakaway should be determined by a normal force balance. As the gas exit velocity reduces further, or as the liquid velocity increases while the gas exit velocity holds constant, the dominant force balance switches to the tangential mode.

In the normal force balance mode, at least two possibilities exist. At the high $\frac{v_{pore}}{U_o}$ end, the Saffman lift force balances the inertial force surrounding an expanding bubble. From (2.5), we find, from the modified Voloshko equation that for this case,

$$\varepsilon_m \left[\frac{\pi\rho\, r_{pore}^4\, v_{pore}^2}{12R^2}\right] = C_L \pi\rho R^2 U_o u_\tau$$

where u_τ is the local friction velocity, and ρ the liquid density. Rearranging, we find,

$$\frac{d_{bubble}}{d_{pore}} = \left[\frac{\varepsilon_m}{12C_L} \left(\frac{v_{pore}}{U_o}\right)^2 \frac{U_o}{u_\tau}\right]^{1/4} \tag{2.9}$$

which does depend upon the injection coefficient to the 1/2 power. For the lower $\frac{v_{pore}}{U_o}$ case, Saffman lift will balance the surface tension between the gas bubble and water on wall, i.e.,

$$2\pi \frac{r_{pore}^2}{R} \sigma = C_L \pi \rho R^2 U_o u_\tau \quad ,$$

or

$$\frac{d_{bubble}}{d_{pore}} = \left(\frac{4\sigma}{\rho U_o^2 C_L d_{pore}} \frac{U_o}{u_\tau}\right)^{1/3} \quad , \tag{2.10}$$

where $\frac{U_o}{u_\tau} = \left(\frac{2}{c_f}\right)^{1/2}$, and c_f is the local skin friction coefficient on the wall. Notice that this result does not depend upon the injection coefficient. Questions may arise as to whether there are any other normal force balances? This question can be addressed by examining the modified-Voloshko equation below for a flowing liquid.

$$\varepsilon_m \frac{\pi\rho\, r_{pore}^4\, v_{pore}^2}{12\, R^2} + 2\pi \frac{r_{pore}^2 \sigma}{R}$$

$$= \frac{4}{3} \pi R^3 (\rho - \rho')g + C_L \pi \rho R^2 U_o u_\tau + \rho' v_{pore}^2\, \pi r_{pore}^2 \, ,$$

where ε_m is an entrainment coefficient, ρ' the gas density, u_τ the friction velocity. Dividing the above equation by $\rho U_o^2 r_{pore}^2$, we find

$$\frac{\pi\varepsilon_m}{12} \left(\frac{r_{pore}}{R}\right)^2 \left(\frac{v_{pore}}{U_o}\right)^2 + 2\pi \frac{\sigma}{U_o \mu} \frac{\nu}{R U_o}$$

$$= \frac{4\pi}{3} \frac{g r_{pore}}{U_o^2} \left(\frac{R}{r_{pore}}\right)^3 + \pi C_L \left(\frac{R}{r_{pore}}\right)^2 \left(\frac{c_f}{2}\right)^{1/2} + \pi \frac{\rho'}{\rho} \left(\frac{v_{pore}}{U_o}\right)^2 .$$

Substituting

$$\varepsilon_m = 32, \quad \frac{\sigma}{U_o \mu} \cong 7, \quad \frac{\nu}{U_o R} \cong 10^{-2}, \quad c_f \cong 3 \times 10^{-3}$$

$$\frac{g r_{pore}}{U_o^2} \cong 10^{-6}, \quad \pi C_L = 6.46, \quad \frac{\rho'}{\rho} \cong 10^{-3}, \text{ for } U_o = 10 \text{ m/sec}, \; r_{pore} = 10 \; \mu\text{m}$$

we can write the above equation as follows,

$$8.3 \left(\frac{r_{pore}}{R}\right)^2 \left(\frac{v_{pore}}{U_o}\right)^2 + 0.4 = 10^{-6} \left(\frac{R}{r_{pore}}\right)^3 + 0.25 \left(\frac{R}{r_{pore}}\right)^2 + 10^{-3} \left(\frac{v_{pore}}{U_o}\right)^2 .$$

From this, we see that we can neglect the buoyancy term (the first term on RHS) as compared with the Saffman's lift term (the second term on RHS). The balance type of force depends upon the magnitude of only two parameters;

$\frac{R}{r_{pore}}$ and $\frac{v_{pore}}{U_o}$. For $\frac{v_{pore}}{U_o} \cong 1$, the first term on the left-hand-side i.e., the water inertia surrounding an expanding bubble, balances the Saffman lift, and gives a result of $\frac{R}{r_{pore}} \cong 2.4$. For $\frac{v_{pore}}{U_o} \cong 0.1$, again the Saffman lift force dominates the right-hand-side of the equation, while the dominant term on the left-hand-side shifts to the surface tension term (the second term), and gives a lower result of $\frac{R}{r_{pore}} \cong 1.2$. No other physically meaningful possibility seems to exist, therefore, we conclude the discussion of the normal force balance.

For the tangential force balance case, only one possible combination exists, i.e., that the water drag on gas bubble, or on a gaseous cap, balances the surface tension between the gas bubble and the wall. As indicated earlier in Fig. 2.2, two bubble breakaway geometries are possible. One is a nearly spherical bubble case and the other one is a nearly spherical cap case. In reality, neither is true. In fact, unstable waves on bubble surface should be present due to the unsteady forcing of the turbulent flow over them, and also due to the liquid local pressure gradient generated by the growing bubble. Whether the bubbles will breakaway as a result of this unstable wave growth is unknown at the present time. A water tunnel experiment on a single bubble formation visualized at close range is a plausible approach to address this issue.

For the spherical bubble case, we have

$$\pi R^2 \, \tau_w = \pi\sigma \frac{r_{pore}^2}{R} \, \delta\Theta \, ,$$

where $\delta\Theta \cong 0.17$ as given before, and τ_w is the local wall shear stress. From this, we find,

$$\frac{d_{bubble}}{d_{pore}} = \left(\frac{2\sigma\delta\Theta}{\tau_w d_{pore}}\right)^{1/3} \qquad (2.11)$$

Alternatively, we can replace τ_w by $\frac{1}{2}\rho U_o{}^2 C_D$, where C_D is the draf coef-coefficient over the bubble. But this also opens questions about the validity of using U_o as the relative liquid velocity over the bubble, and also that C_D is not known until the bubble Reynolds number is known which requires knowledge of the bubble size a priori. As the injection coefficient reduces, we move into the spherical cap regime. For the tangential force balance, the tangential surface tension force is described by (2.3). The shear force is $2\pi R h \tau_w$, where $h = R(1 - (1 - (\frac{r_{pore}}{R})^2)^{1/2})$ or h

$\simeq \frac{1}{2} r_{pore} \frac{r_{pore}}{R}$ where small $\frac{r_{pore}}{R}$ is applied. Equating them, we find $R \simeq \frac{\sigma\delta\Theta}{\tau_w}$. Substituting the values of $\sigma, \delta\Theta$, and $\tau_w = \rho u_\tau^2$ at 10 m/sec, we find $R \simeq 74$ μm which is 7 times greater than the pore size (10 μm.), consistent with our assumption that $\frac{r_{pore}}{R} < 1$. The bubble diameter after breakaway, d_{bubble}, is related to the cap height h by $d^3_{bubble} = h^3 + 3r^2_{pore} h$. Substituting h and carrying out simplications, we have

$$d^3_{bubble} = r^3_{pore} \left(3 \frac{r_{pore}}{2R} + \left(\frac{r_{pore}}{2R}\right)^3\right) \cong \frac{3}{2} r^3_{pore} \frac{r_{pore}}{R}$$

or

$$d_{bubble} = \left(\frac{3}{2} \frac{r_{pore}}{R}\right)^{1/3} r_{pore}$$

i.e.,

$$\frac{d_{bubble}}{d_{pore}} = \frac{1}{2} \left(\frac{3}{4} \frac{\tau_w d_{pore}}{\sigma\delta\Theta}\right)^{1/3},$$

an interesting result in that the shear stress dependence is the reverse of that for the single bubble regime case (2.11). In the spherical cap regime, bubbles will be smaller than the pore size, and increase in size with τ_w, while in the single bubble regime, bubbles will be greater than the pore size, and decrease with the local τ_w.

2.1.3 Estimation of Pore Characteristics of a Porous Surface

Porous shells and flat plates have been used extensively for microbubble drag reduction. These surfaces are typically formed by pressing and sintering powdered metal particles. The void space among the powders provide the required porosity. Estimation of the pore characteristics is essential for predicting and controlling the bubble size. In this subsection, we derive the required relationship among the pore size, number of pores per unit area and percent surface porosity.

Porous titanium shells and plates are traditionally used in the industry for filtration purposes. The pore size is typically specified in terms the size of the smallest particles filtered out, instead of the pore size *per se*. In liquid flows, titanium filters made by Gould will normally retain particles approximately 1/3 of the mean pore size. No generalization to gas flows was made, but the retention size is known to be much finer.

Table 2.1 Gould titanium standard filtration grades.

Grades	Size Range mean pore size (Microns)	Smallest particle filtered out in Liquid (Microns)
Ti-2003	3	1
Ti-2505	5	2
Ti-4010	10	3
Ti-5015	15	5
Ti-6525	25	8

The distribution of pore size is not available, but given the fact that particles 3 times smaller than the mean pore size are filtered out, there must be a sharp drop-off in the large pore size end of spectrum. The standard deviation in pore size σ_{pore} must be smaller than the mean pore size $\bar{d}_{pore}$. Therefore for engineering purposes, one may assume $\sigma_{pore} = \frac{1}{2} \bar{d}_{pore}$, and that the pores follow a Gaussian distribution, i.e.,

$$p(d_{pore}) = \frac{1}{\sqrt{2\pi}\, \sigma_{pore}} e^{-\frac{1}{2}\left(\frac{d_{pore} - \bar{d}_{pore}}{\sigma_{pore}}\right)^2} .$$

This distribution can be used to make engineering estimates of pore size and bubble sizes.

To calculate the gas flow rate through individual pores, it is necessary to know the surface porosity, i.e. the percent area occupied by pores. Typically, a "density" is given which indicates the ratio of the weight of the porous material to that of the solid material of the same nominal volume. Let us denote this "density" by ξ, then it is seen that

$$\xi = \frac{\rho_{solid} V_{solid}}{\rho_{solid} V_{total}} = \left(\frac{\ell_{solid}}{L}\right)^3 ,$$

where ℓ_{solid} and L are lengths characterizing the solid part and the entirety of the porous material respectively. If $\ell^3_{solid} + \ell^3_{void} = L^3$, where ℓ_{void} is a length characterizing the void portion of the porous material, then

$$\frac{\ell_{void}}{L} \simeq (1 - \xi)^{1/3} \quad ,$$

so that the surface porosity η can be estimated to be

$$\eta \simeq \left(\frac{\ell_{void}}{L}\right)^2 = (1 - \xi)^{2/3} \quad . \tag{2.13}$$

Given a 45% "density" we find the surface porosity to be 67%.

It is sometimes useful to relate the mean powder size $\bar{d}_{powder}$ to the mean pore size $\bar{d}_{pore}$. This is possible if we assume that there are as many pores as powder particles. Then we can say

$$\frac{N_{powder} \frac{1}{6} \pi \bar{d}^3_{powder}}{V_{total}} = \xi$$

or

$$\frac{N_{pore} \frac{1}{6} \pi \bar{d}^3_{pore}}{V_{total}} = 1 - \xi$$

so that

$$\frac{\bar{d}_{powder}}{\bar{d}_{pore}} = \left(\frac{\xi}{1 - \xi}\right)^{1/3} \simeq 0.93, \text{ for } \xi = 0.45 \quad . \tag{2.14}$$

Once $\bar{d}_{pore}$ and η are found, the number of pores per unit surface area,

N_{pore}/S

can be calculated to be

$$\frac{N_{pore}}{S} = \frac{(1 - \xi)^{2/3}}{\frac{\pi}{4} \bar{d}^2_{pore}} \quad . \tag{2.15}$$

For $\bar{d}_{pore} = 10\mu m$, and $\xi = 45\%$, we have

$$\frac{N_{pore}}{S} = 0.85 \times 10^6 \text{ pores}/\text{cm}^2 \quad .$$

2.2 Bubble Size Distribution in a Turbulent Boundary Layer

From the analysis of bubble size spectrum discussed in section 2.1, one may realize that, given a flow speed U_o, by keeping one of the two parameters Q, and r_{pore} constant but varying the other one, bubbles of different sizes will be generated as a result of activating different physical mechanisms. For the commonly utilized porous ejectors made of pressed and sintered metal powders, the pore sizes are not uniform and the gas ejector's pore size spectrum will lead to a spectrum of bubble sizes. Furthermore, in a turbulent boundary layer environment, even if both of Q, r_{pore} are held constant, the local shear stress on and near the wall fluctuates. The absolute magnitude of the fluctuating shear stress is greater than twice the mean shear stress i.e., $\tau_{max} \sim \overline{2\tau_w}$. Since the local instantaneous τ_w determines nearly all modes of balances, bubbles generated in a turbulent boundary layer should have a bubble size spectrum with a bandwidth equal to at least two times of the mean bubble size. In this subsection, we discuss the shear stress fluctuations, analyze a bubble size spectrum obtained in the ocean, identify possible water, physics affecting bubble formation in seawater versus that in fresh and point out the possible bubble splitting and coalescence effects on a bubble size spectrum.

2.2.1 Shear Stress Fluctuation in a Turbulent Boundary Layer

The instantaneous τ_w in a turbulent boundary layer is a commonly measured variable, but most of data are presented in terms of the statistics of τ_w without providing its probability density function (PDF's). To name just a few relevant ones, BURTON [10] gave the ratio of the variance $\sigma(\tau_w)$ to the mean $\overline{\tau_w}$, $\frac{\sigma(\tau_w)}{\overline{\tau_w}} = 0.31$, SANDBORN [42] found the same ratio ranges from 0.2 to 0.4, KREPLIN and ECKLEMANN [27] determined the skewness $S(\tau_w) = 0.65$ and the flatness $F(\tau_w) = 3.5$ on the wall. Several sets of PDF of τ_w do exist but were measured at a small distance away from the wall, i.e., $y^+ \cong 1$ to 5. BRODKEY, WALLACE and ECKLEMANN [9] determined individually the PDF's of u as a result of ejection, sweep, inward and outward interactions,and superposed them to obtain the total PDF(u), at $y^+ = 3.4$. Since within $y^+ = 5$,the $P(u) \cong P(\tau_w)$ one can regard this resultant $P(u)$ to be representative of $P(\tau_w)$. BHATIA, DURST and JOVANOVIC [7] obtained $P(\tau_w)$ at $y^+ = 3.7$ to 6, while KREPLIN and ECKLEMANN (1979) obtained it at $y^+ = 1.6$ to 100. The most useful $P(\tau_w)$ for our purpose is given by SANDBORN [42] which was obtained in a wind tunnel with the local length Reynolds number equal to 10^7.

Figure 2.4 shows Sandborn's PDF(τ_w) data versus the normalized $\tau_w/\overline{\tau_w}$. Inspired by the lognormal distribution suggested by NAKAGAWA and NEZU [34] for the streak spacing λ^+, and also by the fact that λ^+ and τ_w are both positive variables, we have attempted to fit the $P(\tau_w)$ data by a lognormal distribution, i.e.,

$$P(\tau_w) = \frac{1}{\sqrt{2\pi}\sigma\tau_w} \exp\left(-\left(\frac{\ell n\tau_w - \mu}{2\sigma^2}\right)^2\right)$$

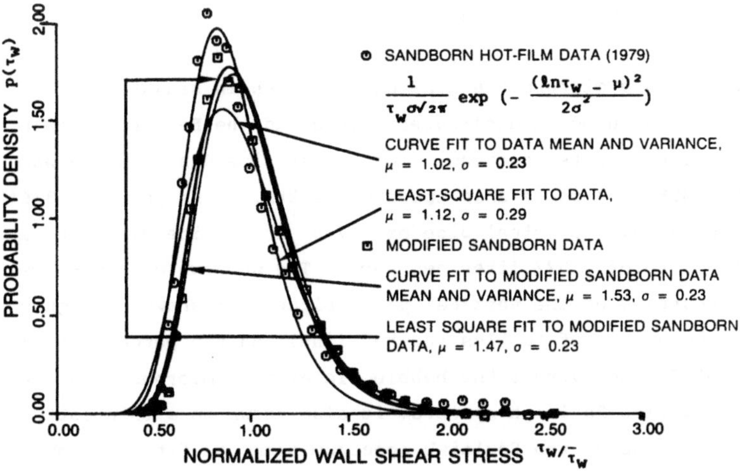

Figure 2.4. Normalized probability density distribution of surface shear stress over a flat plate.

where $\mu = \overline{\ln\tau_W}$ and $\sigma^2 = \overline{(\ln\tau_W-\mu)^2}$. Several possibilities exist. The obvious one is to first obtain μ and σ^2 from the raw data. As is seen, the curve fit to the data mean and variance is poor. As a comparison, the least square fit is also shown. It was indicated by Sandborn that the high shear end of data was not trustworthy, therefore we reduced the value of the highest five data points and refitted by the lognormal distribution and the least square, both of which are also shown in Fig. 2.4. Due to the reduction of τ_W, using the normalized abscissa variable, the curves are shifted to the right. But as we can see, a much lesser difference is seen in the later approach between the lognormal and the least square fits, and in fact, $\bar{\tau}_W$, and variance of τ_W, i.e. $V(\tau_W)$ can be calculated from the μ, σ^2 by:

$$\bar{\tau}_W = e^{\mu + \frac{\sigma^2}{2}}, \quad V(\tau_W) = e^{2\mu}\left[e^{2\sigma^2} - e^{\sigma^2}\right],$$

and from the values given in Fig. 2.4, we find

$$\frac{(V(\tau_W))^{1/2}}{\bar{\tau}_W} = e^{-\frac{\sigma^2}{2}}\left(e^{2\sigma^2} - e^{\sigma^2}\right)^{1/2} = 0.24$$

from the raw data fit, and similarly 0.24 from the modified Sandborn data set, which are both close to the value accepted in the turbulence community.

Given the equation which determines the bubble breakaway diameter from d_{pore} and τ_W, we see that knowledge of the PDFs for the pore size and the wall shear stress will determine a PDF for the bubble diameter.

2.2.2 Gallagher's Bubble Size Distribution

GALLAGHER [18] obtained a microbubble size distribution in seawater at typical ship speeds. He used an underwater camera and the Bete-Fog bubble analyzer to obtain bubble size spectrum. One bubble size distribution was presented by GALLAGHER [18] for a ship speed of 12 kts, using pores of 3/64" in diameter through a vertical steel pipe of 2 7/8" in outside diameter, and pumping air at the rate of 0.027 liter/sec/pore. The underwater camera was located 2 feet downstream of the vertical pipe. Figure 2.5 shows Gallagher's data in terms of the number of bubbles (dashed line) and the percent of occurrence (heavy solid line) versus the bubble diameter in microns. On the same figure, a normalized Rayleigh spectrum is also shown in a thin dotted solid line. As can be seen, the Rayleigh spectrum seems to fit the bubble size spectrum, and is represented by a simple parameter of the bubble diameter of the maximum percent occurrence at 221 μm. Also illustrated are the bubble sizes predicted using the different force balances described in the previous subsection; lift = surface tension and drag = surface tension for the bubble regime, and jet instability. Assuming $\tau_w = \frac{1}{2} \rho U_o^2 C_D$ and using $C_D = 0.2$ at $Re_d \cong 4 \times 10^5$ from SCHLICHTING [43] we find $d_{bubble} = 206\mu m$ from drag (tangential) balance, and $\frac{d_{bubble}}{d_{pore}} = 0.17$, while $d_{bubble} = 317\mu m$ from the lift (normal) balance, and $\frac{d_{bubble}}{d_{pore}} = 0.26$. The former balance gives a result fairly consistent with the observed maximum percent occurrence value (d_{max}) at 221 μm. The bubble size indicated by the splitting will be addressed in the following subsection, suffice it to say that it seems to occur where the bubble size spectrum data deviates the most from the Rayleigh spectrum. The bubble size due to jet instability was 5000 μm.

An attempt to fit Gallagher's data by the lognormal distributions was also carried out. A comparison of the best fits of the Rayleigh and lognormal distributions are shown in Fig. 2.6. This comparison was motivated by the following consideration. Given that $P(\tau_w)$ follows a lognormal distribution $\Lambda(\mu,\sigma^2)$, and (from the force balances) that $P(d_{bubble}) = P\left(\frac{1}{\tau_w^{1/3}}\right)$, then according to the lognormal distribution (AITCHISON and BROWN, pp. 11 [2]) $P(d_{bubble})$ should follow a lognormal distribution of $\Lambda(-\frac{1}{3\mu}, \frac{\sigma^2}{9})$. As is shown in Fig. 2.6 (except for the secondary peak at small bubble diameter, which may be a result of the bubble splitting phenomenon) the lognormal distribution seems to fit the data better. By

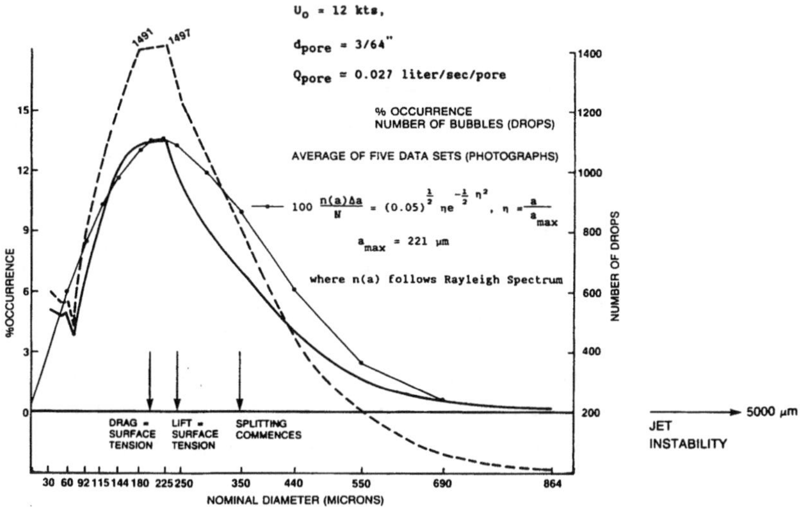

Figure 2.5. Bubble size spectrum obtained by Gallagher (1984) using a M.I.T. camera and bete fog analyzer.

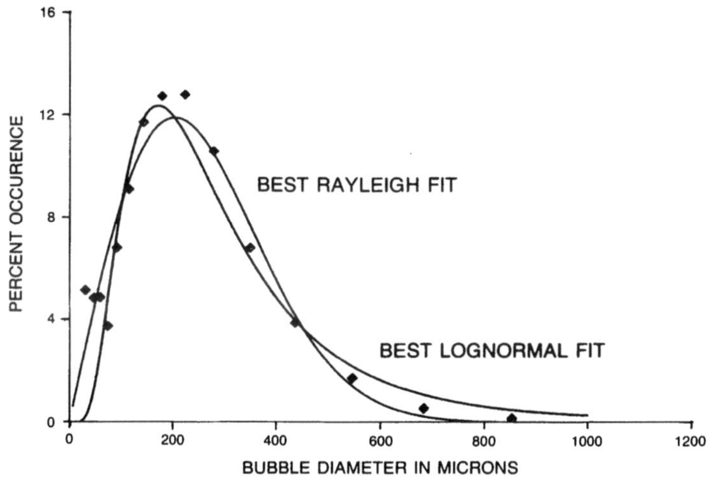

Figure 2.6. Comparison of Rayleigh and Lognormal fits to Gallagher's bubble size spectrum.

combining the force balance equations with the PDF of fluctuating shear stress results, and assuming the $P(d_{bubble}) = P((\frac{\sigma r_{pore}^2 \delta\Theta}{\tau_w})^{1/3})$ from drag balance and $P(d_{bubble}) = P((\frac{4\sigma d_{pore}^2}{\rho U_0^2 C_L})^{1/3}\frac{U_0}{u_\tau})$ from lift balance, we can construct simulated bubble size distributions. These results are shown in Fig. 2.7. The solid histograms are for the drag balance case, while the blank histograms are for the lift balance case. Initially, C_D was assumed to be 0.2, and the simulated PDF's were generated. As are shown in the figure, the peak values are consistent with the observed d_{max} value but the size bandwidths are one order of magnitude smaller than the data indicated. Then, since we did not know the exact geometry of the pores relative to the pressure minimum point of flow over a long cylinder, we reduced the C_D by a factor of 20 and the resulting pairs of bubble size spectrum are also shown in Fig. 2.7. The lift balance gives considerably greater spread in the bubble size but the peak value moved out of the observed bubble size range. The drag balance case increases the bubble size of maximum popula tion by a factor of two but the bandwidth increased very little. This analysis and comparison of data clearly indicated that some more physics are lacking in the modeling presented so far. Two more physical phenomena have been identified, they are: the seawater physics and bubble splitting in a turbulent boundary layer, and are the subjects to be discussed below.

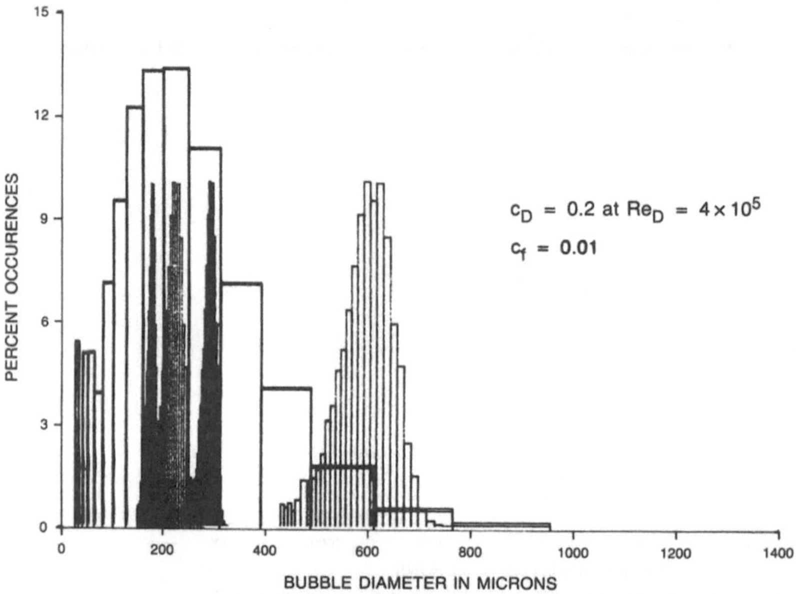

Figure 2.7. Comparison of simulated bubble size spectra with Gallagher's ocean data.

2.2.3 Smaller Microbubbles in Seawater

2.2.3.1 Historical Background

As early as 1954, FOX and HERZFELD [17] found that minute bubbles are stabilized by encapsulation in an organic film. RILEY [39] found a coincidence between stable microbubble formation and the seasonal occurrence of natural organic particles. BLANCHARD [8] found that bubble coalescence is much more rapid in tap water than in seawater, MONAHAN [32] found much smaller bubbles produced in seawater than in tap water.

SCOTT [44] conducted a rigorous experiment designed to determine the effects of salt versus the organic materials, in which he purified the tap water by distillation from potassium permanganate solution and by filtration through activated carbon. Commercial grade salt was heated at red heat sufficient for fusion, driving off any organic material. Then Nitrogen was blown through a porous glass membrane, 3 cm in diameter, with pore sizes in the 5 to 15μm range. He found a rapid decrease in the bubble size as a result of adding the pure salt into the purified water. The bubble distribution in the purified salt water (35g/liter concentration) was also found to be visually identical to that formed in real seawater.

JOHNSON and COOKE [24] set out to prove the existence of and to visualize the surface organic film on bubbles formed in seawater. They formed bubbles in air saturated seawater by the shear produced at the surface of a sintered glass frit through which air was blown. After the bubble dissolution, a small transparent particle, about 5μm in diameter, composed of the material originally present on the bubble remained and was systematically photographed. The organic material was identified by DETWILER [16] to be proteinaceous molecules, such as glycoproteins and protoglycans.

2.2.3.2 Laboratory Visualization

To verify the above facts, we have equipped an axisymmetric model with air ejectors and towed it in the Gould/Newport tow tank in both the tap water and in saltwater. Figure 2.8 shows the bubbles generated in tap water and in saltwater. The bubbles were generated through the shearing motion of the water. The saltwater was prepared by putting 640 lbs. of fine table salt in 3500 gallons of tap water. The salinity was calculated to be 21% by weight, which is about 60% of the typical seawater salinity. As one can see from Fig. 2.8 much smaller bubbles were formed in saltwater, and a much denser bubble cloud was formed. Bubbles were observed to stay at the depth (17") where they were generated and eventually dissolved into the ambient saltwater, while in tap water, the bubbles rose and coalesced to form larger bubbles which rose faster and eventually burst on the surface.

U_o = 8 ft/sec
C_q = 0.001

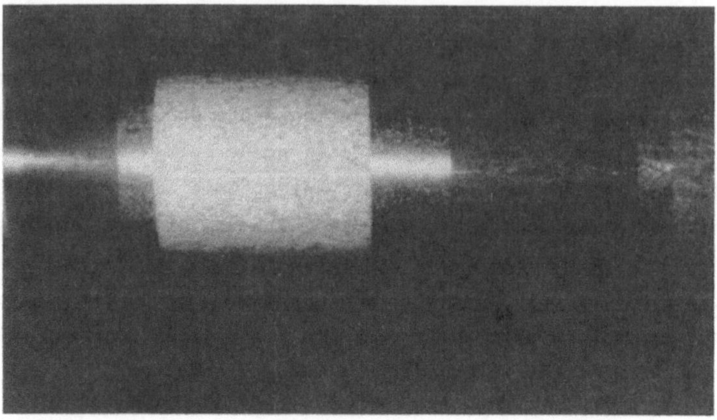

A) TAP WATER

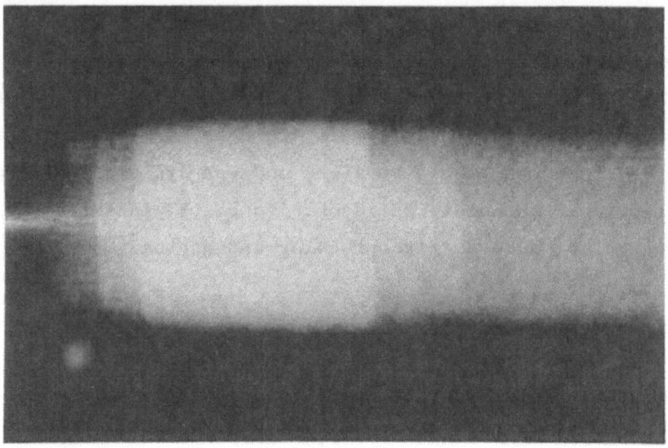

B) SALT WATER

Figure 2.8. Microbubbles ejected through two rings.

2.2.3.3 Possible Physics

Three major basic physic mechanisms can be identified to be the possible origins of smaller bubbles in seawater. They are; the reduced surface tension between the gas bubble and water, the organic film, and the metal ionic film.

The most intuitive physics is the reduction of surface tension between the gas bubble and water. SCOTT [44] stated that the surface activity of salt increases the surface tension of the solution, consistent with the formula given in KRAUS [26], i.e., $\sigma_{seawater}$ = 75.63 (dyne/cm) - 0.144T(^{o}C) + 0.221S ($^{o}/_{oo}$). On the other hand, it is well known BATCHELOR [6] that any absorbed contaminant molecules at an air-water interface will orient themselves and exert on each other a repulsive molecular force that partially balances the surface tension of the pure water. So that the net effect is that in seawater the surface tension value is smaller than that in purified tap water. The decrement is proportional to the amount of surface contaminant, and the gradient of contaminant concentration. DAVIS and ACRIVOS [14] analyzed the effect of varying surface tension on a bubble and found that the drag coefficient of a bubble covered with a surface contaminant are consistent with the experimental data if the maximum surface tension reduction is from 10 to 45%. The only quantitative way to resolve this seemingly complex issue is to directly measure the surface tension value for each seawater experiment.

Effects of the absorbed organic films on gas bubble radial oscillations have been modelled by FOX and HERZFELD [17], AVETISYAN [5] and by GLAZMAN [19]. Fox and Herzfeld modelled the film as an isotropic elastic shell, therefore no surface tension dependence was found. Avetisyan represented the absorption film as a non-Newtonian liquid characterized by hypothetical viscoelastic properties. GLAZMAN [19] derived a new term at the bubble-film-water interface to include the non-uniform surface concentration and the resultant surface tension, and represented the film as a viscoelastic film. He found that, as a result of the film, smaller bubbles will be formed, and an additional restoring force induced by the films dilational elasticity will increase the bubble resonance frequency. The true physics of the organic film on the bubble is far from being completely resolved, but intuitively one can see the following effects on bubble oscillation. One, the organic film will create a surface tension gradient that will resist bubble oscillation, and the viscoelastic nature of the film will likely to damp the high frequency bubble oscillations. In either case, smaller bubble oscillation amplitudes are expected, and the smaller the bubble, the greater the damping effects of the surface film will be on the bubble oscillation. Another possible effect of the seawater surface contaminant on bubbles is to increase the bubble drag coefficient, making them less mobile, and hence reducing the bubble collision speed, the collision frequency and the subsequent coalescence.

Finally, the potential effect that the resinous surface film will be stabilized by metal ion complex was suggested by DEGENS [15]. This would have the net effect of making the bubbles electrically charged, so that bubbles will repel each other, therefore reducing the occurrence of bubble collision and subsequent coalescence.

2.2.4 Bubble Splitting in a Turbulent Boundary Layer

As was shown earlier in Fig. 2.6 Gallagher's data shows a secondary peak which exists at 50μm indicating that some of the bubbles greater than 50μm may have been broken up into smaller bubbles. Another indication of the existence of bubble splitting can be seen in Fig. 2.7 of the simulated bubble size spectrum using the shear stress spectrum. By comparing the simulated bubble size spectrum with Gallagher's data, we found that the peak matches but not the bandwidth, indicating other mechanisms exist to make bubbles larger and smaller than the predicted mean bubble size. Theoretically speaking, bubble splitting should be considered whenever bubble generation is taking place in a turbulent flow. The reason is given below. In a laminar boundary layer, the maximum shear stress occurs right on the wall and is a steady shear stress, so that once the bubbles are generated and transported away from the wall, they will not encounter higher stress. In a turbulent boundary layer, however, the maxima of Reynolds stresses $\simeq \tau_w$ and the turbulent kinetic energy $\tilde{=} 10\tau_w$ both occur away from the wall at about $y^+ = 30$ to 100, so that bubbles must encounter greater turbulence intensity and unsteady stresses than the stresses very near the wall which are responsible for the bubble creation.

In principle, three splitting stresses can be identified, the mean shear stress τ, the turbulent Reynolds stresses $\overline{\rho u_i' u_j'}$, and the turbulent fluctuating pressure $\overline{p'^2}$ which is proportional to the turbulent kinetic energy. There is, however, only one resistance stress, i.e., the surface tension $\frac{\sigma}{d_{bubble}}$. Depending upon the bubble (or droplet) Reynolds number $\frac{d_{bubble}}{\nu_g}\left(\frac{\tau}{\rho_g}\right)^{1/2}$, two different force balances can be realized, where subscript g represents gas phase. At the low Reynolds number limit, the mean shear stress and surface tension balance each other. At the high Reynolds number limit, the discrete phase, i.e., the bubbles (or droplets), feel the influence of turbulence in the continuous phase. The kinetic energy of the turbulent motion in the continuous phase will bring about the breakup of the discrete phase. This is the case in which we are most interested.

Historically, chemical engineers have shown the greatest interest in this problem, since they are concerned with the problems of dispersion of gas-liquid systems for atomization, and froth formation, and liquid-liquid systems for emulsification. This problem had the attention of the most distinguished scientists during the early part of this century. TAYLOR [48] developed theories and experiments to study the "globule deformation" in the Couette and plane hyperbolic flows. The globules are stationary, and the study corresponds to the $Re_{globule} < 1$ limit, and is valid if the drops are small compared with the local regions of viscous flow. These facts, however, were not explicitly stated. Taylor obtained the results of We $= \frac{\tau d_{globule}}{}$ versus the ratio of μ_d, where where d stands for the dispersed phase, to μ_c, c stands for the continuous phase. His remarkable findings that no breakup will take place both at the

low $\frac{\mu_d}{\mu_c} < 3\times10^{-4}$ and high $\frac{\mu_d}{\mu_c} > 20$ limits has inspired many subsequent studies, especially those by ACRIVOS [1] and RALLISON [36]. Unfortunately, none of these results applies to our case of interest.

HINZE [22] laid the foundation of the fundamental physics of droplet breakup in a moving stream. He put Taylor's work ($Re_{globule} < 1$) into perspective relative to a droplet breakup in an air stream, and emulsification in a turbulent flow, for the $Re_{drop} > 1$ case. For droplet breakup in an air stream, Hinze defined a new Weber number

$We = \frac{\rho_c U_o^2 d_{droplet}}{\sigma}$, replacing τ in Taylor's expression by the dynamic pressure of the air stream $\rho_c U_o^2$. From his experiment, he found that the mode of droplet breakup was very different for We greater or less than the $We_{critical}$. For $We \gg We_{critical}$, droplets were stripped off the drops owing to the waves and ripples generated on the drops. This relevant fact may have its counterpart for bubbles when $We \gg We_{critical}$. For emulsification in a turbulent flow, Hinze pointed out that in a turbulent flow, the critical Weber number will not be the same for all the globules, some statistical mean value will determine the average size of the largest globules that can withstand the breakup forces. For an isotropic and homogeneous turbulence case, Hinze worked out the maximum droplet diameter from the experimental data obtained by CLAY [13]. Clay's data was obtained in an apparatus consisting of two concentric cylinders containing two immiscible fluids. The inner cylinder rotated and one of the fluids formed discrete drops. By relating his analytic expression and Clay's data, Hinze found the maximum drop diameter is determined by the following critical

Weber number,

$$We_{critical} = \frac{\rho \overline{u'^2}\, d_{drop}}{\sigma} \cong 1.17 \quad .$$

From Hinze's analysis of the maximum drop size in a turbulent flow, i.e.,

$$d_{drop_{max}} \cong 0.725 \left(\frac{\sigma}{\rho_c}\right)^{3/5} \varepsilon^{-2/5} \quad ,$$

we can obtain an expression for the corresponding drop size in a turbulent boundary layer, if we assume that the isotropic turbulence assumption still applies. From HINZE, pp.647 [23] we find the maximum ε to be

$$\varepsilon_{max} \simeq \frac{30 u_\tau^3}{\delta} \quad ,$$

where δ is the local boundary layer thickness. Expressing δ and ρu_τ^2 in terms of Re_x and substituting into the above relationship, we obtain,

$$d_{drop_{max}} = 1.03 \left(\frac{\sigma}{\rho U_o^2}\right)^{3/5} x^{2/5} Re_x^{0.04}$$

where x is the downstream distance along a flat plate. SLEICHER [47] did turbulent flow experiments in a 48 ft. long Lucite pipe with 1 1/2" inner diameter using two immiscible liquids. He found that the fraction of drops that broke up was very sensitive to flow velocity, and in every case the breakup occurred very close to the pipe wall where the turbulence was the least isotropic and homogeneous. Although Sleicher's data does not deal with bubble breakup, it is highly relevant to our concern here, since it deals with breakup in a turbulent boundary layer. He gave the following results for the critical Weber number,

$$\frac{\rho_c U_o^2\, d_{drop_{max}}}{\sigma} = 38\, \eta^{1/2} (1 + 0.7\eta^{-0.7}) \quad ,$$

where $\eta = \frac{\sigma}{\mu_c U_o}$, μ_c is the vicosity of the continuous phase and U_o its velocity.

SEVIK and PARK [45] conducted a study of air bubbles splitting in a turbulent jet. Their apparatus consisted of a water jet, oriented vertically upwards, with an air nozzle located at the center of the water jet. The water nozzle diameter was 1.5" and the water velocity at the nozzle exit varied from 7 to 16 ft/sec. Air was injected into the center of the water jet, and air nozzle diameters varied from 0.071" to 0.25". Bubbles with diameters varying from 4 to 5.8 mm were generated in the laminar core of the low turbulence jet of water, and subsequently broken up by the jet turbulence downstream. The initial bubble sizes were designed to be greater than the KOLMOGOROV microscale, which was on the order of 25μm, so that the bubble breakup would occur, and that the breakup would be characterized by the single parameter of critical Weber number, according to Kolmogorov (1949). They found that the bubbles remained near the center of the water jet, and that breakup was substantially completed at an axial distance of 9 to 10 water nozzle diameters. Beyond that, no further changes in bubble size took place. The bubbles were also progressively broken into smaller and smaller ones as the jet velocity increased. The most important finding was that by setting a characteristic frequency of the turbulence to the lowest resonant frequency of the bubble oscillation with constant volume, they found that the critical Weber numbers corresponded to both Clay's and their own experimental data. Furthermore, they stated that progressively smaller bubbles were generated when the critical Weber number was exceeded, and when the higher frequencies of the turbulence excited the higher modes of the bubble oscillation.

Briefly, SEVIK and PARK [45] measured $We_{critical}$ of air bubbles generated in a water jet, and found that

$$\frac{\rho \overline{u'^2}\, d_{bubble}}{\sigma} = 2.6 \,. \tag{2.16}$$

While by equating bubble and flow frequencies,

$$\frac{(\overline{u'^2})^{1/2}}{d_{bubble}} = f_{bubble} = \frac{1}{2\pi}\left[\frac{(n+1)(n-1)(n+2)\,\sigma}{\rho r^3_{bubble}}\right]^{1/2}, \text{ at } n = 2$$

they found the following theoretical result,

$$\frac{\rho \overline{u'^2}\, d_{bubble}}{\sigma} = 2.48 \,, \tag{2.17}$$

which is close to the experimental result. Inspired by these results, we consider the balance of the turbulent kinetic energy exerting influence on a bubble to the surface tension energy, which is the only energy resisting the breakup, i.e.,

$$\rho \overline{u'^2}\, \frac{4}{3}\pi \left(\frac{d_{bubble}}{2}\right)^3 = \frac{\pi}{2}\sigma d^2_{bubble}$$

We find

$$\frac{\rho \overline{u'^2}\, d_{bubble}}{\sigma} = 3 \,, \tag{2.18}$$

which is fairly close to the above experimental result. To summarize all these results, Fig. 2.9 is provided to illustrate the maximum bubble diameters versus the flow speed. As one can see that all three curves of

$$\frac{\rho \overline{u'^2}\, d_{bubble}}{\sigma} = C, \text{ with, } C = 2.48,\ 2.6 \text{ and } 3$$

are nearly identical. The $y^+ = 25$ and 100 curves are provided to indicate the boundaries of maximum turbulence production. As the majority of the bubbles become smaller than $y^+ = 100$, we conjecture that more microbubble drag reduction can be expected. Interesting enough, one sees that at 5 m/sec, the splitting just commences, which may have some relevance to the observation that drag reduction begins to emerge at this speed. At 20 m/sec more bubbles will be of the size of $y^+ = 25$, further increase in speed should ensure that most of the bubbles are smaller than $y^+ = 25$, therefore increasing the bubbles opportunity for interacting with the turbulence production that would normally take place in a single phase fluid turbulent flow. It is intuitively convincing that the more bubble splitting is taking place the more the turbulence kinetic energy must be drained from the TBL, therefore, the more skin-friction reduction. Hinze's isotropic turbulence result for droplets is also shown, which is never smaller than $y^+ = 100$. Also shown is the curve for 0.01δ, which shows that at

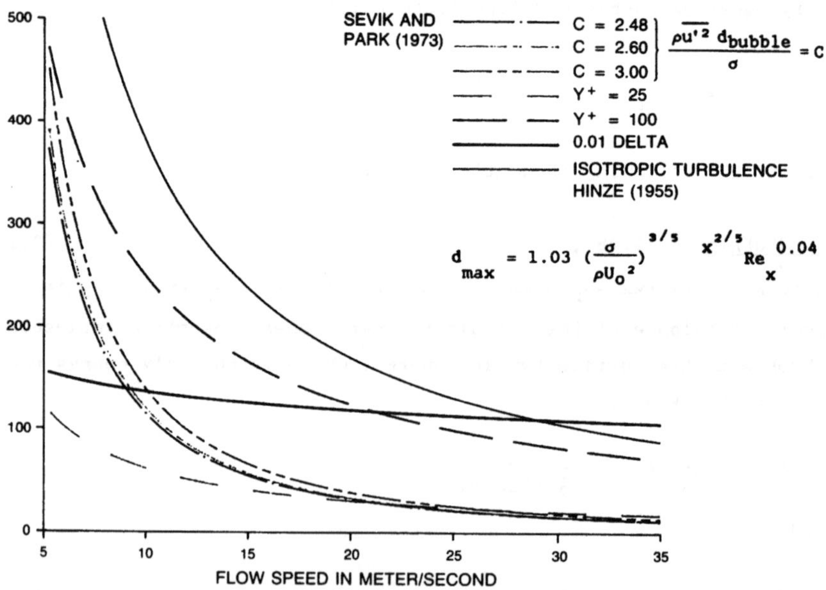

Figure 2.9. Maximum bubble size in a TBL as a result of bubble splitting.

speeds less than 10 m/sec, bubbles larger than 0.01δ would survive the splitting while at speeds greater than 10 m/sec, bubbles will be much smaller than 0.01δ.

2.3 Bubble Transport in TBL

2.3.1 Hydrodynamic Forces on a Microbubble During Transport

One of the tools necessary to investigate microbubble drag reduction is a means of determining the motion of microbubbles in a turbulent boundary layer. The model employed here assumes a spherical microbubble in a mean turbulent boundary layer velocity profile. Any effects due to the rotation of the bubble are ignored. The forces on the microbubble are calculated as if the bubble were actually a rigid particle. This assumption is based on evidence that surface phenomena encountered in the real world environment make this a better approximation than the usual free surface model (see BATCHELOR [6]). The fluid is assumed to be Newtonian and incompressible.

We assume that the time scale, T, over which a bubble alters its "break-away" size by diffusion, splitting or coalescence is such that

$$T \gg \nu/u_\tau^2 \ .$$

The bubble size may then be assumed constant during the calculation of its short time trajectory in a turbulent boundary layer. This is equivalent to stating that the bubble size does not change significantly until the bubble has travelled many, many viscous lengths.

The drag of the microbubble is

$$D = C_D\,(Re)\ .\ \frac{1}{2}\rho\ U_{rel}^{\,2}\ \pi\ R^{\,2}$$

where the drag coefficient, C_D, is a function of the relative velocity bubble-diameter Reynolds number

$$Re = 2\ \frac{U_{rel}R}{\nu}$$

and U_{rel} is the difference between the bubble velocity and the fluid velocity. The drag coefficient is calculated from a series of functional fits given by MORSI and ALEXANDER [33].

A microbubble in a shear flow also experiences a lift force. For small

Reynolds numbers this force is

$$L = 6.46\ \rho\ U_{rel}\ R^2\ (\nu k)^{1/2}$$

where k is the magnitude of the velocity gradient SAFFMAN [40].

All pressure gradients are taken into account, including the pressure gradient due to gravity which causes bouyancy.

All inertial effects are also considered. The mass of the microbubble, the added mass of the microbubble, and the so-called Basset's force.

$$B = 6\ \rho\ R\ (\pi\nu)^{1/2} \int_{-\infty}^{t} \frac{dU_{rel}}{d\tau} \frac{d\tau}{(t-\tau)^{1/2}}\ .$$

The basic equations thus take the form

$$\frac{4}{3}\pi R^3 \frac{d}{dt}\{\rho' u + C_A \rho(u-\tilde{u})\}$$

$$= - C_D \cdot \frac{1}{2}\rho\ |\vec{V}|\ (u-\tilde{u})\ \pi R^2$$

$$- 6.46\ \rho\ (v-\tilde{v})\ R^2\ \nu^{1/2} \left(\frac{\partial\tilde{v}}{\partial x}\right)^{1/2}$$

$$- \frac{4}{3}\pi R^3 \frac{\partial p}{\partial x}$$

$$- C_H\ \rho\ R^2\ (\pi\nu)^{1/2} \int_{-\infty}^{t} \frac{d}{d\tau}(u-\tilde{u}) \frac{d\tau}{(t-\tau)^{1/2}} \qquad (2.19)$$

and

$$\frac{4}{3}\pi R^3 \frac{d}{dt}\{\rho' v + C_A \rho(v-\tilde{v})\}$$

$$= - C_D \cdot \frac{1}{2}\rho\ |\vec{V}|\ (v-\tilde{v})\ \pi R^2$$

$$- 6.46\ \rho\ (u-\tilde{u})\ R^2\ \nu^{1/2} \left(\frac{\partial\tilde{u}}{\partial y}\right)^{1/2}$$

$$- \frac{4}{3}\pi R^3 \frac{\partial p}{\partial y} + \rho' \frac{4}{3}\pi R^3 g$$

$$- C_H\ \rho\ R^2\ (\pi\nu)^{1/2} \int_{-\infty}^{t} \frac{d}{d\tau}(v-\tilde{v}) \frac{d\tau}{(t-\tau)^{1/2}} \qquad (2.20)$$

Where the velocity of the bubble is given by

$$\vec{V} = u\ \vec{i} + v\ \vec{j}$$

and the velocity of the fluid in the absence of the bubble is

$$\vec{\tilde{V}} = \tilde{u}\ \vec{i} + \tilde{v}\ \vec{j}$$

The coefficients C_A and C_H are the added mass and history (or Basset's) force coefficients, respectively. These are functions of the "acceleration number"

$$Ac = \frac{|\vec{V} - \vec{\tilde{V}}|^2}{2R \frac{d}{dt}|\vec{V}-\vec{\tilde{V}}|}$$

as given by ODAR and HAMILTON [35].

The relative importance of the forces on a microbubble may be determined by forming their ratios and calculating the numerical values of the ratios for any case of interest. To simplify and avoid confusion one may use the drag force as the basis of comparison, since even a cursory examination of the physics involved serves to convince one that the drag force must always be significant in any case of interest.

Assuming that the relative velocity bubble Reynolds number is small, the form of the drag force used for comparison with other forces is Hadamard's drag

$$4 \pi \mu R U_{rel} ,$$

where μ is the fluid viscosity, R is the microbubble radius and U_{rel} is the relative velocity between the fluid and the microbubble. The forces that will be compared with the drag are: lift,

$$6.46 \rho U_{rel} R^2 (\nu k)^{1/2}$$

where k is the magnitude of the velocity gradient; Basset's force,

$$C_H \rho R^2 (\pi\nu)^{1/2} \int_{-\infty}^{t} \frac{dU_{rel}}{d\tau} \frac{dt}{(t-\tau)^{1/2}} ;$$

buoyancy,

$$\frac{4}{3} \pi R^3 \rho' g;$$

gravity,

$$\frac{4}{3} \pi R^3 \rho' g;$$

inertia,

$$\frac{4}{3} \pi R^3 \rho' \frac{dU_{rel}}{dt} ;$$

and added mass,

$$C_A \frac{4}{3} \pi R^3 \rho \frac{dU_{rel}}{dt} .$$

Forming ratios of these forces with the drag force one finds;

$$\frac{\text{LIFT}}{\text{DRAG}} = 0.51\, R^+ (k^+)^{1/2} ,$$

$$\frac{\text{BASSET}}{\text{DRAG}} = \frac{C_H}{4\pi^{1/2}} \frac{R^+}{U^+} \int_{-\infty}^{t^+} \frac{dU^+}{d\tau^+} \frac{d\tau^+}{(t^+-\tau^+)^{1/2}} ,$$

$$\frac{\text{BUOYANCY}}{\text{DRAG}} = \frac{1}{3} \frac{R^{+^2} G^+}{U^+} ,$$

$$\frac{\text{GRAVITY}}{\text{DRAG}} = \frac{1}{3} \alpha \frac{R^{+^2} G^+}{U^+} ,$$

$$\frac{\text{INERTIA}}{\text{DRAG}} = \frac{1}{3} \alpha \frac{R^{+^2}}{U^+} \frac{dU^+}{dt^+} ,$$

and

$$\frac{\text{ADDED MASS}}{\text{DRAG}} = \frac{1}{3} C_A \frac{R^{+^2}}{U^+} \frac{dU^+}{dt^+} .$$

In these equations

$$R^+ = \frac{Ru_\tau}{\nu} , \qquad \text{where } u_\tau = \left(\mu \frac{du}{dy}\Big|_{wall} / \rho\right)^{1/2}$$

$$k^+ = \frac{k\nu}{u_\tau^2} ,$$

$$U^+ = \frac{U_{rel}}{u_\tau} ,$$

$$t^+ = \frac{tu_\tau^2}{\nu} ,$$

$$G^+ = \frac{\nu g}{u_\tau^3} ,$$

and

$$\alpha = \rho'/\rho .$$

As an example let us assume

$$u_\tau \simeq 1.44 \text{ m/s}$$

and

$$\nu = 10^{-6} \text{ m}^2/\text{s}$$

one finds

$$G^+ \simeq 3.3 \times 10^{-6}$$

Now assume that for a bubble in ocean

$$\alpha \simeq 2.2 \times 10^{-2}$$

To estimate the nondimensional relative velocity, acceleration and velocity gradient one should consider regions where these quantities are large, that is, regions where the forces will be large. As an order of magnitude estimate one may state that $U^+ \simeq 10.0$, $k^+ \simeq 1.0$, $\frac{dU^+}{dt^+} \simeq 10.0$

In order to estimate Basset's force one may suppose that the bubble has experienced a roughly constant acceleration during the last Δt^+ of time, prior to which it was moving with constant velocity. This supposition yields an order of magnitude estimate for most cases since it may be shown that solutions of the above equations tend to behave exponentially. Hence, the periods of high accelerations occur over short time intervals. With these assumptions one may approximate Basset's force as

$$\int_0^{t+} \frac{dU^+}{d\tau^+} \frac{d\tau^+}{(t^+ - \tau^+)^{1/2}} \simeq 2 \frac{dU^+}{dt^+} (\Delta t^+)^{1/2} .$$

In addition the above arguments show that one may state that

$$\Delta t^+ \simeq 1 .$$

If one now takes the added mass coefficient, C_A, and the history or Basset's force coefficient, C_H, to have their high acceleration values $C_A = 1/2$, $C_H = 6$, then the force ratios may be calculated. Proceeding with the aforementioned example, with microbubble sizes chosen which are typical of those found in engineering situations, one arrives at the results listed below.

	Bubble 10μ	Bubble 100μ	Bubble 1000μ
R+	14.4	144.0	1440.0
Lift / Drag	7.3	73.5	735.
Inertia / Drag	1.5	152.1	15210.0
Added Mass / Drag	34.5	3450.0	3.5×10^5
Basset / Drag	24.4	243.8	2437.5
Basset / Added Mass	0.705	0.07	0.007
Gravity / Drag	5.0×10^{-7}	5.0×10^{-5}	5.0×10^{-3}
Buoyancy / Drag	2.3×10^{-5}	2.3×10^{-3}	2.3×10^{-1}

The results are striking. With the exception of gravity and bouyancy, which may safely be neglected, all other influences on a microbubbles acceleration are at least of the same order of magnitude as the drag and generally dominate the drag during periods of high velocity and acceleration. Of particular interest is the lift force, which can be orders of magnitude greater than the drag force in the high shear flow of a boundary layer. The lift force acts in the direction of increasing relative velocity magnitude. Thus a microbubble released at the wall experiences a strong lift force pulling it away from the wall. Initially the lift force is many times the drag force and the bubble will move almost perpendicularly away from the wall. As the drag force slowly brings the relative velocity between the microbubble and the fluid to zero and the bubble moves to regions of lower shear the lift force diminishes rapidly and the bubble eventually moves parallel to the wall. (see Fig. 2-10 and 2-11).

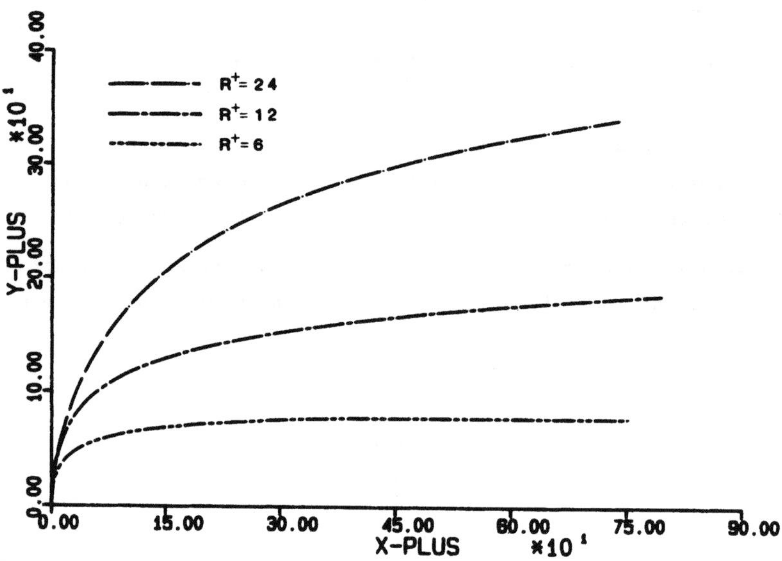

Figure 2.10. Non-dimensional particle trajectories.

Outside of the boundary layer, however, the lift force will generally be negligible. This is due to the comparatively small values of velocity gradient encountered in the external flow. For example, the maximum velocity gradient that exists in the potential flow about a circular cylinder is of the order of U_∞/a. where U_∞ is the free-stream velocity and a is the radius of the cylinder. If we assume that u_τ = 1.44 m/s, and if a = 1 m, then in terms of the non-dimensional quantities introduced earlier, this velocity gradient is

$$\frac{\tilde{U}}{\tilde{a}} = \left(\frac{U_\infty}{u_\tau}\right)/\left(\frac{au_\tau}{\nu}\right) = 1.7 \times 10^{-5}$$

which is very small compared to the value found in the boundary layer where $k^+ \simeq 1$.

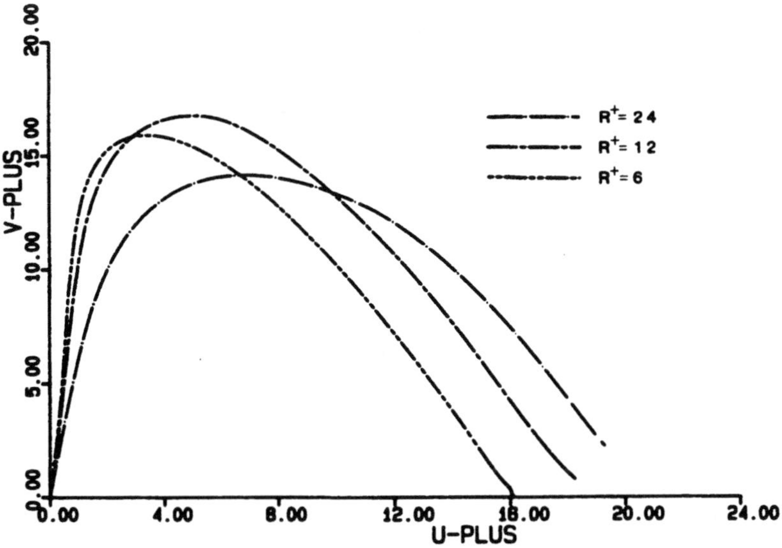

Figure 2.11. Non-dimensional particle velocities.

Hence, outside of the boundary layer the main forces to be taken into account will be the drag force, the pressure force, and the inertial and added mass forces.

A program. APT-1 (for Advected Particle Trajectory), has been developed which takes all of the relevant forces into consideration and thus is capable of calculating the trajectories of the bubbles from the time they depart the wall to their long time asymptotic path. A comparison of the terms included in APT-1 with those included in two other particle transport models is shown in Table 2.2.

2.3.2 Microbubble Trajectories in TBL

Once the microbubble size spectrum has been determined (including bubble splitting and coalescense effects) it is known where, in diameter, the bulk of the microbubble population resides. One may employ the microbubble diameter probability density function (PDF), p(D), to describe the distribution in bubble sizes passing any downstream station. From this PDF one may define the mean bubble diameter μ_D as well as the variance σ_D^2 and skewness σ_D^3 of the distribution. The bulk of the microbubble population may be said to lie between the diameters $\mu_D \pm \sigma_D$. The trajectories followed by bubbles of these sizes, as they pass through the turbulent

TABLE 2.2 APPROXIMATE PROPERTIES OF GASES IN EXHAUST AND EXPERIMENTS (100F, 1 atm)

	H_2	He	H_2O VAPOR (212) °F	CO	N_2	AIR	CO_2
MOLECULAR WEIGHT	2	4	18	28	28	28.97	44
ρ, DENSITY, lbm/ft³, 1 atm	0.0049	0.010	0.0372	0.069	0.069	0.071	0.108
k, THERMAL CONDUCTIVITY Btu/hrft F	0.11	0.092	0.0145	0.015	0.015	0.015	0.010
μ, VISCOSITY (lbm/FT SEC) x 10^5	0.620	1.36	0.870	1.24	1.22	1.29	1.05
ν, KINEMATIC VISCOSITY (ft^2/sec) x 10^3	1.26	1.34	0.234	0.18	0.18	0.18	0.098
c_p, SPECIFIC HEAT, Btu/lbm F	3.42	1.24	0.451	0.25	0.25	0.24	0.20
Pr, PRANDTL NUMBER	0.695	0.68	0.96	0.75	0.71	0.72	0.77
$(\partial T/\partial p)_h$, JOULE-THOMPSON COEFFICIENT °F/atm		~0	0.4	0.36	0.35	~1.8	
σ, SURFACE TENSION, dynes/cm			72				
$\rho_{g, sal}$ SOLUBILITY MOLE FRACTION (x 10^5)	1.2	0.8		1.3	1	1.4	40

boundary layer, serve to bound that portion of the boundary layer which contains most of the microbubbles. For sufficiently small variance and skewness these trajectories also bound the region of greatest void fraction. Hence the magnitude of the local void fraction maximum can be increased by reducing the variance in bubble diameters. Similarly, control of the location of this void fraction maximum can be accomplished by controlling the mean bubble diameter.

In order to examine the phenomenon of microbubble drag reduction further it became necessary to obtain an estimate of the void fraction variation in the turbulent boundary layer. As a first attempt the APT-1 code was employed to calculate trajectories for bubbles of the diameters μ_D and $\mu_D \pm \sigma_D$, for the cases of three microbubble diameter spectra. For these spectra the means and standard deviations (μ_D, σ_D) in microns were (338, 34), (156, 16) and 33,5, 3.5, respectively. The trajectories for these three cases are shown in Fig. 2.12. These trajectories were calculated with a mean turbulent boundary layer profile produced by the axisymmetric TAPS code. The body shape was that of an axisymmetric body with a length Reynolds number of $\frac{U_\infty L}{\nu} = 8.9 \times 10^7$ and a Froude number of $\frac{U_\infty}{\sqrt{gL}} = 2.55$.

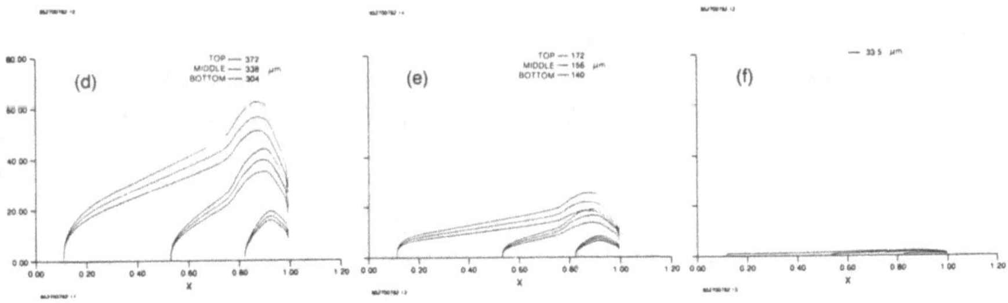

Figure 2.12. Microbubble trajectories.

The trajectories were calculated from three different initial locations in each case. These locations correspond to $\frac{x}{L} = 0.11, 0.53$ and 0.82 .

The trajectories are presented in terms of both $\frac{y}{\delta}$ and y^+ vs $\frac{x}{L}$.

Only the trajectories for the mean bubble size are shown for the case μ_D = 33.5 microns for reasons of clarity.

2.3.3 Void Fraction Profiles

The microbubble trajectories can be utilized to obtain estimates of the void fraction profile. To this end consider the evolution with downstream distance of bubbles with diameters in the interval $(D, D + \Delta D)$ in proximity to a plane or nearly plane wall. Following typical turbulence modelling practice we can consider the flow to consist of a basic steady portion (the mean turbulent flow profile) with a strong spatially varying cross-stream diffusion rate (an eddy viscosity model $\nu_e = \nu_e(y)$). Then if, for the moment, we neglect the diffusion we see that we may easily calculate the trajectories of the bubbles of diameter D and $D + \Delta D$ as though the flow were laminar. For $\frac{\Delta D}{D} << 1$ these trajectories are separated by a very small distance, Δ and all the bubbles with diameters in the interval under discussion have trajectories which lie between these two trajectories.

For convenience we are assuming here that all the bubbles in the initial distribution begin their trajectories at one point on the body with the same initial conditions. Each of the trajectories under discussion may be described by a function of the form $y = f_D(x)$. Thus if $\gamma(D;x,y)\,\Delta D$ describes the void fraction at (x,y) due to bubbles in the interval $(D, D + \Delta D)$ then the above initial condition may be written as

$$\gamma(D;x,y)\,\Delta D = C\,\delta(y - f_D(x))\,p(D)\,\Delta D$$

where C is an as yet undetermined constant and δ is the Dirac delta function. Assuming the diffusion process obeys the linear diffusion equation with diffusion coefficent ν_e the void fraction, γ, will then evolve downstream as

$$\gamma(D,x,y)\,\Delta D = \frac{C\,p(D)\,\Delta D}{2(\pi\nu_e t)^{1/2}} \exp\{\frac{-(y - f_D(x))^2}{4\,\nu_e t}\}$$

where we have assumed that $\frac{\partial\nu_e}{\partial y} << \frac{u\Delta}{L}$ in the regions of interest, and that $\frac{U_\infty y}{\nu_e} >> 1$ so wall effects may be neglected. The coordinates of the mean bubble path (x_D, y_D) are related by $y_D = f_D(x_D)$ and may also be expressed parametrically by

$$x_D = x_D(t)$$
$$y_D = y_D(t) \quad .$$

Hence, by the implicit function theorem, we may write

$$t = t_D(x)$$

so that

$$\gamma(D;x,y)\ \Delta D = \frac{C\ p(D)\ \Delta D}{2\ (\pi\ \nu_e t_D(x))^{1/2}} \exp\{\frac{-(y - f_D(x))^2}{4\ \nu_e t_D(x)}\} \quad .$$

Then the local void fraction can be calculated by:

$$\alpha(x,y) = \int_0^\infty dD\ \gamma(D;x,y)$$

The quantities f_D and t_D in the expression for γ are implicit functions of D.

In order to determine the value for C we require that the total void flux past any downstream plane, perpendicular to an axisymmetric body with diameter D_{body}, be conserved so that

$$Q = \pi\ D_{body} \int_0^\infty dy\ \alpha(x,y)\ u(x,y) \quad .$$

Thus we see that

$$Q = C\ \pi\ D_{body} \int_0^\infty dy\ u(x,y) \int_0^\infty dD\ p(D)\ g(D;x,y)$$

where

$$g(D;x,y) = \frac{1}{2\ (\pi\ \nu_e t_D(x))^{1/2}} \exp\{\frac{-(y - f_D(x))^2}{4\ \nu_e t_D(x)}\} \quad .$$

Interchanging orders of integration we may write

$$Q = C\ \pi\ D_{body} \int_0^\infty dD\ p(D) \int_0^\infty dy\ u(x,y)\ g(D;x,y) \quad .$$

However if u(x,y) varies slowly enough with y, that is if

$$\frac{1}{U_\infty} \frac{\partial u}{\partial y}\Big|_{y_D} << \left(\frac{U_\infty}{\nu_e L}\right)^{1/2}$$

then we may approximate u(x,y) by some $\tilde{u}$ for the purposes of the integration over y to find that

$$\int_0^\infty dy\ u(x,y)\ g(D;x,y) \simeq \tilde{u} \quad .$$

If we further assume that the initial microbubble distribution is sufficiently narrow so that

$$\frac{1}{U_\infty} \frac{\partial u}{\partial y} << \frac{1}{\Delta}$$

then any dependence of $\tilde{u}$ on D may be ignored and the second integration may be performed to arrive at

$$Q \cong C \pi D_{body} \tilde{u} .$$

Therefore the local void fraction can be computed as,

$$\alpha(x,y) \cong \frac{Q}{\pi D_{body} \tilde{u}} \int_0^\infty dD \, p(D) \, g(D,x,y)$$

or

$$\cong \frac{Q}{\pi D_{body} u(x,y)} \int_0^\infty dD \, p(D) \, g(D,x,y) .$$

Thus the information contained in the microbubble trajectories do provide the quantities f_D and t_D necessary to perform the integration. Examples of void fraction profiles and their evolution with downstream distance on an axisymmetric body are shown in Fig. 2.13.

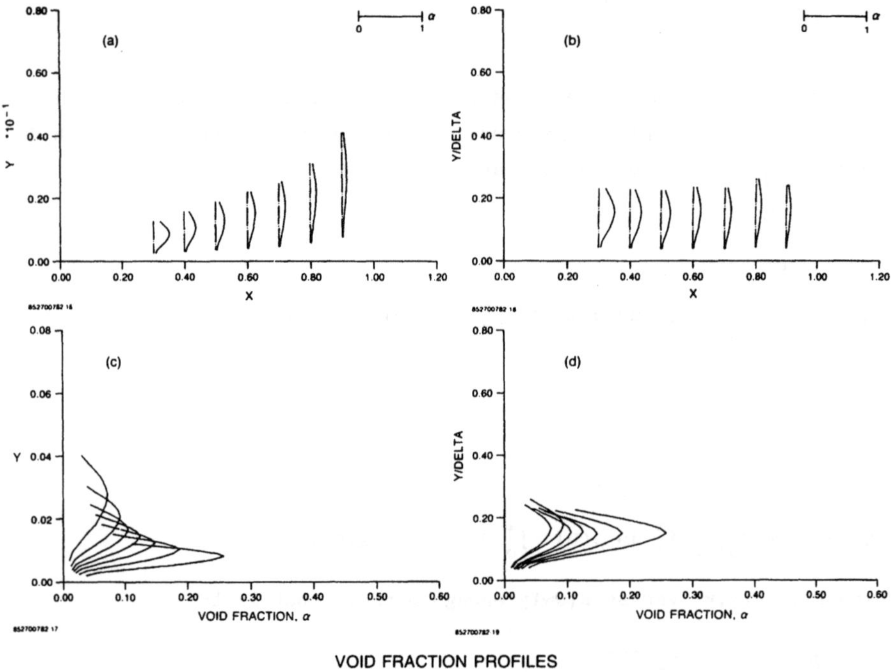

Figure 2.13. Void fraction profiles.

2.4 Microbubble Drag Reduction Mechanism

2.4.1 Persistence of Skin-Friction Reduction

MADAVAN, DEUTSCH and MERKLE [29], abbreviated MDM hereafter, indicated that the ratio $\frac{\Delta c_f}{c_f}$ decays by 50% over a distance of 25δ, where δ is the boundary layer thickness. This observation raises a question about why this happens and how it can be prevented or mitigated in practice. Intuitively, three basic mechanisms can be suggested to be the potential origin of the loss of effectiveness of the microbubble drag reduction. They are:

- o Growth of the turbulent boundary layer thickness reduces the local void fraction,
- o Bubble diffusion away from the wall due to the turbulent eddy diffusivity,
- o Bubble coalescence and subsequent rise away from the wall due to the larger buoyancy.

The last mechanism is expected to be greatly reduced in seawater, especially for higher speeds. The second mechanism depends upon the magnitude of the turbulent eddy diffusivity $u_\tau\delta$, which also depends upon δ, the boundary layer thickness. This fact led us to focus our discussion on the first mechanism, i.e., how the growth of $\delta(x)$ will degenerate the effectiveness of microbubble drag reduction.

The nominal void fraction, i.e., the average void fraction in a microbubble-laden TBL was suggested by MADAVAN, DEUTSCH and MERKLE [29] as,

$$C_V = \frac{Q_{gas}}{Q_{gas} + Q_{boundary\ layer}} ,$$

where $Q_{b.\ell.} = (1 - \frac{\delta^*}{\delta}) U_o b\delta$, and b being the width of gas ejector.

In a single-phase fluid TBL, $1 - \frac{\delta^*}{\delta} = 0.87$, so that

$$C_V = \frac{1}{1 + \frac{0.87\ U_o b\delta}{Q_{gas}}} .$$

If one defines $Q_{gas} = C_q U_o S = C_q U_o bL$, where S is the porous surface area of the gas ejector, $b = \pi D$, where D is the diameter of the axisymmetric body and L the length of the ejector, we then have,

$$C_V = \frac{1}{1 + \frac{0.87\ \delta}{C_q\ L}} . \qquad (2.22)$$

The rate of change of C_V and δ are then related by,

$$\frac{dC_V}{dx} = -\frac{0.87}{C_q L} C_V^2 \frac{d\delta}{dx} , \tag{2.23}$$

so that if $\frac{d\delta}{dx}$ is reduced, so will be the $\frac{dC_V}{dx}$, hence hopefully, the decay of the microbubble drag reduction can be reduced. The following discussion will be focused on whether this basic mechanism can be identified as the primary one and then how much reduction of $\frac{d\delta}{dx}$ is needed to slow down the decay of C_V and how, in practice, such a mitigation can be implemented.

First, let us compare the decay rate of C_V with the MDM data on the persistence of skin friction reduction over the distance downstream. According to MDM, $\frac{\Delta c_f}{c_f}$ decays by about a factor of two over a distance of 30δ. Since their data shows that $\frac{\Delta c_f}{c_f}$ is proportional to the C_V, we expect that C_V should be reduced by a factor of 2 over a distance of 30δ. Such an assumption can be compared with (2.22). MDM [29], Fig. 17, gave the following key parameters;

- $L = 7"$, $C_q \simeq 0.02$ for $\frac{\Delta c_f}{c_f}$ at speed $U_o = 10$ m/sec,

- $C_{V_o} = 0.48$ at $x_o = 409$ mm will yield a $\left.\frac{\Delta c_f}{c_f}\right|_o = 0.85$,

- at $x_1 = 575$ mm, $\left.\frac{\Delta c_f}{c_f}\right|_1 = 0.4$ with an equivalent $C_{V_1} = 0.20$.

Namely, a nearly 50% reduction in $\frac{\Delta c_f}{c_f}$ for a 50% reduction in C_V. Using $\delta = 0.37 \frac{x}{Re_x^{0.2}}$ and $U_o = 10.8$ m/sec, we find $\delta(x_o) = 0.71$ cm and $\delta(x_1) = 0.93$ cm, substituting into (2.22), we find

$$\frac{C_{V_1}}{C_{V_o}} \simeq 0.8 ,$$

certainly not sufficient to explain the 50% reduction observed in MDM's data.

From MIGIRENKO and EVSEEV's [31] void fraction profile data, we see a distinct peak of the void fraction at $\frac{y}{\delta} \simeq 0.1$, and theoretically, the local void fraction should follow the diffusion equation, if one neglects the bubble-bubble interactions. Namely, the void fraction should follow $c(y) \propto e^{-\frac{y^2}{4\nu_e t}}$, where y is the distance from the void fraction peak, and $\nu_e \simeq 0.04\, u_\tau \delta$ HINZE, pp. 645 [23]. To find one e-fold time, or equivalently the distance downstream, for c(y) to reduce by e^{-1}, we set

$4\nu_e t = (1/2\ \delta)^2$. Substituting ν_e into above, we obtain

$\Delta x \cong \frac{1}{2} U_o t \simeq \frac{1}{1.28} (\frac{2}{c_f})^{1/2} \delta \simeq 35\delta$ for $c_f = 0.001$. The above Δx is not very different from the MDM data of 30δ. Therefore indicating that diffusion of bubbles instead of growth of the TBL thickness may be the dominant mechanism for the decay of microbubble drag reduction.

The next relevant question is then, how much growth is there in $\delta(x)$ over a distance of 30δ, and how much suction is required to reduce its growth? From the expression for $\delta(x)$ we find at the same speed,

$$\frac{\Delta\delta(x)}{\delta(x)} = 0.8\,\frac{\Delta x}{x} = 24\,\frac{\delta(x)}{x} = \frac{8.88}{Re_x^{0.2}} \quad .$$

so that at $Re_x = 10^7$, $\frac{\Delta\delta}{\delta} \simeq 0.3$.

The required suction coefficient is then defined as

$$C_{q_s} = \frac{\pi D \int_0^{0.3\,\delta(x)} u\,dy}{U_o A} \quad ,$$

where A is the suction area and D the diameter of the axisymmetric body. Equivalently, the above can be rewritten as

$$C_{qs} = \frac{\delta}{L} \int_0^{0.3} [1-(1-\frac{y}{\delta})^{1/7}]\ d(\frac{y}{\delta}) \quad ,$$

where $A = \pi DL$, and $\frac{u}{U_o} = 1 - (1 - \frac{y}{\delta})^{1/7}$ HINZE, pp. 632 [23] were applied.

2.4.2 Bubble Splitting as a Turbulence Reduction Mechanism

From discussion in Section 2.2 about bubble splitting, one can make an estimate of the total amount of turbulent kinetic energy needed to split bubbles in a turbulent boundary layer and compare that with the total available energy in a TBL. From this estimation, one can identify whether the bubble splitting is a plausible basic mechanism for reducing the turbulence. Furthermore, if it is indeed plausible, can one identify what can be done to maximize its effects?

From the energy balance point of view it takes a Reynolds stress $\overline{\rho u'^2} = 3\,\frac{\sigma}{d_{bubble}}$ to initiate bubble splitting. Assuming that bubbles stay spherical in shape, the work needed to split a bubble from a diameter of d_{bubble} into bubbles of diameter $\frac{1}{2}d_{bubble}$ is, abbreviating d_{bubble} by d_b from here on,

$$\int_{\frac{d}{2}}^{d} \frac{3}{d_b} 4\pi \left(\frac{d_b}{2}\right)^2 dd_b = \frac{9}{8} \pi\sigma d_b^2 ,$$

and the energy needed to split N bubbles will be $\frac{9}{8}\pi\sigma d_b^2 N$.

Assuming the bubble size spectrum follows a Rayleigh spectrum as shown in Figure 2.5, i.e,

$$n(d_b) = N \frac{d_b}{d_{b_o}^2} e^{-\frac{1}{2}\left(\frac{d_b}{d_{b_o}}\right)^2} ,$$

the total amount of turbulence energy required to split all N bubbles, into bubbles with diameters half of their original sizes, can be calculated from,

$$N \int_0^\infty \frac{9}{8} \pi\sigma \frac{d_b^3}{d_{b_o}^2} e^{-\frac{1}{2}\left(\frac{d_b}{d_{b_o}}\right)^2} dd_b ,$$

$$= \frac{9}{4} \Gamma(2) \pi\sigma N d_{b_o}^2$$, where d_{b_o} is the bubble diameter of maximum

population in the Rayleigh spectrum. The power P_s needed to split all N bubbles is then $\frac{9}{4}\Gamma(2)\pi\sigma\dot{N}d_{b_o}^2$. To relate the bubble generation rate N to the gas ejection rate Q, we first integrate the Rayleigh spectrum to find the total gas volume, then differentiate it with respect to time to obtain the volumetric gas flow rate.

Namely, total gas volume $= N \int \frac{4}{3} \pi \left(\frac{d_b}{2}\right)^3 \frac{d_b}{d_{b_o}^2} e^{-\frac{1}{2}\left(\frac{d_b}{d_{b_o}}\right)^2} dd_b ,$

$$= \frac{\sqrt{2}}{3} \pi N \Gamma\left(\frac{5}{2}\right) d_{b_o}^3 = \left(\frac{\pi}{2}\right)^{3/2} N d_{b_o}^3 .$$

Therefore, the volumetric gas flow rate Q can be expressed as,

$$Q = \left(\frac{\pi}{2}\right)^{3/2} \dot{N} d_{b_o}^3 , \qquad (2.24)$$

where $\dot{N}$ is the number of bubbles generated per time. Expressing $\dot{N}$ in terms of Q, we have the rate of turbulence energy required to split bubbles to be,

$$P_s = \frac{9}{4} \pi\sigma d_{b_o}^2 \left(\frac{2}{\pi}\right)^{3/2} \frac{Q}{d_{b_o}^3} ,$$

or

$$P_s = \frac{9}{(2\pi)^{1/2}} \frac{\sigma}{d_{b_o}} Q \quad . \tag{2.25}$$

This power should be compared with the total turbulence power available in a turbulent boundary layer over an axisymmetric body, which can be derived as follows. The power per unit width is

$$P'_{TBL} = \int_o^\delta \rho\overline{q^2}dy, \quad \text{where } \overline{q^2} = \overline{u'^2} + \overline{v'^2} + \overline{w'^2}$$

and $\sim \overline{q^2}^2 = 8\, u_\tau\, (1-\frac{Y}{\delta})$ from HINZE, pp.642, Fig. 7-20 [23], so that

$P'_{TBL} = 4\rho u_\tau^2\delta$, where δ is boundary layer thickness. Multipying P'_{TBL} by πDU_o, where D is the diameter of an axisymmetric body and U_o the body speed, we have

$$P_{TBL} = 4\pi\rho u_\tau^2 \delta U_o D \quad . \tag{2.26}$$

The ratio of P_S to P_{TBL} is then,

$$\frac{P_s}{P_{TBL}} = \frac{9}{2(2\pi)^{1/2}} \left(\frac{\sigma}{\rho u_\tau^2 d_{b_o}}\right) \left(\frac{Q}{\pi D\delta U_o}\right) \quad . \tag{2.27}$$

The first term in parenthesis represents the ratio of surface tension stress to turbulent wall shear stress for a bubble with a diameter of d_{b_o}, the second parenthesis represents the ratio of gas flow rate to that of the TBL axisymmetric body. To estimate the first parenthesis, we apply the following facts,

$$\rho\overline{u'^2_{max}} \cong 9\, \rho u_\tau^2 \ , \quad d_{b_{max}} \cong 3\, d_{b_o} \quad ,$$

so that,

$$\frac{\rho\overline{u'^2_{max}}\, d_{b_{max}}}{\sigma} = 3 = 27 \frac{\rho u_\tau^2\, d_{b_o}}{\sigma}, \text{ and } \frac{\sigma}{\rho u_\tau{}^2 d_{b_o}} \cong 10 \quad .$$

(2.27) represents the fraction of energy expended to split all bubbles from d_b into $\frac{1}{2}\, d_b$. This ratio should be equal to the drag reduction ratio in practice. Although this analysis is preliminary in nature, it does indicate that bubble splitting can be a primary candidate responsible for the effectiveness of microbubbles to reduce turbulence. To increase the effectiveness, one can increase the surface tension between the water and gas, therefore injecting certain liquid or using a gas having higher value of σ may be useful. Since d_{b_0} decreases with flow speed U_0, by increasing U_0 higher reduction may also be possible, however, we must also increase the gas flow rate Q proportionally.

3. Conclusion and Future Plans

The following conclusions have been obtained:

- Hydrodynamic forces, both tangential and normal to the wall, have been identified. Quantitative relationship with gas injection induced bubble formation on a wall are obtained. But fundamental unknowns are the bubble shape, drag force over it and lift force on a bubble in a TBL.

- Bubble size regimes have been identified, they are: the spherical cap, single bubble, intermediate and jet disintegration regimes. The controlling parameter has been found to be the ratio of gas exit velocity to the external flow velocity. At very low injection rate, the tangential force balance mode determines bubble size, i.e., water drag and surface tension balances each other. At higher injection rates, the normal force balance mode takes over, i.e., the lift force balances the surface tension or the water inertia surrounding a bubble. At even higher injection rate, jet instability determines the bubble size. There are still unknown regions between the gas jet and bubble regimes.

- Effects of seawater have been identified, but not quantified.

- Effects of bubble splitting have been identified to be very significant, and may well be the key origin of microbubble drag reduction mechanisms.

- Transport of bubbles in a TBL has been simulated. The smaller the bubble, the more likely the bubbles will stay near the wall. The unknowns are the quantitative lift force formulation, bubble coalescence and splitting effects.

4. References

1. A. Acrivos: The Breakup of Small Drops and Bubbles in Shear Flows, New York Academy of Sciences, p. 1. (1983)

2. J. Aitchison and J.A.C. Brown: The Lognormal Distribution, (Cambridge at the University Press, University of Cambridge)(1957).

3. Al-Hayes and Winterton: Bubble Growth in Flowing Liquids, Int. J. Heat Mass Transfer, 24, pp. 213-221 (1981a).

4. Al-Hayes and Winterton: Bubble Diameter on Detachment in Flowing Liquids, Int. J. Heat Mass Transfer, 24, pp. 223-230 (1981a).

5. I. A. Avetisyan: Sov. Phys. Acoust., 23, pp. 285-288 (1977).

6. G. K. Batchelor: An Introduction to Fluid Dynamics, (Cambridge at the University Press, University of Cambridge) (1967).

7. J. C. Bhatia, F. Durst and J. Jovanovic: Corrections of Hot-Wire Anemometer Measurements Near Walls, J. Fluid Mech., 122, pp. 411-431. (1982)

8. D. C. Blanchard: The Electrification of the Atmosphere by Particles from Bubbles in the Sea, Progress in Oceanography, 1, pp. 72-202 (1963).

9. R. S. Brodkey, J.M. Wallace and H. Eckelmann: J. Fluid Mech. 63, p. 209 (1974).

10. T. E. Burton: The Connection Between Intermittent Turbulent Activity Near the Wall of a Turbulent Boundary Layer with Pressure Fluctuations at the Wall, Report No. 70208-10, Acoustics and Vibration Laboratory, (Massachusetts Institute of Technology) (1974).

11. C. P. Chen, Y. Goland and E. Reshotko: Generation Rate of Turbulent Patches in the Laminar Boundary-Layer of a Submersible, in Hough, Ed. G.R. Hough, Viscous Flow Drag Reduction, Progress in Astronautics and Aeronautics, pp. 73-89 (1980).

12. C. I. Christov and P.K. Volkov: Numerical Investigation of the Steady Viscous Flow Past a Stationary Deformable Bubble. J. Fluid Mech., 158, pp. 341-363 (1985).

13. P. H. Clay: Proc. Roy. Acad. Sci., (Amsterdam), 43, pp. 852-979 (1940).

14. R. E. Davis and A. Acrivos: The Influence of Surfactants of the Creeping Motion of Bubbles, Chemical Engineering Science, 21, pp. 681-685 (Pergamen Press Ltd., Oxford) (1966).

15. E. T. Degens: In the Global Carbon Cycle: SCOPE Report 13, pp. 57-77 B. Bolin et al., Eds., (Wiley, New York)(1979).

16. A. Detwiler: Surface-Active Contamination on Air Bubbles in Water, in Surface Contamination: Genesis, Detection, and Control, Ed. K.L. Mittal (Plenum, New York, 1979), 2, pp. 993-1007 (1979).

17. F. E. Fox and K.F. Herzfeld: J. Acoust. Soc. AM-26, pp. 984-989 (1954).

18. J. J. Gallagher: Microbubble Size Distributions Data Collection and Analyses (presented in the OCEAN 1984 conference) (1984).

19. R. E. Glazman: Effects of Absorbed Films on Gas Bubble Radial Oscillations, J. Acoust. Soc. Am., 74, p. 980 (1983).

20. I. S. Gradshteyn and I.M. Ryzhik: Table of Integrals, Series, and Products, Editor Alan Jeffrey, Academic Press (1980).

21. L. G. Guerri, L. G. Lucca and A. Prosperetti: A Numerical Method for the Dynamics of Non-Spherical Cavitation Bubbles. Proc. 2nd Int. Colloquium on Drops and Bubbles. (Editor Dennis H. LeCroissette, Monterey, California, Nov. 19-21). NASA JPL Publication 82-7 (1981).

22. J. O. Hinze: Fundamentals of the Hydrodynamic Mechanism of Splitting in Dispersion Processes, A.I.Ch.E. Journal, 1, No. 3, p. 280 (1955).

23. J. O. Hinze: Turbulence, Second Edition, McGraw-Hill, New York, New York (1975).

24. B. D. Johnson and R.C. Cooke: Generation of Stabilized Microbubbles in Seawater, SCIENCE, 213, p. 209 (1981).

25. A. N. Kolmogorov: On the Disintegration of Drops in a Turbulent Flow, Doklady Akad. Nauk., SSSR, 66, p. 825 (1949).

26. E. B. Kraus: Atmosphere-Ocean Interaction, Clarendon Press, Oxford (1972).

27. H.R. Kreplin and H. Ecklemann: Bearer of the Three Fluctuating Velocity Components in the Wall Region of a Turbulent Channel Flow, Physics of Fluids, 22, No. 7, p. 1233 (1979).

28. F. MacIntyre: Bubble Generation on a Heated Body, Graduate School of Oceanography (University of Rhode Island) personal notes (1981).

29. N. K. Madavan, S. Deutsch and C.L. Merkle: Measurements of Local Skin Friction in a Microbubble Modified Turbulent Boundary Layer, TM. No. 84-136, Penn. State Univ., Applied Research Laboratory (1984).

30. M. R. Maxey and J.J. Riley: Equation of Motion for a Small Rigid Sphere in a Nonuniform Flow, Phys. Fluids, 26, (4), pp. 883-889 (1983).

31. G.S. Migirenko and A.R. Evseev: Turbulent Boundary Layer with Gas Saturation, Problems of Thermophysics and Physical Hydrodynamics (in Russian) (Novosibirsk) Nauka (1974).

32. E. C. Monahan: Sea Spray and its Relationship to Low Elevation Wind Speed. Ph.D. Thesis (Mass. Institute of Technology, Cambridge, MA)(1966).

33. S. A. Morsi and A.J. Alexander: An Investigation of Particle Trajectories in Two-Phase Flow Systems, J. Fluid mech., 55, pp. 193 (1972).

34. H. Nakagawa and I. Nezu: J. Fluid Mech., 104, p. 1 (1981).

35. F. Odar and W.S. Hamilton: Forces on a Sphere in a Viscous Fluid, J. Fluid Mech., 18, pp. 302 (1964).

36. J. M. Rallison: J. Fluid Mech., 109, p. 456 (1981).

37. Lord Rayleigh: On the Stability of Cylindrical Fluid Surfaces, Phil. Mag. 34, pp. 177-180 (1892).

38. M. M. Reischman and J.M. Holzmann: Digital Image Analysis of Two Phase Flow Data, NOSC Tech. Report 502 (San Diego, California)(1980).

39. G. A. Riley: Limnology Oceanography, 8, p. 372 (1963).

40. P. G. Saffman: The Lift on a Small Sphere in a Slow Water Flow, J. Fluid Mech., 22, pp. 385 (1965).

41. P. G. Saffman: Corrigendum, J. Fluid Mechanics, 31, p. 625 (1968).

42. V. A. Sanborn: Evaluation of the Time Dependent Surface Shear Stress in Turbulent Flows, ASME Publication 79-WA/FE-17 (1979).

43. H. Schlichting: Boundary Layer Theory, McGraw Hill Book Co., New York, New York (1960).

44. J. C. Scott: The Role of Salt in White Cap Persistence, Deep Sea Research, 22, pp. 653-657 (1975).

45. M. Sevik and S.H. Park: The Splitting of Drops and Bubbles by Turbulent Fluid Flow, J. Fluid Engineering, p. 53 (1973).

46. E. Silberman: Gas Jets in Liquids, Production of Bubbles By the Disintegration of Gas Jets in Liquid, 5th Midwestern Conference on Fluid Mechanics, p. 263 (University of Minnesota) (1957).

47. C. A. Sleicher, Jr.: Maximum Stable Drop Size in Turbulent Flow, A.I.Ch.E. Journal, 8, No. 4, p. 471 (1962).

48. G. I. Taylor: The Formation of Emulsion in Definable Field of Flow, Proceeding of the Royal Society, London, Series A, 146, p. 501 (1934).

49. Voloshko, et al,: Inzhenerno-Fizicheskii Zhurnal, 35, No. 6, pp. 1066-1071 (1978).

Some Remarks on the Kutta Condition

Koichi Oshima

1. Introduction

Applicability of the Kutta-Joukowski condition to the steady, two-dimensional, inviscid flow around an airfoil with a sharp trailing edge has been well established experimentally as well as theoretically. It says at the trailing edge that:

(1) the pressure is continuous, that is, there is no tail loading,

(2) the departing velocities from both sides are the same, that is, there is no vorticity shedding,

(3) the departing stream line bisects the trailing edge angle.

This flow model is also applicable to the medium for high Reynolds number flow around a thin airfoil with a sharp trailing edge and with no separation zone, if one accounts for the displacement thickness of the boundary layer along the airfoil surface and applies one of these conditions (usually condition (3)) at this modified body surface. This concept is called the global Kutta condition. This outer potential flow is calculated and gives physically plausible flow characteristics.

For the flows around an airfoil with a round trailing edge, the numerical solutions of the Navier-Stokes equations with the no-slip condition at the body surface are not unique and give an infinite number of physically acceptable solutions. One of these Kutta conditions, usually that the departing streamline bisects the trailing part of the airfoil surface, is applied in order to make the solution unique. This corresponds to the one having the pre-specified, bound circulation. In other words, the flows around the two-dimensional body with a round separation part are uniquely determined only when the global flow condition, such as bound vorticity or far field vorticity distribution, is given. The global Kutta condition can be used in these cases.

W.S.VORUS [1] proposed a new high Reynolds number approximation, which is different from the combination of the boundary layer theory and the potential flow theory. He unified the viscous and potential regimes of the high Reynolds number flow into the single solution domain. His theory can predict the characteristics of separated flows approximately. It recovers thin airfoil theory in the limit of infinite Reynolds number, and concludes that, for high Reynolds number flows, positions of separation and reattachment points are independent of Reynolds number. The separated flow collapses, along with all other viscous-flow characteristics, into the infinitesimally thin body-surface vortex sheet as vorticity convection overwhelms vorticity diffusion in the infinite Reynolds number limit. This gives a theoretical foundation of the concept of the global Kutta condition.

2. Unsteady Flows

For the cases of unsteady flows, such as starting flows or flows around an oscillating airfoil, many questions are still left open to further investigation. A potential flow around a flat plate with zero incidence, which is heaving in uniform flow, was numerically simulated based on the Euler equations by A. NATSUME [2]. Flow patterns are shown in Fig.1. It is noted that a point vortex is located at the trailing edge and that the separated streamline starts at a point on the flat plate and leaves from it perpendicularly. The experimentally observed wake patterns, shown in Fig.2, also confirms this fact.

Flow separating at a sharp wedge has been studied extensively in connection with flows around a moving two-dimensional body with a sharp corner, (for example see [3],[4]). Impulsively started flows past an airfoil with a sharp trailing edge, particularly the vorticity layer shed behind the airfoil, have been investigated theoretically as well as experimentally by C.-Y.CHOW and M.-K.HUANG [5], J.M.R.GRAHAM [6] and D.AUERBACH [7]. An infinitely thin vortex sheet shed from the trailing edge surrounded by an otherwise potential flow, assumed in [6], follows the self-similar solution of the local boundary condition and is straight. On the other hand, Chow and Huang consider the distant geometry to be important to the flow at the edge, and the flow can never be considered self-similar and the initial path of the vortex center must be curved. Auerbach concluded that the initial flow is not self-similar. That is, the flow field around the trailing edge is dependent upon the far field including the whole airfoil, supporting the global Kutta condition concept.

Numerical simulation of the high Reynolds number flow around a two-dimensional oscillating or rotating elliptic airfoil with various time constants has been carried out based on the Navier-Stokes equations (for example, K.OSHIMA, Y.OSHIMA and Y.KURIYAMA [8]) in which the far field boundary conditions are taken from the asymptotic solution of the vorticity equations. Some of the results are presented in Fig.3. It was found that the rear separation point does not stay at the trailing edge and has a time hysterisis character and that the aerodynamic forces acting on the airfoil show loop characteristics. No separation zone appears and the leaving streamline bisects the body surface curvature. That is, the Kutta condition is not locally applicable at the trailing edge and the flow character around there is influenced by the whole flow field characters.

3. Discrete Vortex Methods

Discrete vortex methods have been successfully applied for various flows around steady as well as unsteady, two-dimensional bodies, as seen in K.OSHIMA and Y.OSHIMA [9]. Since the discrete vortex approximation is considered as the high Reynolds number limit of viscous, two-dimensional flows in which vorticity concentrates into discrete points and does not decay, the generation processes of the vorticity has to be modeled based on, for example, the Kutta condition. One of commonly used methods is: Discretize the body surface into a finite number of elements, assign a singularity (usually a point vortex), at the center of each element, apply the condition of solid boundary to each element and determine the strength of the assigned vortex. In order to fulfill the conservation law of circulation, one free vortex is assumed in the neighborhood of the trailing edge, which drift due to the induced velocity caused by all the other vortices. The solid boundary condition is taken as no perpendicular velocity to the solid surface and no mean slip velocity along the element surface (the latter of which is called the no-slip condition).

This method has been successfully applied to many two-dimensional flow problems with a sharp trailing edge, (for example see [10]). In the case when the element including the trailing edge tends to be infinitely close to the trailing edge, this no-slip condition becomes exactly the same as the Kutta condition discussed in Session 1. The case when this element is taken as half of the leading or the trailing part of the symmetric airfoil gives the potential flow solution, that is, no net drag is derived. N. IZUTSU [11] applied this method to a rotating elliptic airfoil, and found that the both solutions do not give downwash of the wake. This is contrary to the experimental observation as shown in Fig. 4, which are the visualized flow pattern around a rotating elliptic airfoil using a smoke wire method. Therefore, he chose this element between such two cases. The boundary of this element is taken so that the potential flow solution along the rest of the surface fulfills the no-slip condition. Some of his results are shown in Fig. 5. Thus we can determine the shed vorticity from the unsteady, two-dimensional airfoil on the solid foundation. Unsteady forces acting on a rotating elliptic cylinder have finite lift and, then, a downwash of the trailing vortex street can be derived, which agrees with the experimental results.

The "global Kutta condition" has been proven to give the most reasonable results, as verified by the results of the experimental observation which were extensively carried out using the wind tunnels and water channels of ISAS and others.

4. Three-Dimensional Flows

Difficulties in extending the discrete vortex method to three-dimensional flow problems were theoretically confirmed by P. G. SAFFMAN and D. I. MEIRON [12], in which inconsistency between vorticity and material coordinates due to stretching of the vortex filament is found to cause the difficulty. Also, vorticity generation processes on three-dimensional bodies in the infinite Reynolds number limit should not be modeled by simple phenomenological means. Experimentally, it is invariably observed that two-dimensional vortical flows always develop a three-dimensional instability. Theoretically, two-dimensional shear layers suffer a Karman-Helmholtz type instability, and eventually turbulence transition takes place. These three-dimensional flows and turbulence are one of the most difficult, unsolved problems in fluid dynamics.

References

1. William S. Vorus: A theory for flow separation, J.Fluid Mech. vol.132 pp.163-183 (1983)
2. Akiko Natsume: Formation and development of the vortex street behind an oscillating airfoil, Dr. Thesis, University of Tokyo (1982) in Japanese
3. J.P.Giesing: Vorticity and Kutta condition for unsteady multienergy flows, J.Appl.Mech. pp.608-613 (1969)
4. P.W.Bearman and J.M.R.Graham: Vortex shedding from bluff bodies in oscillatory flow: A report on Euromech 119, J.Fluid Mech. vol.99 pp.225-245 (1980)
5. Chuen-Yen Chow and Ming-Ke Huang: The initial lift and drag of an impulsively started airfoil of finite thickness, J.Fluid Mech. vol.118 pp.393-409 (1982)
6. J.M.R.Graham: The lift on an aerofoil in starting flow, J.Fluid Mech. vol.133 pp.413-425 (1983)
7. David Auerbach: Experiments on the trajectory and circulation of the starting vortex, J.Fluid Mech. vol.183 pp.185-198 (1987)
8. K.Oshima, Y.Oshima and Y.Kuriyama: Finite element analysis of viscous incompressible flow around an oscillating airfoil, Proc. 6th ICNMFD, pp.433-438 (1978)
9. K.Oshima and Y.Oshima: Flow simulation by discrete vortex method, Proc. 8th ICNMFD, pp.94-106 (1982)
10. K.Ono: Investigation of two-dimensional oscillating airfoil, Dr.Thesis of University of Tokyo (1979)
11. N.Izutsu: Numerical and experimental study of unsteady vortical flow, Dr.Thesis of University of Tokyo (1986)
12. P.G.Saffman and D.I.Meiron: Difficulties with three-dimensional weak solutions for inviscid incompressible flow, Physics Fluid vol.29 pp.2373-2375 (1986)

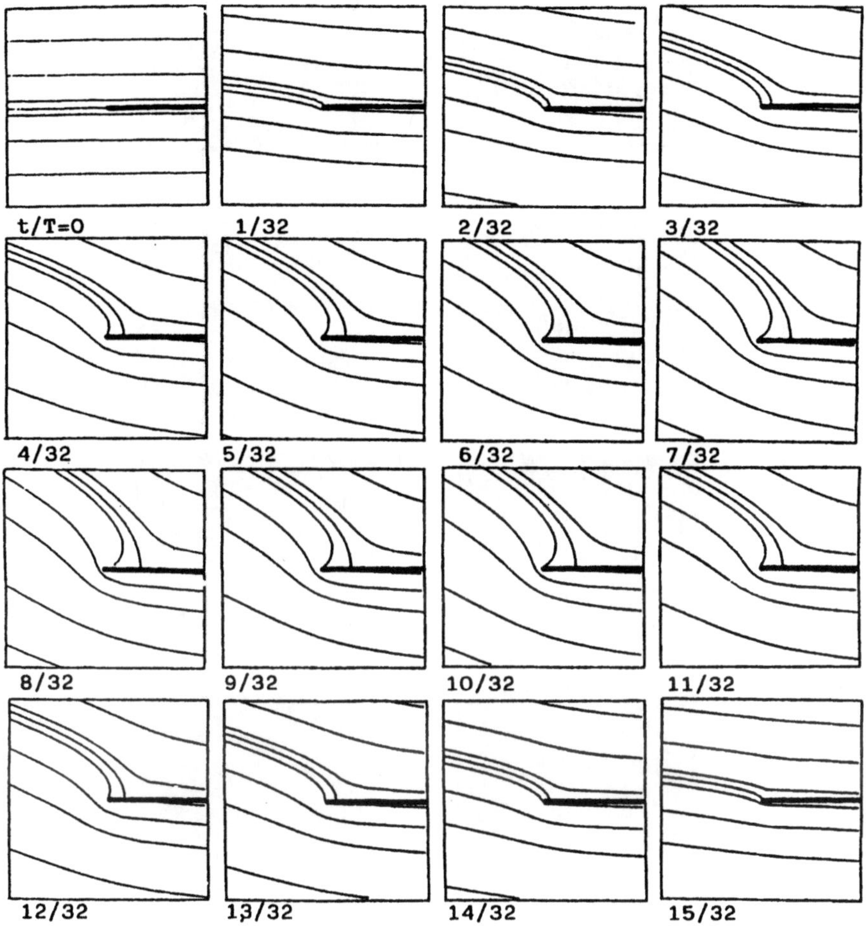

Fig. 1 Numerically simulated flow around trailing edge of a heaving flat plate, which is based on the Euler equations and assume a finite point vortex located at the trailing edge. The value of t/T corresponds to phase time of the oscillation. Note that the leaving streamline start at a upstream portion of the plate.

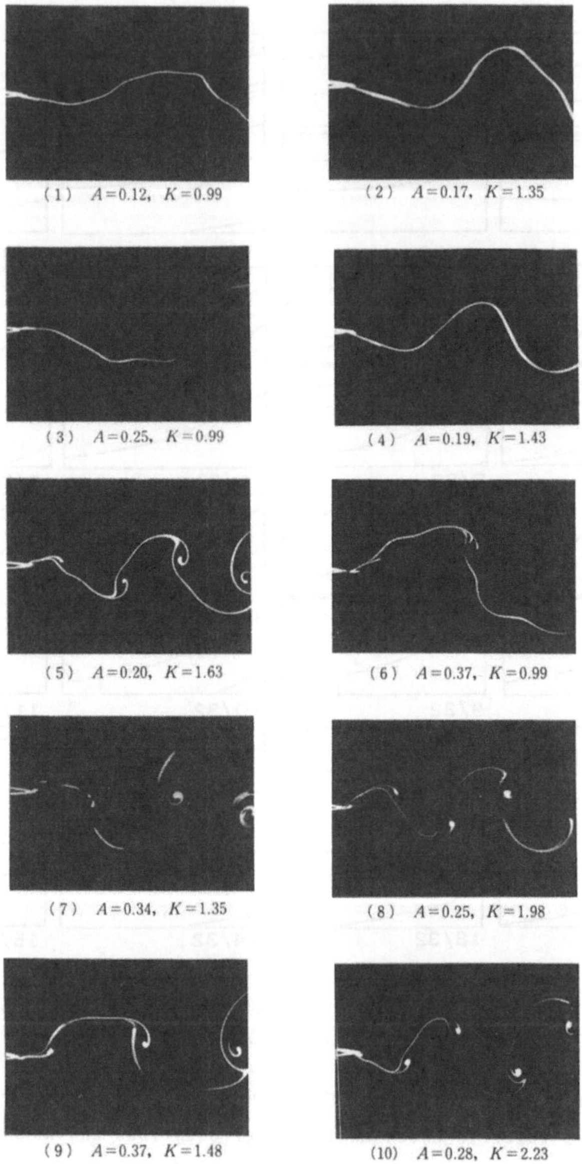

Fig. 2 Visualized wake pattern shed from an oscillating NACA 0012 airfoil using electrolysis method, where A and K are the nondimensionalized amplitude and the nondimensionalized frequency of the airfoil. Note that the wake does not bisect the trailing edge angle, especially for strong oscillation.

(a) t=6.4, K=2.0

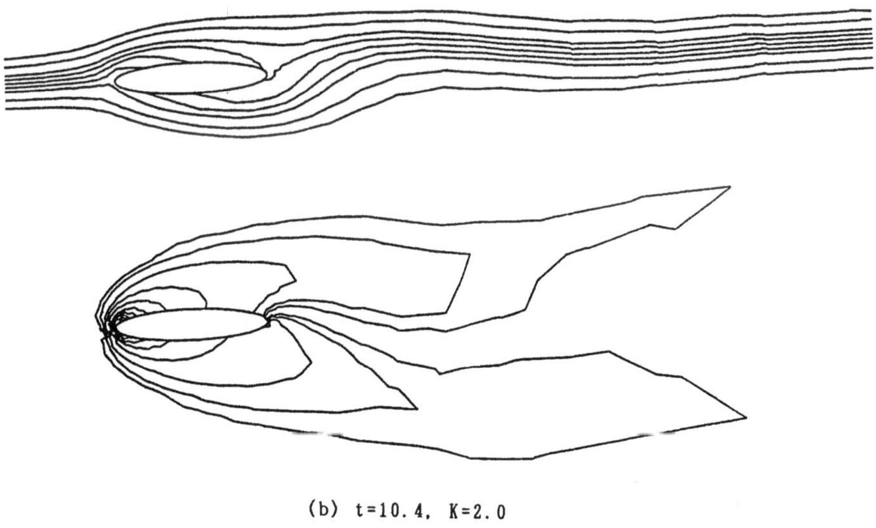

(b) t=10.4, K=2.0

Fig. 3 Numerical solution of flow fields around an oscillating elliptical airfoil based on finite element analysis of the Navier-Stokes equations. Upper is the streamlines and lower is the equi-vorticity contours. Note that no separation zone appears and the leaving streamline starts at an upstream part of the airfoil surface.

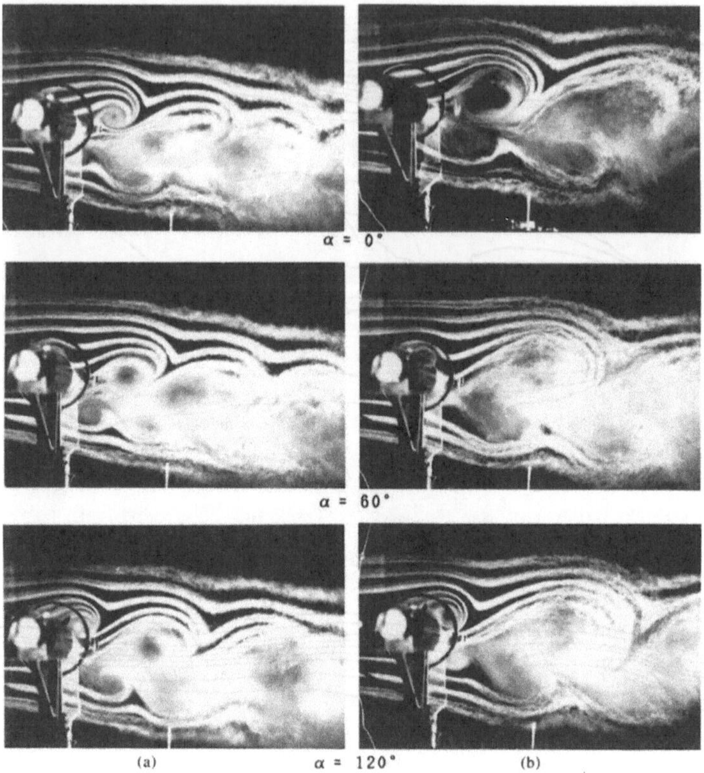

Fig. 4 Visualized flow pattern around a rotating elliptic airfoil. Smoke wire method was applied and the photographs were taken by synchronized strobo-flash lights with the rotation. Note that the vortices are shed regularly and their movement shows downwash clearly.

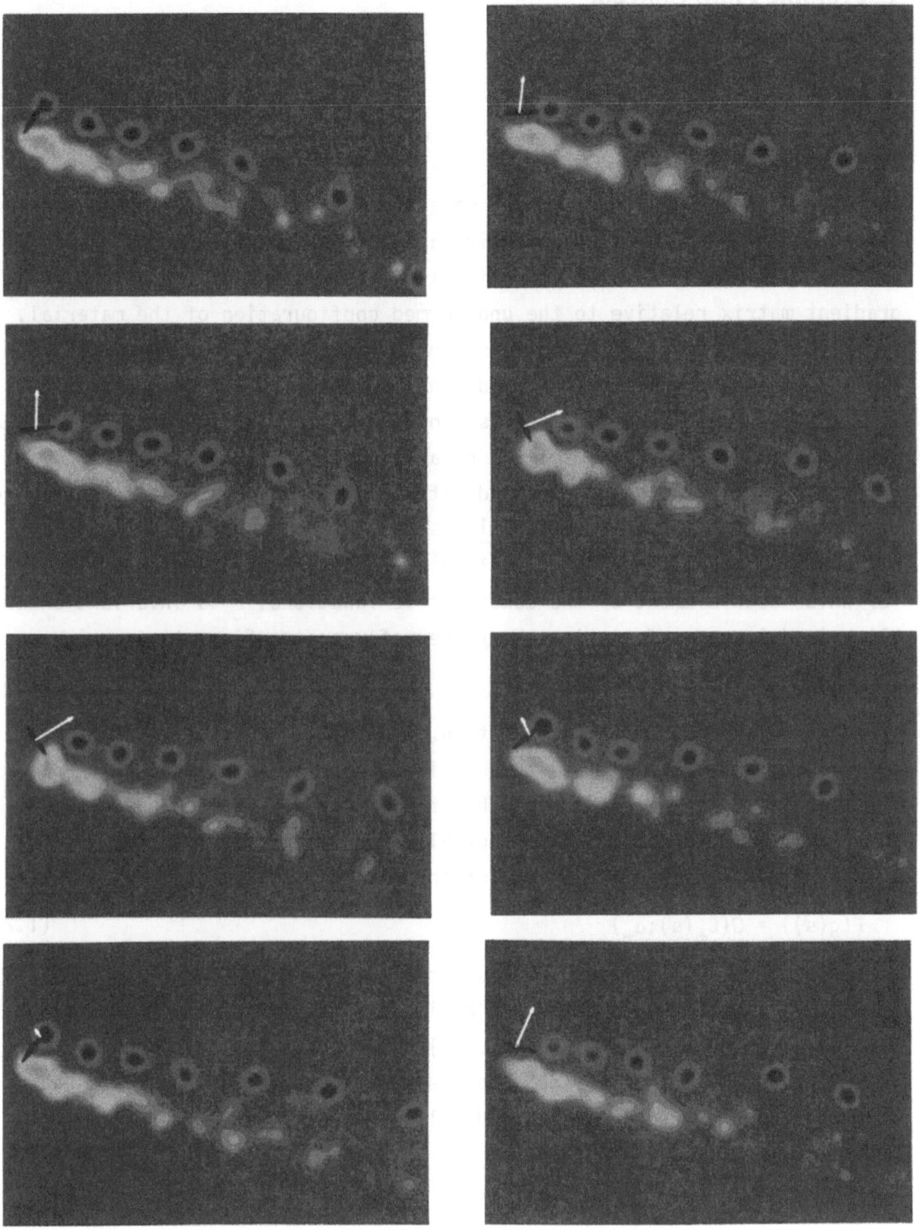

Fig. 5 Flows around the rotating elliptic airfoil were simulated by discrete vortex method. CLearly the downwash of the shed vortex street is visible.

The Simple Fluid Concept

R.S. Rivlin and G.F. Smith

1. Introduction

In [1], Green and Rivlin discussed constitutive equations for materials with memory on the basis of a constitutive assumption that the Cauchy stress $\underset{\sim}{\sigma}(t)$ at time t is a matrix-valued functional of the history of the deformation gradient matrix relative to the undeformed configuration of the material. In [2] Noll, restricting his discussion to the special case when the material is of the hereditary type, made the constitutive assumption that $\underset{\sim}{\sigma}(t)$ is a matrix-valued functional $\underset{\sim}{F}\{\underset{\sim}{g}(s)\}$ of the history of the deformation gradient matrix $\underset{\sim}{g}(s)$ at lapsed time s relative to an arbitrarily chosen fixed reference configuration. He called materials for which such a constitutive assumption is valid <u>simple materials</u> (see §3). He sought to distinguish between simple solids and simple fluids by defining the latter as materials for which each element of the constitutive functional $\underset{\sim}{F}$ is invariant under all unimodular transformations $\underset{\sim}{H}$ of the reference configuration:

$$\underset{\sim}{F}\{\underset{\sim}{g}(s)\} = \underset{\sim}{F}\{\underset{\sim}{g}(s)\underset{\sim}{H}\}. \tag{1.1}$$

Noll concluded from this assumption, together with material frame indifference, that $\underset{\sim}{F}\{\underset{\sim}{g}(s)\}$ must necessarily be expressible as an <u>isotropic</u> functional $\hat{\underset{\sim}{G}}\{\underset{\sim}{C}_t(s);\rho_t\}$ of the Cauchy strain matrix $\underset{\sim}{C}_t(s)$ at lapsed time s, relative to the configuration actually adopted by the material at time t, parametrized by the density ρ_t of the material at time t:

$$\underset{\sim}{F}\{\underset{\sim}{g}(s)\} = \hat{\underset{\sim}{G}}\{\underset{\sim}{C}_t(s);\rho_t\} \tag{1.2}$$

Furthermore, Noll stated, without proof, that any such functional satisfies (1.1) and therefore describes a simple fluid. A proof of this statement has been advanced by TRUESDELL [3]. These results have been called by TRUESDELL and NOLL [4,§32] the "fundamental theorem on simple fluids" and have been much quoted in the periodical and hard-cover literature. In this paper we shall discuss this theorem critically.

Noll's theory has been criticized previously by FAHY and SMITH [5,6] and, more simply, in a brief note, by RIVLIN [7]. They argued, by applying theorems in the theory of invariants, that the relation (1.1) implies restrictions on the constitutive functional far more stringent than those derived by Noll. Following the publication of [7], MURDOCH [8] attempted to defend Noll's theory and his note was followed by further notes by RIVLIN and SMITH [9], MURDOCH [10], TRUESDELL [3], and RIVLIN [11].

It was maintained by MURDOCH [8] and by TRUESDELL [3], as well as by others in private communications, that what Noll really meant was that $\tilde{F}$ is expressible as an isotropic functional of the Cauchy strain matrix relative to a fixed configuration which instantaneously coincides with the configuration actually adopted by the material at time t.

We see no evidence in [2], nor in the exposition by Truesdell and Noll in [4,§32], nor indeed, in the mathematical statement in [3] as distinct from the verbal statement, that this is indeed the case. However, it will be seen that the main conclusions of the present paper are valid for either interpretation.

In §4 of the present paper we show that Noll's defining conditions for a simple fluid, if fully exploited, place restrictions on the form of the constitutive functional more severe than those expressed by (1.2). In order to avoid the accusation that we are misinterpreting Noll's statements, we follow the argument presented by Truesdell and Noll in [4,§32] step by step, changing only the notation. After each step we comment on its validity.

In §5 we show that Truesdell's argument in [3], that any constitutive functional of the form (1.2) satisfies Noll's defining conditions for a simple fluid, could equally well be used to prove that a much wider class of constitutive functionals also satisfies these conditions and can lead to a reductio ad absurdum. Indeed, in [12] we have shown that the form for the constitutive functional for a simple fluid given by Noll's theorem is merely that for any isotropic simple material if the spatial (i.e. Eulerian) description of the deformation history is used.

The origin of these paradoxes which result from Noll's analysis resides in the manner in which the invariance statement (1.1) is used (see §6).

2. Notation

We shall be concerned in this paper with deformations of a body of homogeneous material which take place in the interval $(-\infty,t]$ of the time τ. Let $\tilde{X}$ be the vector position of a generic particle of the body in some fixed reference configuration X, which may or may not coincide with a configuration actually adopted by the body at some time during the deformation. We shall denote the configuration actually adopted by the body at time t by x_t. We define the <u>lapsed time</u> s by

$$s = t - \tau. \tag{2.1}$$

Since we shall be concerned only with materials which are hereditary, we shall describe the deformation by the dependence of the vector position $\underset{\sim}{x}(s)$ of the generic particle at lapsed time s on $\underset{\sim}{X}$ and s:

$$\underset{\sim}{x}(s) = \underset{\sim}{x}(\underset{\sim}{X},s). \tag{2.2}$$

Let X_A (A=1,2,3) and $x_i(s)$ (i=1,2,3) be the components of the vectors $\underset{\sim}{X}$ and $\underset{\sim}{x}(s)$ in a fixed rectangular cartesian coordinate system x.* Then (2.2) may be written in indicial notation as

$$x_i(s) = x_i(X_A,s). \tag{2.3}$$

The deformation gradient matrix, at lapsed time s, relative to the configuration X and referred to the rectangular cartesian coordinate system x, is denoted $\underset{\sim}{g}(s)$ and defined by

$$\underset{\sim}{g}(s) = \left\| g_{iA}(s) \right\| = \left\| \frac{\partial x_i(s)}{\partial X_A} \right\|. \tag{2.4}$$

Correspondingly, the deformation gradient matrix, at lapsed time s, relative to the configuration x_t and referred to the rectangular cartesian coordinate system x, is denoted $\underset{\sim}{g}_t(s)$ and defined by

$$\underset{\sim}{g}_t(s) = \left\| g^t_{ij}(s) \right\| = \left\| \frac{\partial x_i(s)}{\partial x_j(0)} \right\|. \tag{2.5}$$

The Cauchy strain matrices, at lapsed time s, relative to the configurations X and x_t, and referred to the coordinate system x, are denoted $\underset{\sim}{C}(s)$ and $\underset{\sim}{C}_t(s)$ respectively and are given by

$$\underset{\sim}{C}(s) = \underset{\sim}{g}^{\dagger}(s)\underset{\sim}{g}(s), \quad \underset{\sim}{C}_t(s) = \underset{\sim}{g}_t^{\dagger}(s)\underset{\sim}{g}_t(s), \tag{2.6}$$

where here and throughout this paper the dagger denotes the transpose.

We note from (2.4) and (2.5) that

$$\underset{\sim}{g}_t(s) = \underset{\sim}{g}(s)\underset{\sim}{g}^{-1}(0). \tag{2.7}$$

Equations (2.6) then yield

$$\underset{\sim}{C}_t(s) = [\underset{\sim}{g}^{-1}(0)]^{\dagger}\underset{\sim}{C}(s)\underset{\sim}{g}^{-1}(0). \tag{2.8}$$

---

*Throughout this paper, when we speak of a vector we shall mean the column matrix formed by the components of the vector in the coordinate system x.

3. Simple Materials

Let $\underset{\sim}{\sigma}(t) = \| \sigma_{ij}(t) \|$ be the Cauchy stress matrix at time t, i.e. at s = 0, referred to the coordinate system x. We make the constitutive assumption that $\underset{\sim}{\sigma}(t)$ is a functional of the history of the deformation gradient matrix $\underset{\sim}{g}(s)$ for $s = [0,\infty)$, i.e. for all times up to and including t:

$$\underset{\sim}{\sigma}(t) = \underset{\sim}{F}\{\underset{\sim}{g}(s)\}. \tag{3.1}$$

In the particular case when the material is incompressible the constitutive assumption (3.1) is replaced by

$$\underset{\sim}{\sigma}(t) = \underset{\sim}{F}\{\underset{\sim}{g}(s)\} - p\underset{\sim}{I}, \tag{3.2}$$

where p is an arbitrary hydrostatic pressure and $\underset{\sim}{I}$ is the unit matrix.

In indicial notation equations (3.1) and (3.2) may be rewritten as

$$\sigma_{ij}(t) = F_{ij}\{g_{kA}(s)\} \tag{3.3}$$

and

$$\sigma_{ij}(t) = F_{ij}\{g_{kA}(s)\} - p\delta_{ij} \tag{3.4}$$

respectively, where δ_{ij} is the Kronecker delta.

In [2] Noll defines a simple material as a material for which the constitutive assumption (3.1) or (3.2) is valid. He defines the isotropy group of the material as the group H of all unimodular transformations $\underset{\sim}{H}$ for which the constitutive functional $\underset{\sim}{F}$ satisfies the relation ([4] eq. (31.2))

$$\underset{\sim}{F}\{\underset{\sim}{g}(s)\} = \underset{\sim}{F}\{\underset{\sim}{g}(s)\underset{\sim}{H}\}. \tag{3.5}$$

He defines a simple fluid as a simple material for which the isotropy group H is the full unimodular group and a simple solid as a simple material for which H is the full orthogonal group or a sub-group of it. If the group H is the full orthogonal group the material is isotropic. Since the full orthogonal group is a sub-group of the full unimodular group it follows from Noll's definition that a simple fluid is necessarily isotropic.

According to the polar decomposition theorem $\underset{\sim}{g}(s)$ may be expressed in the form

$$\underset{\sim}{g}(s) = \underset{\sim}{R}(s)\underset{\sim}{U}(s), \tag{3.6}$$

where $\underset{\sim}{U}(s)$ is a symmetric matrix with positive proper values and $\underset{\sim}{R}(s)$ is a proper orthogonal matrix. $\underset{\sim}{U}(s)$ is the right-stretch matrix at lapsed time s relative to the configuration X referred to the coordinate system x. If $\underset{\sim}{g}(s)$ is given, $\underset{\sim}{U}(s)$ and $\underset{\sim}{R}(s)$ are determined uniquely by the relations

$$\underset{\sim}{U}(s) = \underset{\sim}{C}^{1/2}(s), \quad \underset{\sim}{R}(s) = \underset{\sim}{g}(s)\underset{\sim}{C}^{-1/2}(s), \tag{3.7}$$

where $\mathbf{C}^{1/2}(s)$ denotes the square root of the matrix $\mathbf{C}(s)$ all of whose proper values are positive.

The Principle of Material Frame Indifference places restrictions on the manner in which the constitutive functional $\mathbf{F}$ can depend on the argument function $\mathbf{g}(s)$. It implies that $\mathbf{F}\{g(s)\}$ must be expressible in the form

$$\mathbf{F}\{\mathbf{g}(s)\} = \mathbf{R}(0)\mathbf{G}\{\mathbf{U}(s)\}\mathbf{R}^{+}(0). \tag{3.8}$$

In [2] Noll obtains in explicit form the restrictions on the form of $\mathbf{F}$ for a simple fluid which, he claims, are implied by (3.8) and by the validity of (3.5) for all unimodular $\mathbf{H}$. He finds that $\mathbf{F}\{g(s)\}$ must be expressible in the form

$$\mathbf{F}\{\mathbf{g}(s)\} = \hat{\mathbf{G}}\{\mathbf{C}_t(s);\rho_t\}. \tag{3.9}$$

Here $\mathbf{C}_t(s)$ is defined by $(2.6)_2$, ρ_t is the density of the material at time t, and $\hat{\mathbf{G}}$ is a matrix-valued functional of $\mathbf{C}_t(s)$ and an ordinary function of ρ_t which satisfies the relation

$$\mathbf{Q}\hat{\mathbf{G}}\{\mathbf{C}_t(s);\rho_t\}\mathbf{Q}^{+} = \hat{\mathbf{G}}\{\mathbf{Q}\mathbf{C}_t(s)\mathbf{Q}^{+};\rho_t\} \tag{3.10}$$

for all orthogonal $\mathbf{Q}$.

We shall discuss in §4 the argument which leads to this conclusion, as it is presented in the article by TRUESDELL and NOLL [4,§32].

Noll further asserted in [2] that any simple material whose constitutive functional $\mathbf{F}$ satisfies (3.9) and (3.10) is a simple fluid. Noll did not give a proof of this statement in [2], nor did Truesdell and Noll in [4]. However, in [3] Truesdell has given such a proof which, he claims, "any qualified beginner can follow step by step." We shall discuss this proof in §5.

4. The Argument of Truesdell and Noll

In this section we present, in our notation, the argument advanced by TRUESDELL and NOLL [4,§32] to reach the conclusion that for a simple fluid the constitutive functional is necessarily of the form given by (3.9) and (3.10). Following each step of the argument we shall present our comments with respect to its validity.

Step 1. Truesdell and Noll take as their starting point the condition that if the simple material considered is a simple fluid the constitutive functional $\mathbf{F}$ in (3.1) must satisfy the relation ([4] eq. (32.1))

$$\mathbf{F}\{\mathbf{g}(s)\} = \mathbf{F}\{\bar{\mathbf{g}}(s)\}, \tag{4.1}$$

where $\bar{\mathbf{g}}(s)$ is the deformation gradient matrix with respect to an arbitrary configuration $\bar{X}$ which is related to the configuration X by a unimodular transformation $\underset{\sim}{H}^{-1}$ say; i.e.

$$\bar{\underset{\sim}{X}} = \underset{\sim}{H}^{-1}\underset{\sim}{X}, \quad \bar{\underset{\sim}{g}}(s) = \| \partial x_i(s)/\partial \bar{X}_A \| = \underset{\sim}{g}(s)\underset{\sim}{H}. \tag{4.2}$$

The relation (4.1) is, of course, equivalent to the relation (3.5) with $\underset{\sim}{H}$ unimodular. If $\bar{X}$ is any configuration, related to X by a transformation $\underset{\sim}{H}^{-1}$ which is not necessarily unimodular, and ρ_X and $\rho_{\bar{X}}$ are the densities of the material in the configurations X and $\bar{X}$ respectively, then ([4] eq. (32.3))*

$$\rho_{\bar{X}} = \rho_X \det \underset{\sim}{H}. \tag{4.3}$$

Truesdell and Noll then argue that (4.1) holds whenever the configurations X and $\bar{X}$ are such that ([4] eq. (32.4))

$$\rho_{\bar{X}} = \rho_X. \tag{4.4}$$

They assert, without further justification, that for a simple fluid the functional $\underset{\sim}{F}$ in (3.1) depends only on the density ρ_X and not on any other properties of the configuration X, and therefore the general constitutive functional in (3.8) reduces to ([4] eq. (32.5))

$$\underset{\sim}{\sigma}(t) = \underset{\sim}{F}\{\underset{\sim}{g}(s)\} = \underset{\sim}{R}(0)\underset{\sim}{K}\{\underset{\sim}{U}(s);\rho_X\}\underset{\sim}{R}^{\dagger}(0), \tag{4.5}$$

where $\underset{\sim}{K}$ is a functional of the history $\underset{\sim}{U}(s)$ and a function of the scalar parameter ρ_X. $\underset{\sim}{K}$ is independent of the choice of the reference configuration, whether or not these are related by unimodular transformations.

<u>Comment</u>. It is not at all clear that this assertion is a valid deduction from (4.1) and (3.8). However, in the case when the material considered is incompressible the question becomes moot, since all configurations which can be adopted by an incompressible material are related by unimodular transformations. Then with (4.1) and (3.8) we obtain from (4.5)

$$\underset{\sim}{\sigma}(t) + p\underset{\sim}{I} = \underset{\sim}{F}\{\underset{\sim}{g}(s)\} = \underset{\sim}{R}(0)\underset{\sim}{K}\{\underset{\sim}{U}(s)\}\underset{\sim}{R}^{\dagger}(0), \tag{4.6}$$

where the functional $\underset{\sim}{K}$ is independent of the choice of the reference configuration.

Notwithstanding the questionable validity of (4.5) we shall, in the remainder of this paper, assume that it is correct.

<u>Step 2</u>. From (4.5) Truesdell and Noll argue that since the functional $\underset{\sim}{K}$ is independent of the reference configuration, "If the configuration at time t is taken as the reference configuration," then $\underset{\sim}{R}(0) = \underset{\sim}{I}$ and (4.5) becomes ([4] eq. (32.7))

---

*Our $\underset{\sim}{H}$ is $\underset{\sim}{P}^{-1}$ in the notation of Truesdell and Noll.

$$\underset{\sim}{\sigma}(t) = \underset{\sim}{F}\{\underset{\sim}{g}(s)\} = \underset{\sim}{K}\{\underset{\sim}{U}_t(s);\rho_t\}, \tag{4.7}$$

where $\underset{\sim}{U}_t(s)$ is the right stretch matrix at lapsed time s relative to the configuration at time t and ρ_t is the density at time t.

Comment. If $\bar{X}$ is any configuration other than X, then from (4.5) it follows that

$$\underset{\sim}{\sigma}(t) = \underset{\sim}{\bar{R}}(0)\underset{\sim}{K}\{\underset{\sim}{\bar{U}}(s);\rho_{\bar{X}}\}\underset{\sim}{\bar{R}}^{\dagger}(0), \tag{4.8}$$

where $\underset{\sim}{\bar{U}}(s)$ is the right stretch matrix at lapsed time s relative to the configuration $\bar{X}$ and $\underset{\sim}{\bar{R}}(0)$ is the corresponding rotation matrix at time t in the polar decomposition of the deformation gradient matrix $\underset{\sim}{\bar{g}}(s)$ relative to the configuration $\bar{X}$:

$$\underset{\sim}{\bar{g}}(s) = \underset{\sim}{\bar{R}}(s)\underset{\sim}{\bar{U}}(s). \tag{4.9}$$

From (4.7) and (4.8)

$$\underset{\sim}{K}\{\underset{\sim}{U}_t(s);\rho_t\} = \underset{\sim}{\bar{R}}(0)\underset{\sim}{K}\{\underset{\sim}{\bar{U}}(s);\rho_{\bar{X}}\}\underset{\sim}{\bar{R}}^{\dagger}(0) \tag{4.10}$$

for all choices of the configuration $\bar{X}$. It is evident that not all functionals $\underset{\sim}{K}$ satisfy this relation.

Essentially the same criticism was made by us in [9]. In responding to this TRUESDELL [3] states that "Noll's definition of a simple fluid is framed within the theory of simple materials...and hence presumes a fixed reference placement. The definition restricts the peer group of the response; the peer group itself is defined...by comparing the responses with respect to different fixed reference placements." As is seen in the quotation in Step 2 the choice of the reference configuration which leads to (4.7) does not satisfy this condition. A similar situation is also evident in Noll's original paper (see, for example, Theorems 4(c) and 4(d) and Theorem 7(c) in [2]). However, if we interpret the choice of the reference configuration to be the fixed configuration which instantaneously coincides with the configuration at time t our criticism remains valid.

Step 3. Truesdell and Noll assert that since the orthogonal group is a sub-group of the unimodular group, $\underset{\sim}{K}$ in (4.7) must satisfy the relation ([4] eq. (32.6))

$$\underset{\sim}{Q}\underset{\sim}{K}\{\underset{\sim}{U}_t(s);\rho_t\}\underset{\sim}{Q}^{\dagger} = \underset{\sim}{K}\{\underset{\sim}{Q}\underset{\sim}{U}_t(s)\underset{\sim}{Q}^{\dagger};\rho_t\} \tag{4.11}$$

for all orthogonal $\underset{\sim}{Q}$.

Comment. There is an evident inconsistency in this argument. Since the constitutive functional $\underset{\sim}{F}$ given by (4.7) satisfies both material frame indifference and the condition (3.5) for all unimodular $\underset{\sim}{H}$, and hence for all orthogonal $\underset{\sim}{H}$, any constitutive functional of the form (4.7) should be isotropic without further restriction.

Step 4. Finally Truesdell and Noll rewrite (4.7) in the form ([4] eqs. (32.8) and (32.9))

$$\sigma(t) = \hat{K}\{G_t(s);\rho_t\} - p(\rho_t)I, \tag{4.12}$$

where p is a scalar function of ρ_t and $\hat{K}$ is an isotropic functional of the matrix $G_t(s)$ defined by

$$G_t(s) = C_t(s) - I. \tag{4.13}$$

Comment. This result is equivalent to the statement of Noll's theorem in §3. It follows directly from (4.7) and (4.11). However, it is obtained by Truesdell and Noll in a manner which seems unnecessarily indirect.

5. Truesdell's Sufficiency Argument

Neither in [2] nor in [4] has any proof been given that any constitutive functional of the form (3.9) which satisfies the isotropy condition (3.10) satisfies the defining condition (3.5) for a simple fluid. However, in [3], Truesdell has given a proof which, he claims, "any qualified beginner can follow step by step."

Truesdell proves that if

$$F\{g(s)\} = \hat{G}\{C_t(s);\ \det g(0)\}, \tag{5.1}$$

where

$$C_t(s) = g_t^{\dagger}(s)g_t(s), \tag{5.2}$$

then the relation (cf. (3.5))

$$F\{g(s)\} = F\{g(s)H\} \tag{5.3}$$

is satisfied for all unimodular H.

His argument is essentially that since (cf. (2.7))

$$g_t(s) = g(s)g^{-1}(0) \tag{5.4}$$

replacement of $g(s)$ by $g(s)H$ and $g(0)$ by $g(0)H$ leaves $g_t(s)$ unaltered for any nonsingular H. Also, if H is unimodular

$$\det g(0)H = \det g(0). \tag{5.5}$$

Truesdell's conclusion follows immediately. We note that in this argument no use is made of the prescription that $\hat{G}$ is an isotropic functional, so that in fact Truesdell has proved that any constitutive functional of the form (5.1) satisfies the simple fluid condition, whether or not $\hat{G}$ is an isotropic functional (cf. our comment in §4 on Step 3).

This is, indeed, an argument which "any qualified beginner can follow step by step." However, a perceptive beginner might observe the following peculiarity which emerges from Truesdell's argument.

Let $\tilde{X}$ be any configuration of the material other than the configuration X and let $\tilde{\underset{\sim}{X}}$ be the vector position in this configuration of the particle which has vector position $\underset{\sim}{X}$ in the configuration X. The deformation gradient matrix $\tilde{\underset{\sim}{g}}(s)$ at lapsed time s relative to the configuration $\tilde{X}$ is defined by

$$\tilde{\underset{\sim}{g}}(s) = \left\| \tilde{g}_{iA}(s) \right\| = \left\| \frac{\partial x_i(s)}{\partial \tilde{X}_A} \right\| . \qquad (5.6)$$

Let $\tilde{\underset{\sim}{C}}(s)$ denote the corresponding Cauchy strain matrix defined by

$$\tilde{\underset{\sim}{C}}(s) = \tilde{\underset{\sim}{g}}^{\dagger}(s)\tilde{\underset{\sim}{g}}(s). \qquad (5.7)$$

Let $\tilde{\underset{\sim}{R}}(0)$ be the rotation matrix in the polar decomposition of $\tilde{\underset{\sim}{g}}(0)$; i.e.

$$\tilde{\underset{\sim}{R}}(0) = \tilde{\underset{\sim}{g}}(0)\tilde{\underset{\sim}{C}}^{-1/2}(0). \qquad (5.8)$$

Let $\underset{\sim}{h}^{-1}$ be the transformation, not in general unimodular, which carries the configuration X to $\tilde{X}$. Then,

$$\underset{\sim}{h}^{-1} = \left\| \frac{\partial \tilde{X}_A}{\partial X_B} \right\| . \qquad (5.9)$$

It follows that

$$\tilde{\underset{\sim}{g}}(s) = \underset{\sim}{g}(s)\underset{\sim}{h}. \qquad (5.10)$$

Replacement of $\underset{\sim}{g}(s)$ by $\underset{\sim}{g}(s)\underset{\sim}{H}$ and $\underset{\sim}{h}^{-1}$ by $\underset{\sim}{h}^{-1}\underset{\sim}{H}$ leaves $\tilde{\underset{\sim}{g}}(s)$ unaltered for any nonsingular $\underset{\sim}{H}$. Therefore, any functional $\underset{\sim}{F}\{\underset{\sim}{g}(s)\}$ of the form

$$\underset{\sim}{F}\{\underset{\sim}{g}(s)\} = \underset{\sim}{H}\{\underset{\sim}{g}(s)\underset{\sim}{h}\} = \underset{\sim}{H}\{\tilde{\underset{\sim}{g}}(s)\} \qquad (5.11)$$

is unchanged by this replacement. Also if $\underset{\sim}{H}$ is unimodular

$$\det \tilde{\underset{\sim}{g}}(0)\underset{\sim}{H} = \det \tilde{\underset{\sim}{g}}(0). \qquad (5.12)$$

It follows that, in particular, any constitutive functional of the form

$$\underset{\sim}{F}\{\underset{\sim}{g}(s)\} = \tilde{\underset{\sim}{R}}(0)\underset{\sim}{K}\{\tilde{\underset{\sim}{C}}(s) \; ; \; \det \tilde{\underset{\sim}{g}}(0)\}\tilde{\underset{\sim}{R}}^{\dagger}(0) \qquad (5.13)$$

is unchanged by the replacement, provided that $\underset{\sim}{H}$ is unimodular. We note that the constitutive functional (5.13) satisfies material frame indifference. Since it also is unchanged by unimodular transformation of the initially chosen reference configuration X, it satisfies all of Noll's defining conditions for a simple fluid.

We have thus reached an evident paradox since $\tilde{X}$ might well have been chosen, instead of X, as the configuration with respect to which the deformation gradient matrix history, in the initial constitutive assumption (3.1) for a simple material, is defined.

In particular, as was pointed out in [9], if we consider deformation histories which take place in the time interval $[t_o,t]$ we could choose both X and $\tilde{X}$ to be the identical (undeformed) configurations of the material at two different times prior to t_o. (t_o may, of course, be chosen as far in the past as we please.) It then follows that

$$\underset{\sim}{R}(0)\underset{\sim}{K}\{\underset{\sim}{C}(s)\ ;\ \det \underset{\sim}{g}(0)\}\underset{\sim}{R}^{\dagger}(0) = \tilde{\underset{\sim}{R}}(0)\underset{\sim}{K}\{\tilde{\underset{\sim}{C}}(s)\ ;\ \det \tilde{\underset{\sim}{g}}(0)\}\tilde{\underset{\sim}{R}}^{\dagger}(0) \tag{5.14}$$

and, consequently, any constitutive functional of the form

$$\underset{\sim}{F}\{\underset{\sim}{g}(s)\} = \underset{\sim}{R}(0)\underset{\sim}{K}\{\underset{\sim}{C}(s)\ ;\ \det \underset{\sim}{g}(0)\}\underset{\sim}{R}^{\dagger}(0) \tag{5.15}$$

would describe a simple fluid. Since $\det \underset{\sim}{g}(0) = [\det \underset{\sim}{C}(0)]^{1/2}$ and $\underset{\sim}{C}(0)$ is a functional of $\underset{\sim}{C}(s)$, we can rewrite (5.15) as

$$\underset{\sim}{F}\{\underset{\sim}{g}(s)\} = \underset{\sim}{R}(0)\hat{\underset{\sim}{K}}\{\underset{\sim}{C}(s)\}R^{\dagger}(0). \tag{5.16}$$

If we introduce the same inconsistency as was introduced by Noll, then $\underset{\sim}{K}$ and $\hat{\underset{\sim}{K}}$ are isotropic functionals and (5.16) is then the expression for the constitutive functional for <u>any</u> isotropic simple material [12]. An attempted refutation of this argument by Murdoch [10] is based on an algebraic error, as was pointed out in [11].

6. Origin of the Paradox

Our conclusions from the discussions in §§4 and 5 are in evident conflict. On the one hand it emerges in §4 that the restrictions on the constitutive functional which result from Noll's defining condition for a simple fluid are far greater than those derived by Noll and embodied in his theorem. On the other hand it appears from the discussion in §5 that a far wider range of constitutive functionals satisfies Noll's defining condition.

We can illustrate the origin of this contradiction most simply by considering the case when the material is incompressible, so that all possible configurations are related by unimodular transformations. This enables us to avoid the questionable step 1 in §4. We then have from (3.2) and (3.5)

$$\underset{\sim}{\sigma} + p\underset{\sim}{I} = \underset{\sim}{F}\{\underset{\sim}{g}(s)\} = \underset{\sim}{F}\{\underset{\sim}{g}(s)\underset{\sim}{H}\} \tag{6.1}$$

for all unimodular $\underset{\sim}{H}$, where $\underset{\sim}{F}$ has the form (cf. (3.8))

$$\underset{\sim}{F}\{\underset{\sim}{g}(s)\} = \underset{\sim}{R}(0)\underset{\sim}{G}\{\underset{\sim}{U}(s)\}\underset{\sim}{R}^{\dagger}(0), \tag{6.2}$$

so that material frame indifference is satisfied. We shall in the following argument suppose that $\underset{\sim}{F}$ has this form.

Since the condition (6.1) on $\underset{\sim}{F}$ is valid for all unimodular $\underset{\sim}{H}$ we have

$$\underset{\sim}{F}\{\underset{\sim}{g}(s)\} = \underset{\sim}{F}\{\underset{\sim}{g}(s)\underset{\sim}{H}_1\} \tag{6.3}$$

and

$$\underset{\sim}{F}\{\underset{\sim}{g}(s)\} = \underset{\sim}{F}\{[\underset{\sim}{g}(s)\underset{\sim}{H}_1]\underset{\sim}{H}_2\} = \underset{\sim}{F}\{\underset{\sim}{g}(s)[\underset{\sim}{H}_1\underset{\sim}{H}_2]\} \tag{6.4}$$

for all unimodular $\underset{\sim}{H}_1$ and $\underset{\sim}{H}_2$. These relations express the transitivity of invariance under a group of transformations. It is this transitivity which is not ensured in the arguments of Truesdell and Noll.

In effect they take

$$\underset{\sim}{H}_1 = \underset{\sim}{g}^{-1}(0) \tag{6.5}$$

in (6.3) and obtain

$$\underset{\sim}{F}\{\underset{\sim}{g}(s)\} = \underset{\sim}{F}\{\underset{\sim}{g}(s)\underset{\sim}{g}^{-1}(0)\} = \underset{\sim}{F}\{\underset{\sim}{g}_t(s)\}, \tag{6.6}$$

which, with (6.2), is the equivalent of (4.7) for an incompressible material. TRUESDELL [3] then argues that, with this expression for $\underset{\sim}{H}_1$,

$$\underset{\sim}{F}\{\underset{\sim}{g}(s)\underset{\sim}{H}_1\} = \underset{\sim}{F}\{\underset{\sim}{g}(s)\underset{\sim}{H}_2(\underset{\sim}{H}_2^{-1}\underset{\sim}{H}_1)\}. \tag{6.7}$$

for all unimodular $\underset{\sim}{H}_2$. This is indeed true, but is seen to be vacuous in that it is valid for <u>any</u> choice of $\underset{\sim}{H}_1$, and indeed for any non-singular choice of $\underset{\sim}{H}_2$ whether unimodular or not.

On the other hand (6.3) and (6.4) yield

$$\underset{\sim}{F}\{\underset{\sim}{g}(s)\underset{\sim}{H}_1\} = \underset{\sim}{F}\{\underset{\sim}{g}(s)\underset{\sim}{H}_1\underset{\sim}{H}_2\} \tag{6.8}$$

and with (6.5) we obtain the restriction

$$\underset{\sim}{F}\{\underset{\sim}{g}_t(s)\} = \underset{\sim}{F}\{\underset{\sim}{g}_t(s)\underset{\sim}{H}_2\} \tag{6.9}$$

for all unimodular $\underset{\sim}{H}_2$, which corresponds to the restriction (4.10) in our comments on Step 2 in §4.

The fallacy in Truesdell's argument evidently arises from the fact that, having introduced the transformation $\underset{\sim}{H}_1$, given in (6.5), he then regards it as part of the entity transformed in testing whether the functional is invariant under unimodular transformations of the reference configuration. In doing so he neglects the fact, previously pointed out in [9], that the reference configuration has now become the configuration at time t (or a fixed configuration which instantaneously coincides with it).

<u>References</u>

1. A. E. Green and R. S. Rivlin, Arch. Rat'l Mech. Anal. <u>1</u>, 1 (1957).

2. W. Noll, Arch. Rat'l Mech. Anal. 2, 197 (1958)
3. C. Truesdell, J. Non-Newtonian Fluid Mech. 15, 249 (1984)
4. C. Truesdell and W. Noll, The Non-Linear Field Theories of Mechanics, in Handbuch der Physik III/3, ed. S. Flügge, Springer, Berlin (1965).
5. E. J. Fahy and G. F. Smith, Mech. Res. Comm. 5, 167 (1978).
6. E. J. Fahy and G. F. Smith, J. Non-Newtonian Fluid Mech. 7, 33 (1980).
7. R. S. Rivlin, J. Non-Newtonian Fluid Mech. 11, 209 (1982).
8. A. I. Murdoch, J. Non-Newtonian Fluid Mech. 12, 387 (1983).
9. R. S. Rivlin and G. F. Smith, J. Non-Newtonian Fluid Mech. 12, 393 (1983).
10. A. I. Murdoch, J. Non-Newtonian Fluid Mech. 15, 247 (1984).
11. R. S. Rivlin, J. Non-Newtonian Fluid Mech. 15, 253 (1984).
12. R. S. Rivlin and G. F. Smith, Int'l J. Solids Structures 23, 325 (1987).

On the Uniqueness of the Discrete Shock Waves

V.V. Rusanov and I.V. Bezmenov

1. Introduction

Let us consider the hyperbolic system

$$w_t + \{F(w)\}_x = 0, \quad w = \{w_1, .., w_q\}^T, F = \{F_1, .., F_q\}^T, \tag{1.1}$$

where $A = dF/dw$ has d real eigenvalues $\lambda_1 < \lambda_2 < .. < \lambda_d$ and $PAQ = \Lambda$, where Λ is a diagonal matrix. The rows P_k of P and columns Q_k of Q are, respectively, the left and right eigenvectors of A; the eigenvalue λ_k is called nonlinear if $(\text{grad}_w \lambda_k, Q_k) \neq 0$.

Let for some w_1 and D the system

$$F(w) - Dw = F(w_1) - Dw_1 \tag{1.2}$$

have, besides w_1, one and only one solution $w = w_2 \neq w_1$, such that

$$\lambda_k(w_2) > D > \lambda_k(w_1),$$

$$\lambda_{k+1}(w_1) > D > \lambda_{k-1}(w_2), \tag{1.3}$$

where $\lambda_k(w)$ may be nonlinear. Then the system (1.1) has the generalized solution $w(x,t) = v(y); \quad y = x - Dt$

$$v(y) \begin{cases} w_- = w_2, & y < x_o \\ w_+ = w_1, & y \geq x_o \end{cases} \qquad w_+ - w_- = \Delta w, \tag{1.4}$$

which is called the shock wave of the k-type.

Consider the stable difference scheme approximating (1.1) on the floating grid with steps $\Delta x = h$ and $\Delta t = \tau$, $\kappa = \tau/h$:

$$w_x^{n+1} = G(w_{x-\nu_1 h}^n, ..., w_{x-\nu_2 h}^n), \quad \nu_1, \nu_2 \geq 0, |\nu_1| + |\nu_2| \neq 0, \tag{1.5}$$

where $w_x^n = w^n(x) = w(x, t^n) = w(x, t^\circ + n\tau)$. If (1.5) is conservative then

$$G(\varphi_{-\nu_1}, ..., \varphi_{\nu 2}) = \varphi_0 + H(\varphi_{-\nu_1+1}, ..., \varphi_{\nu 2}) - H(\varphi_{-\nu 1}, ..., \varphi_{\nu_2 - 1}). \tag{1.6}$$

Analogously to (1.4) we search for the solution of (1.5) in the form

$$w_x^n = v(x - n\tau D) = v(x - n\delta h) = v_{x-n\delta h}, \quad \delta = \kappa D.$$

Inserting it into (1.5) we get

$$v_{x-\delta h} = G(v_{x-\nu_1 h}, \ldots, v_{x+\nu_2 h}), \qquad |x| < +\infty,$$

$$v_{\pm\infty} \equiv \lim_{x \to \pm\infty} v_x = w_\pm. \tag{1.7}$$

The function v_x satisfying (1.7) is called the discrete shock wave on the floating grid. Because of the stability of (1.5) we get from (1.3)

$$|\delta| < \kappa \max_k \{|\lambda_k(w_-)|, |\lambda_k(w_+)|\} \leq 1,$$

Without loss of generality we could suppose that $0 \leq \delta < 1$.

In the case $d = 1$ the existence of v_x is proved for the monotone [1,2] and TVD schemes [3]. For the monotone schemes only, it is proven also [2,6] that the iterative process

$$v_{x-\delta h}^{n+1} = G(v_{x-\nu_1 h}^n, \ldots, v_{x+\nu_2 h}^n) \tag{1.8}$$

converges to the unique function v_x if a special condition [5] for the initial function $v^\circ(x)$ is fulfilled.

For $d > 1$ the existence of v_x is proven for some class of difference schemes in the case of rational $\delta = p/q$ and if $||w_+ - w_-||$ is sufficiently small, i.e. for the weak shocks [4].

The proofs mentioned are based on general methods and give no means to implement the restrictions used in theorems that are essential for the uniqueness. In order to clear the question we made investigations of some concrete problems. The results obtained allow one to understand more clearly the structure of v_x in the nonmonotone case and may be useful for the subsequent theoretical investigations.

2. Some Properties of the Function v_x

Let us consider the case $d = 1$. Let v_x be a solution of (1.7) such that

$$|v_x - w_\pm| = 0(|X|^{-N}) \qquad \text{as} \quad x \to \pm\infty, \tag{2.1}$$

the solution has the property (2.1).

Define the function $X_v(\varsigma)$ as follows:

$$X_v(\varsigma) = (\Delta w)^{-1} \sum_{m=-\infty}^{\infty} (\varsigma + mh + \frac{h}{2})(v_{\varsigma+(m+1)h} - v_{\varsigma+mh}). \tag{2.2}$$

It is easy to see that $X_v(\varsigma + h) = X_v(\varsigma)$ and

$$h^{-1} \int_0^h X_v(\varsigma) d\varsigma = (\Delta w)^{-1} \int_{-\infty}^{\infty} x\, dw(x) = Z_v. \tag{2.3}$$

The Z_v is the "mean point" of the distribution $v(x)$. For the step-wise function $v(y)$ determined by (1.4), $Z_v = X_o$, i.e. the point of the discontinuity.

It was proven [5] that for the solution satisfying (2.1), $X_v(\varsigma + \delta h) = X_v(\varsigma)$. Because of this we have two possibilities:

1. δ is irrational, then $X_v(\varsigma) = \text{const}$;

2. $\delta = p/q$ is rational, then $X_v(\xi)$ has the period h/q.

It was proven in [6] that for any monotone scheme and any δ there exists one and only one solution v_x of (1.7) satisfying (2.1) with $X_v(\xi) = \text{const}$. Evidently in this case $X_v(\xi) = Z_v$. Let us note that for the step-wise function (1.4), $X_v(\varsigma)$ is nonconstant and discontinuous. Namely $X(\varsigma) = x_0 + h(s - \frac{1}{2})$ where s is fractional portion of $(\varsigma - x_0)/h$.

Consider now the case $\varsigma = p/q$. Setting $h = q$ we get that period of $X_v(\varsigma)$ is equal one and

$$v_{x-p} = G(v_{x-\nu_1 q}, ..., \nu_{x+\nu_2 q}), \quad v_{\pm\infty} = w_{\pm}. \tag{2.4}$$

Let $x = j + \xi$ and denote $v_x = v_j(\xi)$. Evidently $v_j(\xi + 1) = v_{j+1}(\xi)$. For the discrete function $v_j(\xi)$ depending on parameter ξ we have the equation

$$v_{j-p} = G(v_{j-\nu_1 q}, ..., \nu_{j+\nu_2 q}), \quad v_{\pm\infty} = w_{\pm}, \tag{2.5}$$

which is the same for all $v_j(\xi)$ on the grid with the space step $h/q = 1$. If $v_x \neq \text{const}$, then there exist ξ', ξ'', $|\xi - \xi''| < 1$ such that $v_j(\xi^x)$ and $v_j(\xi')$ are essentially different. Due to this, any solution v_x of (2.4) (discrete shock wave on the floating grid) induce the one-parametric family of solutions $v_j(\xi)$ of (2.5). Each $v_j(\xi)$ $(-\infty < j < +\infty)$ for given ξ is called the discrete shock wave on the fixed grid

$$\alpha_v(\xi) = (\Delta\, w)^{-1} \sum_{m=-\infty}^{\infty} (\lambda + mq + q/2)\{v_{\lambda+(s+1)q}(\xi) - v_{\lambda+sq}(\xi)\}. \tag{2.6}$$

Setting $\xi = \lambda + \xi, \lambda = 0, 1, ..., q-1, 0 \leq \xi < 1$, we get from (2.2) and (2.6) that $\alpha_v(\xi)$ does not depend on λ and

$$X_v(\lambda + \xi) = X_v(\xi) = \alpha_v(\xi) + \xi.$$

If $X_v(\xi) = X_v = \text{const}$ then $\alpha_v(\xi) = X_v - \xi$ determines the shift of the fixed grid, chosen by means of ξ with respect to constant "mean point" of the profile. For the monotone scheme the solution of (2.5) on a fixed grid is uniquely determined by α_v [5,6]. The example below shows that for the nonmonotone scheme, the same α_v could correspond different functions v_j.

3. The Example of Nonuniqueness of v_j

Consider the three-point scheme

$$w_m^{n+1} = w_m^n - (\kappa/2)\{F(w_{m+1}^n) - F(w_m^n)\} + (\omega/2)\{w_{m+1}^n - 2w_m^n + w_{m-1}^n\}, \quad (3.1)$$

where $F''(w) < 0$.

Let $D = 0$, $\delta = p = 0$, $q = 1$, $F(w_+) = F(w) = F$. Then from (1.2) and (1.6) we get

$$v_j = v_j + H_{j+1/2} - H_{j-1/2}, \quad H_{j+1/2} = \text{const};$$

$$H_{j+1/2} = -\frac{\kappa}{2}\{F(v_{j+1}) + F(v_j)\} + \frac{\omega}{2}\{v_{j+1} - v_j\} = H_\pm = -\kappa F.$$

The connection between v_j and v_{j+1} is determined by the formula

$$\Phi^-(v_{j+1}) = \Phi^+(v_j), \quad (3.2)$$

where

$$\Phi^\pm(v) = v \pm \frac{\kappa}{\omega}(F(v) - F).$$

For the monotone scheme the functions $\Phi^\pm(v)$ are monotonic in $[w_-, w_+]$.

Consider the solution of (3.1) and take some value of v_{j_o}. The value of v_{j_o+1} is the root of the equation

$$\Phi^-(v) = \Phi^+(v_{j_o}). \quad (3.3)$$

If $\Phi^-(v)$ is monotonic then v_{j_o+1} exists and is unique for any $v_{j_o} \epsilon [w_-, w_+]$. It is easy to show that $v_{j_o+1} > v_{j_o}$. So if the scheme is monotonic the sequence v_j increases monotonically and $\lim_{j\to\pm\infty} v_j = w_\pm$.

If the scheme is nonmonotonic then $\Phi^-(v)$ has a minimum in (w_-, w_+) and for v_{j_o+1} two different solutions can exist.

One of them is located on the ascending branch of $\Phi^-(v)$, where $[\Phi^-(v)]' > 0$, and the other on the descending branch, where $[\Phi^-(v)]', 0$. Consider the pair v_j, v_{j+1} satisfying (3.2). We define the pair as irregular when one or both of the following conditions 1, 2, are satisfied:

(1) $$[\Phi^-(v)]'_{v_j} > 0 \quad \text{and} \quad [\Phi^-(v)]'_{v_{j+1}} < 0;$$

(2) $$[\Phi^+(v)]'_{v_j} < 0 \quad \text{and} \quad [\Phi^+(v)]'_{v_{j+1}} > 0.$$

In all the other cases we define the pair as regular. We call the solution v_j of (3.2) regular if all it pairs are regular. Any other solution we call irregular.

In the monotone case only regular solutions exist and each solution is determined uniquely by its a_v. There are two possibilities only for two solutions $v_j^{(1)}$ and $v_j^{(2)}$. Namely, either $v_j^{(1)} = v_j^{(2)}$ or $v_j^{(1)} \neq v_j^{(2)}$ for all j [7].

To investigate the nonmonotone case consider $F(w) = (1-w^2)/2, w^{\pm} = \pm 1$. The direct computation for some ω and κ (for example $\omega = 0.005$, $\kappa = 0.05$) shows that two different regular solutions can coincide partially, namely for an infinite number of j, but not for all j. Different regular solutions have different α_v and the floating grid solution constructed out of regular solutions by means of the rule $\xi = \alpha_v + const$ is continuous and smooth but not monotonic. There are also irregular solutions with different α_v. For some α_v, one or several irregular solutions may exist beside one regular solution. Thus the nonuniqueness of the stationary solution of (1.7) is proven for $D = 0$ and the nonmonotone scheme.

4. The convergence of iterations

The stationary solution of (2.5) can be computed as the limit of the iterative process given by the formula analogous to (1.8):

$$v_{j-p}^{n+1} = G(v_{j-\nu_1 q}^n, \ldots, v_{j+\nu_2 q}^n). \tag{4.1}$$

For numerical study of the convergence, a special model on the finite grid was developed. The nonlinear operator transforming v^n

$$\begin{aligned} v_j^{n+1} &= v_j^n + H_{j+1/2}^n - H_{j-1/2}^n, \quad -M^- + \leq j \leq M^+ - 1, \\ v_{M^+}^{n+1} &= v_{M^+}^n + H^+ - H_{M-1/2}^n, \\ v_{-M^-}^{n+1} &= v_{-M^-}^n + H_{-M^-+1/2}^n - H^-. \end{aligned} \tag{4.2}$$

In numerical experiments made for the same $F(w), \omega, \kappa$ as before the following results were obtained:

1. Any stationary regular solution is stable with respect to the iterative process (4.2).
2. For any initial function v_j° with some α_v the iterative process (4.2) converges to the stationary regular solution with the same α_v.
3. The paragraph 2 is valid also for the v_j° as the irregular solution. Thus the irregular solution is unstable with respect to (4.2). The word "any" in (1) and (2) should be understood as "any of those checked numerically".

Let us consider linearized operator $\mathcal{L}$ for (4.2) with some solution $\hat{\nu}_j$ of (2.5) chosen as basic. The operator $\mathcal{L}$ is defined as follows:

$$\begin{aligned} u^{n+1} &= \mathcal{L}u^n, \quad u = \{u_{-M^-}, ..., v_{M^+}\} \\ u^{n+1}_{-M^-} &= (1 - C_{-M})u^n_{-M^-} + A_{-M^-_{+1}}u^n_{-M^-+1} \\ u^{n+1}_j &= C_{j-1}u^n_{j-1} + B_j u^n_j + A_{j+1}u^n_{j+1}; \quad -M + 1 \leq j \leq M^+ - 1, \\ u^{n+1}_{M^+} &= C_{M^+-1}u^n_{M^+-1} + (1 - A_{M^+})u^n_{M^+}, \end{aligned} \tag{4.3}$$

where

$$\begin{aligned} A_j &= \tfrac{\omega}{2}[\Phi^-(\nu)]'_{\hat{\nu}_j} = (\omega + \kappa F'(\hat{\nu}_j))/2, \quad B_j = 1 - \omega, \\ C_j &= \tfrac{\omega}{2}[\Phi^+(\nu)]'_{\hat{\nu}_j} = (\omega - \kappa F'(\hat{\nu}_j))/2. \end{aligned} \tag{4.4}$$

For some regular and irregular solutions $\hat{\nu}_j$, the eigenvalues of linearized operator for (4.3) were computed. For all regular solutions they were situated inside of the unit circle. For irregular solutions there are always several eigenvalues outside the unit circle.

The number of points in (4.2) $(M^+ + M^-)$ in experiments were between 10 and 25.

5. Conclusion

The basic result of this paper is the nonuniqueness of the solution constructing the example in which two solutions, namely regular and irregular have the same α_v. It was proven before [7] that in the monotone case, for different α_v there corresponds different solutions which are all regular and vice versa. On the grounds of numerical experiments it is possible to suppose that when $D = 0$ and $F''(\omega) \neq 0$ the situation is the same but it is not yet rigorously proven.

The three statements (1), (2), (3) which are formulated in section 4 also have to be considered as a hypothesis strongly supported by numerical experiments.

References

1. G. Jennings: Discrete Shocks. Comm. on Pure and Appl. Math. 1974 vol. 37, p. 25-37.

2. V. V. Rusanov, I. V. Bezmenov: The Asymptotic of the Profile of a Discrete Discontinuity in Computing a Shock Wave by Difference Schemes. Doclady AC. Sci. USSR, 1981, vol. 261, N4, p. 817-820; English transl. in Soviet Math Dokl. Vol. 24 (1981), No. 3.

3. V. V. Rusanov, I. V. Bezmenov: The Existence of the Limiting profile Shock Wave Type for TVD Schemes. Preprint no. 177, Keldysh Inst. Appl. Math. Acad. Sci. USSR, Moscow, 1986 (Russian).

4. A. Maida, J. Ralstone: Discrete Shock Profiles for System of Conservation Lays. Comm. on Pure and Appl. Math, 1979, vol. 32, p. 446-482.

5. V. V. Rusanov: Nonlinear Analysis of the Shock Profile in Difference Schemes. Proc. of the Second ICNMFD, Lecture notes in Physics, 1970, vol. 8, p. 270-278.

6. I.V. Bezmenov: The Generation of a Limit Profile of a Discrete Discontinuity in Computing a Shock Wave by Difference Schemes. Preprint no. 60, Keldysh Inst. Appl. Math., Acad. Sci. USSR, Moscow, 1981. (Russian)

7. V. V. Rusanov, I. V. Bezmenov: Asymptotics of a Solution of Shock-wave Type For a Finite Difference Equation. Trudy Mat. Inst. Steklov 157(1981), 178-190; English transl. in Proc. Steklov Inst. Math. 1983, no. 3 (157).

Computational Fluid Mechanics in R&D on Uranium Enrichment

Soubbaramayer

1 Introduction

The contributions of MAURICE HOLT to Computational Fluid Mechanics represent a gigantic monument. This symposium is a natural occasion to every participant to focus the light on some particular stone of that architecture. In the present paper, I want to make a flashback on one among his thousand contributions and to indicate the future evolution of the topics.

Maurice HOLT has actively participated in several workshops on the mechanics of strongly rotating gases held respectively in France (1977), Italy (1979), Great Britain (1981) and USA (1983). He was the co-author of two papers / 1, 2 /. At these meetings the subject was investigated mainly in view of its applications to uranium enrichment by gas centrifuge / 3 /. Uranium enrichment is a multi-billion dollar industry and represents an essential link in the cycle of nuclear fuels for power production. There are many processes of uranium enrichment but three of them dominate the nuclear history in the earlier days (gaseous diffusion in the fifties), then centrifugation in the seventies, up to the present trend (laser process). I want to take this opportunity to point out the important role played by the computational fluid mechanics in the three processes. The type of problem raised respectively in each of the processes is the following : channel flow with wall suction in gaseous diffusion, mechanics of strongly rotating gases in centrifugation and two-phase flow in laser process.

2 Gaseous Diffusion : Channel Flow with Wall Suction

This process, having reached very early the full commercial and industrial maturity, is used for more than 70 % of the present world production of enriched uranium. An excellent review of the problem is / 4 /. The fluid mechanics problem involved is the channel flow (cylindrical channel) with suction by lateral wall. This is an old problem where the computer methods have not yet played a role as intensive as in to day's problems. Moreover the flow being turbulent, semi-empirical formula were heavily used, particularly in mass transfer problems / 5, 6 and 7 /. As this is not my topics, I will go on to the two next problems.

3 Centrifugation : Mechanics of Strongly Rotating Gases

The principle of uranium enrichment by strongly rotating centrifuge is developped in / 3 /. The fluid mechanics problem involved in this process is formulated below.

A cylinder of radius a and height h (Fig. 1), closed at both ends by caps, contains a gas (uranium hexafluoride of

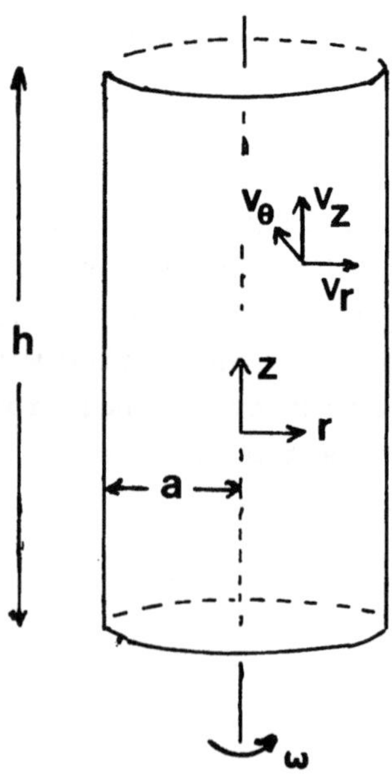

Fig. 1 - Schematic diagram of a centrifuge.

molar mass M = 352 g). The cylinder rotates very rapidly about its axis of symmetry with a constant angular velocity ω. The rotation speed is such that the centrifugal acceleration near the periphery of the cylinder is practically several hundred thousands of times greater than gravity. If no other constraint is imposed, the motion of the gas is a very simple one, termed in the literature by "rigid body rotation" or "wheel flow" : the gas rotates as a rigid body with the same angular velocity as the cylinder. The velocity components and the thermodynamic variables of the gas are easily derived :

$$V_\theta = \omega r \qquad V_z = 0 \qquad V_r = 0 \qquad T = T_o$$
$$p(r)/p(o) = \rho(r)/\rho(o) = \exp\,(M\,\omega^2\,r^2/2\,RT_o) \tag{1}$$

A practical typical case is ω a = 500 $m.s^{-1}$, T_o = 310 K.

Eq. (1) yields

$$p(a)/p(o) = \rho(a)/\rho(o) = 2.10^7$$

The gas in a centrifuge is strongly stratified in the radial direction : there are seven orders of magnitude between the densities at the axis and the periphery. This will cause the first difficulty in the numerical treatment for the choice of meshes.

The flow in centrifuge is more complex than the wheel flow. As a matter of fact, in practical devices, a small axial circulation of the gas is superimposed on the wheel flow. Such axial motion is termed in the literature by "countercurrent flow". The countercurrent flow can be generated by thermal method or mechanical method. The thermal method can work if the rotor wall and end caps are made with a good heat conducting material. The method consists in imposing a proper temperature distribution on the solid boundaries of the cylinder : the bottom end cap is slightly heated, the top end cap is slightly cooled and, along the side wall, a linear profile of temperature joins the two end cap temperatures. The mechanical method of generating the countercurrent is of different nature. A stationary object (a scoop) is placed in the gas stream near the bottom end cap. In practice both methods are used. Investigating the countercurrent flow is one of the important objectives of the fluid mechanics analysis of the centrifuge.

3.1 Model

The gas flow in a centrifuge is governed by the conservation laws of continuum fluid mechanics (mass, momentum and energy) and the thermodynamic gas state law. In cylindrical coordinates with axial symmetry and at steady state the set of six equations is written here. In this system, (z, r) are the axial and radial coordinates, (V_z, V_r, V_θ) the components of the velocity in the axial radial and azimuthal directions, (p, ρ, T) the thermodynamic variables (pressure, mass density and temperature) ; the viscosity μ, the thermal conductivity κ and the specific heat at constant volume C_v are assumed to be constant. The last equation is the perfect gas law.

$$-.\ \frac{\partial}{\partial z}(\rho\, V_z) + \frac{1}{r}\frac{\partial}{\partial r}(\rho\, rV_r) = 0$$

$$-.\ \rho(V_z \frac{\partial V_z}{\partial z} + V_r \frac{\partial V_z}{\partial r}) = -\frac{\partial p}{\partial z} + \mu\,(\nabla^2 V_z + \frac{1}{3}\frac{\partial}{\partial z}\,\mathrm{div}\,V)$$

$$-.\ \rho(V_z \frac{\partial V_r}{\partial z} + V_r \frac{\partial V_r}{\partial r} - \frac{V_\theta^2}{r}) = -\frac{\partial p}{\partial r} + \mu\,((\nabla^2 - \frac{1}{r^2})\,V_r + \frac{1}{3}\frac{\partial}{\partial r}\,\mathrm{div}\,V)$$

$$-.\ \rho(V_z \frac{\partial V_\theta}{\partial z} + V_r \frac{\partial V_\theta}{\partial r} + \frac{V_r V_\theta}{r}) = \mu\,(\nabla^2 - \frac{1}{r^2})\,V_\theta \tag{2}$$

$$-.\ \rho\, C_v\,(V_z \frac{\partial T}{\partial z} + V_r \frac{\partial T}{\partial r}) + p\,\mathrm{div}\,V = k\,\nabla^2 T + \phi_{visc}$$

$$\phi_{visc} = 2\,\mu\,((\frac{\partial V_z}{\partial z})^2 + (\frac{\partial V_r}{\partial r})^2 + \frac{V_r^2}{r^2}$$

$$+ \frac{1}{2}\,((\frac{\partial V_\theta}{\partial z})^2 + (\frac{\partial V_r}{\partial z} + \frac{\partial V_z}{\partial r})^2 + (\frac{\partial V_\theta}{\partial r} + \frac{V_\theta}{r})^2) - \frac{1}{3}\,(\mathrm{div}\,V)^2)$$

$$\mathrm{div}\,V = \frac{\partial V_z}{\partial z} + \frac{1}{r}\frac{\partial}{\partial r}(r\,V_r), \quad \nabla^2 = \frac{\partial^2}{\partial z^2} + \frac{\partial^2}{\partial r^2} + \frac{1}{r}\frac{\partial}{\partial r}$$

$$-.\ P = \frac{R}{M}\,\rho\,T\ .$$

Note that in momentum equations the viscous stresses take into account the compressibility of the gas through the term $(\frac{1}{3}\text{ div V})$ and note also that gravity is neglected. The term $(\frac{V_\theta^2}{r})$ in the radial momentum equation and the term $(\frac{V_r V_\theta}{r})$ in the tangential momentum equation represent the centrifugal and CORIOLIS forces respectively. In the energy equation, the term (P div V) represents the reversible work done on the fluid due to compression or expansion and the term $\phi_{visc.}$ represents the viscous dissipation. The system (2) should be completed by appropriate boundary conditions. It is through the boundary conditions that is introduced the cause of the countercurrent flow. For the thermal drive, the boundary condition is obvious : the temperature has a prescribed profile on the horizontal and vertical boundaries. The mathematical simulation of the scoop effect is more complex. Many authors have roughly simulated the scoop effect by a disk rotating slightly slower than the rotor. In actual fact, the scoop is an aerodynamic object in the gas stream and the flow around is three dimensional and transonic. Maurice HOLT has guided the French works on this point, through one of his students Marc HITTINGER / 1, 2 /.

Determining the countercurrent flow field in the centrifuge is achieved by solving the set of eq. (2) together with boundary conditions.

3.2 Numerical Solution of the Model

The preceding model was solved numerically by the finite element method with a computer code named CENTAURE, whose main features have been published in / 8 /. The meshing was carried out not only by taking into account the very strong radial stratification of the gas density in the centrifuge but also by taking advantage of mathematical solutions that many authors have obtained previously by linearizing the problem. The physical assumption underlying the linearization is that the countercurrent flow is a small perturbation about the wheel flow. The system of linearized equations has a singular perturbation character. Solution has been obtained by using the method of boundary layers in contained rotating fluids (see for instance GREENSPAN / 9 /). Fluids rotating about an axis give rise to a variety of boundary layers near the solid walls bounding the fluid : the structure of these boundary layers is quite different according to whether the solid wall is perpendicular to the axis of rotation (EKMAN layer) or parallel to the axis (STEWARTSON layers). The, now classical, methods of investigation of such boundary layers were adapted to the centrifuge and have shown that the flow field is divided into three regions (near the lateral wall, near the end caps and the central core) as depicted in Fig. 2.

Using these results, the meshing was done in quadrangular elements quite different in size according to the region of the flow : Extremely thin meshes in the EKMAN layers and STEWARTSON layers and much larger meshes in the central core. The ratio between the sizes of the biggest and the smallest meshes was 10^4.

As an illustration, we present one result of computation for the following input data :

$a = 25$ cm, $h = 250$ cm, $\omega a = 600$ m.s^{-1},

$p_{wall} = 100$ torr, $T_o = 320$ K.

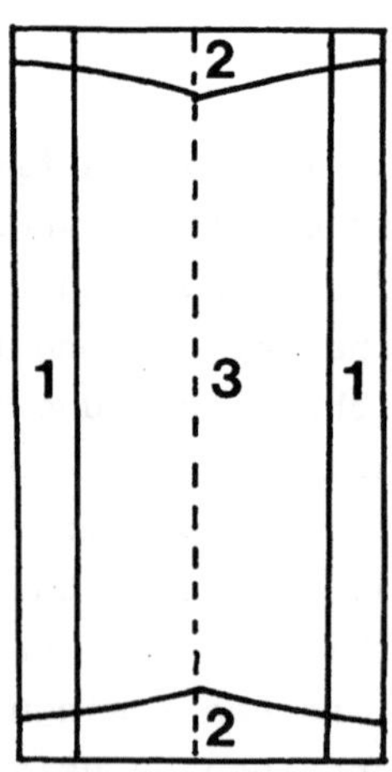

Fig. 2 - Boundary layers in the centrifuge.
1 - Stewartson layer
2 - Ekman layer
3 - Inner core

Fig. 3 represents the axial mass velocity versus the stretched radial coordinate ($A^2 = M\,\omega^2 a^2/2\,RT_o = 24$) for a thermal drive in the lower half of the centrifuge (results are symmetrical with respect to the midplane). The figure shows that the flow is confined in a very thin layer near the side wall, the thickness of the layer being approximately 5 % of the radius.

CENTAURE code has been intensively exploited and used also for optimizing the performance of a given centrifuge / 10 /. Research on fluid mechanics in centrifuges is not fully closed and works are still continuing to be published / 11 /.

4 Laser Process : Thermocapillary-Buoyancy Flow of Molten Uranium in an Evaporation Crucible

The physical aspects of uranium enrichment by the laser process and the major components comprised in a practical device are reviewed in / 12 /. The fluid mechanics problem formulated below is met while modelling the flow in the evaporator.

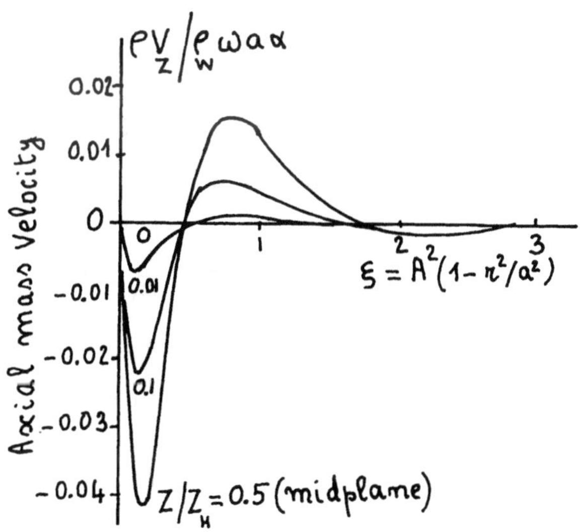

Fig. 3 - Radial profile of the axial mass velocity in the case of a wall thermal drive.

The schema of the problem is depicted in Fig. 4. A crucible with rectangular cross section contains molten uranium. The bottom and the side walls of the crucible are maintained at a constant temperature equal to the melting point T_o of the metallic uranium. The free surface of the liquid is heated from above by focusing an energetic electron beam. The power density deposited by the beam on the free surface is not generally spatially uniform, but decreasing from a high value at the center of the beam to a low value at the edge. The heating of the liquid by the beam should be modelled by the data, as boundary condition, of energy flux on the free surface. For sake of simplicity we choose a rather different approach : A temperature profile is specified on the free surface.

Due to the heating, the liquid is evaporated and the vapor emitted from the free surface expands upwards streaming through to a cold plate where it condenses. The whole device is enclosed in a vacuum casing. The thermal boundary conditions of the crucible induces in the liquid a convection flow driven by three forces : viscosity, thermocapillarity (MARANGONI effect) and thermogravitation (buoyancy effect). We want to investigate this flow in the pool.

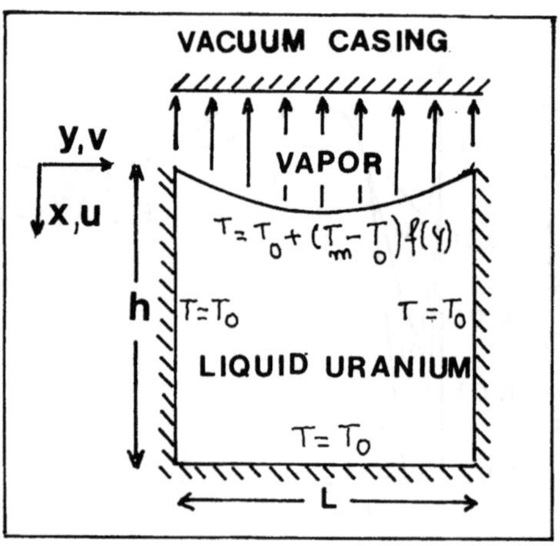

Fig. 4 - Schematic diagram of an uranium evaporator.

4.1 Model

The flow is assumed to be laminar and in steady state. The model includes the hydrodynamics equations in BOUSSINESQ approximation,the boundary conditions on the solid walls and the boundary conditions on the liquid-vapor interface / 13 /. Notations and their meanings are listed below :

x, y	space coordinates
u, v	velocity components
T	temperature
ρ	density, $\rho_o = 17.9\ g.cm^{-3}$; $\rho = \rho_o\ \ 1 - \beta(T-T_o)$
p	pressure
g	acceleration due to gravity, $g = 980\ cm.s^{-2}$
β	coefficient of volumetric expansion : for liquid uranium $\beta = 0.9\ 10^{-4}\ K^{-1}$
T_o	melting temperature of uranium, $T_o = 1405$ K
μ	dynamic viscosity of liquid uranium = $0.06\ g.cm^{-1}.s^{-1}$
ν	kinematic viscosity, $\nu = \mu.\rho^{-1} = 0;0035\ cm^2.s^{-1}$
χ	thermal diffusivity = $0.14\ cm^2.s^{-1}$
σ	surface tension, σ dynes.cm^{-1} = 1.550 - 0.14 T(K)
h	height of the cavity
L	width of the cavity.

The model follows :

$$\frac{\partial u}{\partial x} + \frac{\partial v}{\partial y} = 0 \tag{3}$$

$$u\frac{\partial u}{\partial x} + v\frac{\partial u}{\partial y} = g\ (1 - \beta(T-T_o)) - \frac{1}{\rho_o}\ \frac{\partial p}{\partial x} + \nu\Delta u \tag{4}$$

$$u\frac{\partial v}{\partial x} + v\frac{\partial v}{\partial y} = - \frac{1}{\rho_o}\frac{\partial p}{\partial y} + \nu\Delta v \tag{5}$$

$$u\frac{\partial T}{\partial x} + v\frac{\partial T}{\partial y} = \chi\Delta T \tag{6}$$

$$\Delta = \frac{\partial^2}{\partial x^2} + \frac{\partial^2}{\partial y^2}$$

Boundary conditions on solid walls ($x = h$, $y = 0$, $y = L$) :

$$u = 0 \qquad\qquad v = 0 \qquad\qquad T = T_o \tag{7}$$

On the liquid-vapor interface $x = H(y)$:

$$u - V H_y = 0 \tag{8}$$

$$T = T_o + (T_m - T_o)\, f(y) \qquad f(y) \text{ specified} \tag{9}$$

$$\mu \left(2 H_y \left(\frac{\partial u}{\partial x} - \frac{\partial v}{\partial y}\right) + (1 - H_y^2)\left(\frac{\partial u}{\partial y} + \frac{\partial v}{\partial x}\right)\right)$$

$$= - \frac{d\sigma}{dT}\left(\frac{\partial T}{\partial y} + \frac{\partial T}{\partial x} H_y\right)(1 + H_y^2)^{1/2} \tag{10}$$

$$\frac{\sigma}{R} = p - p_{vap} - \rho_{vap}\, u_{vap}^2 \quad ; \quad \frac{1}{R} = H_{yy}\,(1 + H_y^2)^{-3/2} \tag{11}$$

In eq. (11) the quantities p_{vap}, ρ_{vap} and u_{vap} designate respectively the pressure, density and normal velovity of the vapor adjacent to the interface. This equation is a dynamic generalization of the usual YOUNG-LAPLACE equation, as the third term in the RHS is an additional term due to the vapor thrust. From equations (3) to (11) we have to determine not only the thermo-hydrodynamic field but also the function H(y) representing the interface curve and this is the delicate problem of capturing the free surface.

4.2 Calculation of the Vapor Characteristics near the Interface

The system of equation (3) to (11) constitutes the model of our problem but it is closed only if the characteristics p_{vap}, ρ_{vap} and u_{vap} of the vapor adjacent to the interface and involved in eq. (11) are known. To calculate them we have to determine the vapor flow between the liquid free surface and the condensation cold plate. The topics pertains to the kinetic theory of evaporation-condensation, developped by many authors actively since two decades in the series of rarefied gas dynamics symposia (see for instance / 14 /). More precisely we consider the "two surface problem", depicted in Fig. 5, which consists in investigating the motion of the vapor emitted from one hot plane and streaming through to another parallel cold plane where it condenses. We use the notation : $T_{a,b}$ the temperature of respectively the evaporation surface and condensation surface, $P_{a,b}$ and $\rho_{a,b}$ the pressure and density of the saturated vapor at temperature $T_{a,b}$, l_a the mean free path in the reference state (T_a, P_a) and L' the interplate distance. The KNUDSEN number l_a/L' is small. The behaviour of the vapor is described by the BGK equation in the steady state one dimensional case :

$$\xi_1 \frac{\partial f}{\partial X} = A_{col}\, \rho\, (f_e - f) \qquad (12)$$

with :

$$f_e = \frac{\rho}{(2\pi RT)^{3/2}} \exp\left(- \frac{(\xi_1 - u_1)^2 + \xi_2^2 + \xi_3^2}{2\, RT}\right) ,$$

$$\rho = \int f \, d\xi_1\, d\xi_2\, d\xi_3 \quad , \quad u_1 = \frac{1}{\rho} \int \xi_1\, f\, d\xi_1\, d\xi_2\, d\xi_3 \quad ,$$

$$T = \frac{1}{3R\rho} \int \left((\xi_1 - u_1)^2 + \xi_2^2 + \xi_2^2\right) f\, d\xi_1\, d\xi_2\, d\xi_3 \quad ,$$

where f denotes the velocity distribution function, ξ_i - the molecular velocity, ρ - the density, u_1 - velocity and T - the temperature, $A^*_{col}\cdot\rho$ is the collision frequency and we take $R = \frac{\text{Boltzmann constant}}{\text{molecular mass}}$.

We put : $\xi_1^2 = \xi_1^2 + \xi_2^2 + \xi_3^2$.

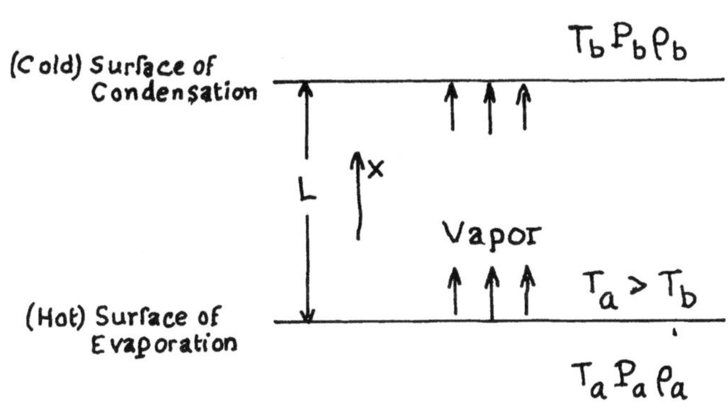

Fig. 5 - Schematic diagram of the two-surface problem.

For the boundary conditions we assume that on each plate the gas molecules leaving the condensed phase have the Maxwellian distribution corresponding to the satured gas at the surface temperature. So the boundary conditions can be written as follows :

$$f = \frac{\rho_a}{(2\pi RT_a)^{3/2}} \exp\left(-\frac{\xi_i^2}{2RT_a}\right) \quad , \quad \text{for } \xi_1 > 0 \quad \text{in } X = 0$$

$$f = \frac{\rho_b}{(2\pi RT_b)^{3/2}} \exp\left(-\frac{\xi_i^2}{2RT_b}\right) \quad , \quad \text{for } \xi_1 < 0 \quad \text{in } X = L' \qquad (13)$$

A computer code has been built by Frederik BOUJOT / 15 / to solve the equations (12) and (13). An example of results is displayed in Fig. 6 where we have represented the spatial distribution of the temperature, pressue and velocity of the vapor. Across a short distance of a few mean free paths from the liquid surface, the vapor characteristics undergo a very strong variation : this is the KNUDSEN kinetic layer. Moreover the computations yield also (at X = 0 in the figure) the characteristics of the vapor in contact with the interface in terms of the temperature T_a of the liquid :

$$T_{vap} = 0.84\ T_a \quad , \quad P_{vap} = 0.54\ P_a \quad , \quad \rho_{vap} = 0.64\ \rho_a \quad ,$$

$$U_{vap} = 0.35 \sqrt{2RT_a} \qquad (14)$$

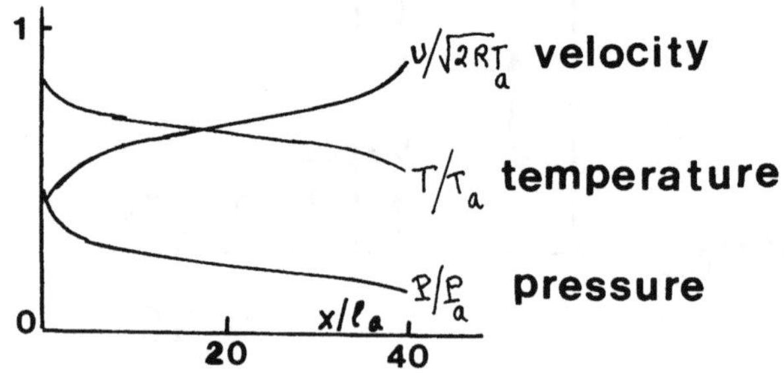

Fig. 6 - Structure of the vapor flow in the 2-surface problem.

In eq. (14) we recall that P_a and ρ_a are the pressure and density of the saturated vapor at the temperature T_a. For uranium, the saturation law is a physical data :
$P_a = \alpha \exp(-C/T_a)$; $\alpha = 0.833\ 10^{12}$ C.G.S. ; $C = 6.033\ 10^4$ K
and the density ρ_a derived from the perfect gas law

$$\rho_a = P_a/RT_a$$

Inserting eq. (14) in eq. (11) closes the model (formed by the equations (3) to (11)) for the flow of molten uranium in the crucible.

An extension of the preceding kinetic computations has been carried out by DONEDDU / 16 / to take into account the presence of energetic electrons in a narrow layer over the interface. This layer is the intersection of the electron beam heating the melt and the vapor beam. The RHS of the BGK equation (12) includes not only the collision atom-atom of uranium but also a second term representing the collision atom-electron

$$\xi_1 \frac{\partial f}{\partial x} = \nu_1 (f_1 - f) + \nu_2 (f_2 - f)$$

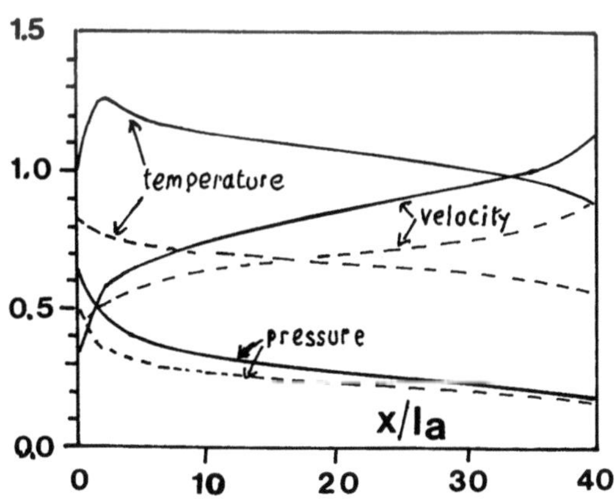

Fig. 7 - Structure of the vapor flow :
—— with collisions atom-electron (F. DONEDDU)
---- without collisions atom-electron (F. BOUJOT)
(Due to the courtesy of F. DONEDDU).

The contribution of this second term is to transfer energy (kinetic and thermal) to uranium atoms. As illustration, an example of computation is presented in Fig. 7 in the case where the width of electron layer is 2.5 mean free paths. It is obvious that for engineering purpose, computations should be rather carried out with this extended model.

4.3 Numerical Solution of the Model

Numerical code named BACCHUS has been used by Ph. RAVIER (from the French computer company CISI) to solve numerically the system of equations (3) to (11). With the only purpose of illustration, we present in figures 8 to 11 results concerning a very simple case of a weak heating ($(T_m-T_o)/T_o < 1$). In that case, we have found / 17 / that the interface has a negligible curvature and is almost flat.

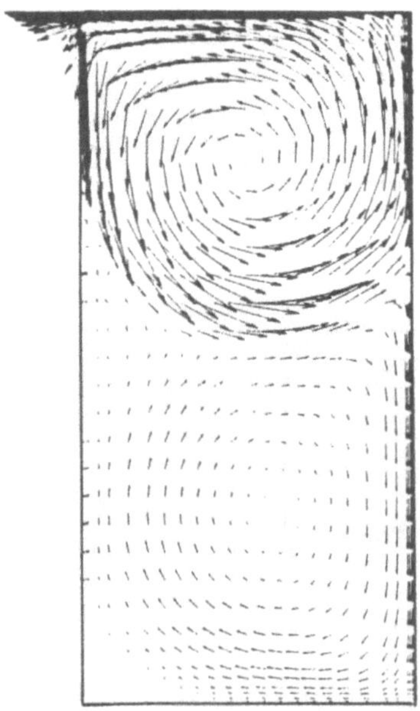

Fig. 8 - Velocity field in the liquid pool.

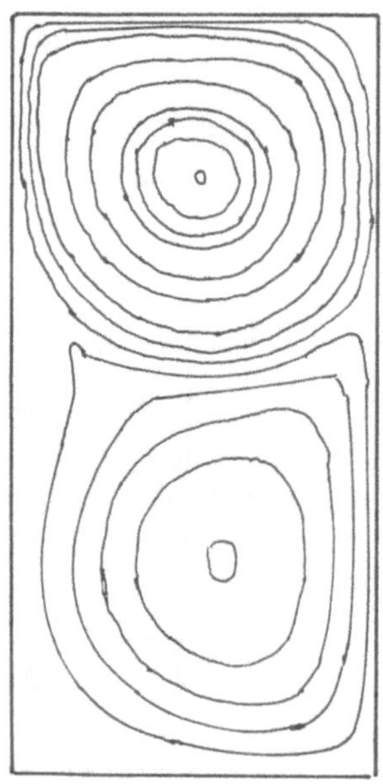

Fig. 9 Streamlines in the liquid pool.

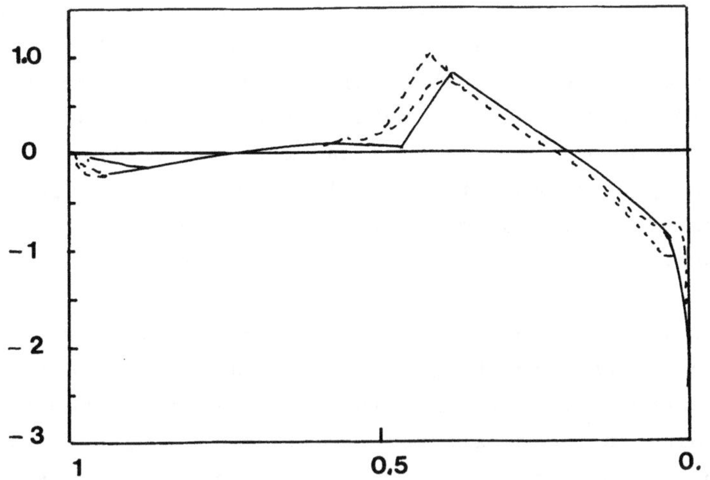

Fig. 10 - Horizontal velocity versus vertical distance.

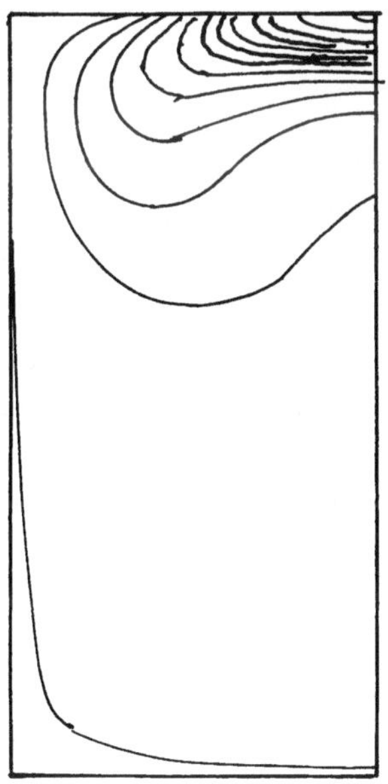

Fig. 11 - Isotherms in the liquid pool.

Fig. 8 represents the velocity field in the left half of the cavity (results are symmetric with respect to the mid vertical axis of the cavity. There are 2 distinct convective cells rotating in opposite sense. The cell near the free surface rotates with higher velocity whereas the cell in the deeper part of the pool is slower. The existence of two distinct cells is also clearly observable in Fig. 9 where we have plotted the streamlines.

Fig. 10 represents the horizontal velocity versus the vertical coordinate in three cross sections. Near the free surface (extreme right in the figure) the velocity has the highest value and decreases very strongly in a very small distance, confering thus a boundary layer character to a thin sheet of liquid just below the free surface : This is the MARANGONI layer investigated by NAPOLITANO / 18 /. In the rest of the cross section the velocity is much smaller and varies more smoothly.

In Fig. 11 we have plotted the isotherms : the effect of convection is clearly in evidence.

5 Conclusion

Research on uranium enrichment processes raises a wide variety of problems of fluid mechanics. The progress in the computational methods in fluid mechanics together with the progress in computers make possible the solution of these difficult problems. Such solutions are also a necessity to get an economically optimal design of industrial plants. In uranium enrichment industry, we should keep in mind that even a few percents of improvement represent a substantial amount of money.

6 References

1. M. Holt, M. Hittinger, Soubbaramayer : Proc. 4th Workshop. Gases in strong rotation, Oxford, 1981. Ed. E. Ratz, Urenco, Centec MBH, Bergisch Gladbach 1. West Germany.

2. C. Cortet, M. Holt, M. Hittinger, Soubbaramayer : Proc. 5th Workshop. Gases in strong rotation, Charlottesville, 1983. Ed. H. Wood, University of Virginia, Charlottesville, USA.
3. Soubbaramayer : in Uranium Enrichment, ed. S. Villani, vol. 35, Springer Verlag, 1979.
4. D. Massignon : in Uranium Enrichment, Ed. S. Villani, vol. 35, Springer Verlag, 1979.
5. Soubbaramayer : C.R. Acad. Sci. Paris, 267A, 904, 1969.
6. R. Jacques, D. Massignon, A. Rosengard, Soubbaramayer : C.R. Acad. Sci. Paris, 268A, 113, 1969.
7. D. Massignon, A. Rosengard, Soubbaramayer : Int. J. Heat Mass Transfer, 13, 879, 1970.
8. J.P. Lahargue, Soubbaramayer : Comp. Meth. Appl. Mech. Eng., 15, 2, 259-273, 1978.
9. H.P. Greenspan : Theory of rotating fluids, Cambridge University Press, London, 1969.
10. Soubbaramayer, J. Billet : Comp. Meth. Appl. Mech. Eng., 24, 165-186, 1980.
11. Proc. Workshop Separation phenomena on liquids and gases, Darmstadt (West Germany), Ed. K. Roesner and E. Rätz, Urenco Centec GMBH, Bergisch Gladbach 1, 1987.
12. Soubbaramayer : Rapport CEA-Conf 9309, C.E.A. France, 1988.
13. V.G. Levich, V.S. Krylov : Ann. Rev. Fluid Mech., 1, 293-316, 1969.
14. Y. Sone : Rarefied gas dynamics, Proc. 14th Symposium, Ed. Haguro Oguchi, University of Tokyo Press, 1984.
15. F. Boujot, Soubbaramayer : C.R. Acad. Sci. Paris, t. 307, Série II, 323-327, 1988.
16. F. Doneddu : Private communication.
17. Soubbaramayer : Annales de Physique, Colloque n° 2, suppl. au n° 3, vol. 13, Juin 1988.
18. L.G. Napolitano, C. Golia : Acta Astronautica, vol. 8, n° 5-6, 417-434, 1981.

Numerical Study of Spherical Taylor-Couette Flow

R.-J. Yang

1. Introduction

Flow in the gap between a rotating inner and a stationary outer sphere can generate Taylor vortices. For the gap width of $\sigma = 0.18$ ($\sigma = (R_2 - R_1)/R_1$, R_1 and R_2 are the radii of inner and outer spheres), SAWATZKI et al. [1] have observed nonunique flow modes at a supercritical Reynolds number ($Re = \Omega_1 R_1^2/\nu$, Ω_1 is the angular velocity of the inner rotating sphere and ν is the kinematic viscosity), namely, three steady axisymmetric and two unsteady nonaxisymmetric modes. Steady modes contain either 0, 1, or 2 vortices per hemisphere in the vicinity of the equator; unsteady modes also contain either 1 or 2 vortices. WIMMER [2] showed in subsequent experiments that the flow modes can be produced by different accelerations of the inner sphere. Each mode requires a certain acceleration of the inner sphere to an angular velocity corresponding to the critical values of the Reynolds number.

Some numerical studies on the three steady axisymmetric flow modes have been made in recent years (BARTELS [3], MARCUS et al. [4], and SCHRAUF [5]). BARTLES [3] solved the unsteady axisymmetric Navier-Stokes equations in stream-function vorticity formulation by means of a finite-difference approximation. He used a quadrantal annulus ($0 \le \theta \le \pi/2$, equator at $\theta = 0$, pole at $\pi/2$) as a computation domain and imposed equatorially symmetric boundary conditions. For the gap width of $\sigma = 0.18$,

Work reported herein was sponsored by NASA/Marshall Space Flight Center, Huntsville, Alabama, under Contract NAS8-40000. Computations were carried out at NASA Ames Research Center.

he could not obtain 1-vortex flow without arbitrary disturbances of the symmetry near the equator. The other two steady modes, 0- and 2-vortex flows, were simulated through straightforward calculations. MARCUS et al. [4] used the same formulation while the equations were solved by a pseudo-spectral method. Their numerical computations were carried out in the whole domain ($-\pi/2 \leq \theta \leq \pi/2$). By controlling the accelerations of the inner sphere, they were the first to reproduce 1-vortex flow as a solution of an initial-value problem directly from the basic flow. As with the long development time for experiments, a long computation time was also needed for this mode. SHRAUF [5] used the continuation method to investigate how the instability of the spherical Couette flow depends on the gap size. Numerical results obtained by these authors agree quite well with the experiments.

The state attained by the flow depends on its history, especially on the transient acceleration of the inner sphere from rest to its final rotation rate, and on the kind of disturbances permitted by experimental setup or numerical schemes. By closely mimicking laboratory experiments, the three steady modes can be simulated numerically by solving the full time-dependent Navier-Stokes equations. Although the three flow modes are axisymmetric and symmetric with respect to the equator at steady states, the whole domain ($-\pi/2 \leq \theta \leq \pi/2$) is required in the case of simulating 1-vortex flow since the transition process of the flow as found to be breaking symmetry at the equator [4]. In this paper, a technique is proposed to simulate the three steady axisymmetric flow modes. In the numerical simulation, only half domain ($0 \leq \theta \leq \pi/2$) is required. Matrix preconditioned technique is applied to the Navier-Stokes equations for a faster convergent rate to the steady state. A method defining routes to each flow mode is introduced. The present solutions, in terms of friction torques and vortex sizes, show excellent agreement with the experiments.

2. Governing Equations and Numerical Methods

The three-dimensional, incompressible, Navier-Stokes equations are modified to the following set of governing equations:

$$\frac{1}{\beta}\frac{\partial p}{\partial t} + \frac{\partial u_i}{\partial x_i} = 0 \tag{1}$$

$$\frac{(\alpha+1)u_i}{\beta}\frac{\partial p}{\partial t} + \frac{\partial u_i}{\partial t} + \frac{\partial u_i u_j}{\partial x_j} = -\frac{\partial p}{\partial x_i} + \frac{\partial \tau_{ij}}{\partial x_j} \tag{2}$$

where t is time; x_i are the Cartesian coordinates; u_i are corresponding velocity components; p is the pressure; and τ_{ij} are the stress tensors. The parameters α and β, according to TURKEL [6], are introduced to precondition the Navier-Stokes equations and allow for a faster convergence to the steady state. The equations are then transformed to a general coordinate system in which the general coordinates ξ_i are related to the Cartesian coordinate x_i by

$$\xi_i = \xi_i(x, y, z) \tag{3}$$

In the generalized curvilinear coordinates, the governing equations in conservation law form are expressed as

$$M\frac{\partial}{\partial t}\hat{D} + \frac{\partial}{\partial \xi_i}(\hat{E}_i - \hat{\Gamma}_i) = 0 \tag{4}$$

Preconditioned matrix M, pressure and velocity vector $\hat{D}$, flux vector $\hat{E}$, and the viscous diffusion vector $\hat{\Gamma}$ are described by

$$M = \begin{bmatrix} \beta^{-1} & 0 & 0 & 0 \\ (\alpha+1)u\beta^{-1} & 1 & 0 & 0 \\ (\alpha+1)v\beta^{-1} & 0 & 1 & 0 \\ (\alpha+1)w\beta^{-1} & 0 & 0 & 1 \end{bmatrix}, \quad \hat{D} = \frac{D}{J} = \frac{1}{J}\begin{bmatrix} p \\ u \\ v \\ w \end{bmatrix}, \quad \hat{E}_i = \frac{1}{J}\begin{bmatrix} U_i \\ uU_i + L_{i1}p \\ vU_i + L_{i2}p \\ wU_i + L_{i3}p \end{bmatrix}$$

$$\hat{\Gamma}_i = \frac{\nu}{J}(\nabla\xi_i \cdot \nabla\xi_j)\frac{\partial}{\partial \xi_j}(0, u, v, w)^T$$

where J is the Jacobian of coordinate transformation, and

$$U_i = L_{i1}u + L_{i2}v + L_{i3}w,$$

$$L_{i1} = (\xi_i)_x,\; L_{i2} = (\xi_i)_y,\; L_{i3} = (\xi_i)_z$$

are the contravariant velocities and the metrics of transformation, respectively.

The governing equations are replaced by an implicit time difference approximation (backward-difference). Nonlinear numerical fluxes at the implicit time level are linearized by Taylor expansion; then, spatial difference approximations (central-difference) are introduced. The result is a system of multidimensional, linear, coupled difference equations for the dependent variables at the implicit time level. To solve these equations, the approximate-factorization or ADI scheme (KWAK et al. [7] and BRILEY et al. [8]) is used. This technique leads to the following system of coupled linear difference equations

$$(M + \Delta t\mathcal{L}_{\xi})\Delta D^{*} = RHS$$
$$(M + \Delta t\mathcal{L}_{\eta})\Delta D^{**} = M\Delta D^{*} \tag{5}$$
$$(M + \Delta t\mathcal{L}_{\zeta})\Delta D = M\Delta D^{**}$$

where

$$\mathcal{L}_{\xi_i} = J\delta_{\xi_i}(\hat{A}_i - \hat{\Gamma}_i),\ RHS = -\Delta t J\Sigma\delta_{\xi_i}(\hat{E}_i - \hat{\Gamma}_i)$$
$$\hat{A}_i = \partial\hat{E}_i/\partial D,\ \Delta D = D^{n+1} - D^{n}$$

ΔD^* and ΔD^{**} are intermediate solutions. The system of Eq. (5) can be written in narrow block-banded matrix structures and be solved by a standard LU decomposition method.

3. Initial and Boundary Conditions

At a supercritical Reynolds number, the final steady states of the spherical Taylor-Couette flow depend on the accelerating history of the inner sphere and initial conditions. The present numerical algorithm uses a preconditioned method and thus transient history has no physical meaning. Disregarding the transient accuracy, the current work is to find routes leading to steady 0-, 1- and 2-vortex flows with a faster convergent rate to the steady states. The computation is conducted in a quadrantal annulus $0 \leq \theta \leq \pi/2$. The initial conditions for all calculations (except some special cases) are given as follows: linear velocity profile across the gap in the circumferential direction, zero velocity components in the meridional plane, and zero static pressure in the entire domain.

The modified Navier-Stokes equations have a symmetric hyperbolic system. The preconditioned matrix is positive definite ($\beta > 0$). The appropriate number of boundary conditions is the same as that for the original Navier-Stokes equations. To make the problem well-posed, the following boundary conditions are used: reflection-symmetric conditon is applied at equator and pole, no-slip condition is used on inner and outer sphere surfaces, normal pressure gradient on both sphere surfaces is balanced by centrifugal and viscous forces, and a reference pressure is set at a specified point in the computation domain.

4. Computation Procedure and Results

Spherical Taylor-Couette flow is characterized by two parameters: the Reynolds number (Re), and the gap width, (σ). The current computations are performed for $\sigma = 0.18$ and $Re \leq 1500$. Preconditioned parameters

($\alpha = 0$ and $\beta = 2$) are used. Numerical convergency is checked by monitoring the friction torques acting on inner and outer sphere surfaces. At a steady state, the friction torques are equal in magnitude. To make sure the steady-state solutions are independent of mesh distributions, two grid systems were tested for the case of Re = 1500: 111 x 51 and 81 x 41 (J x K, J and K are number of mesh points in azimuthal and radial direction, respectively.) Both computed solutions were nearly identical to each other. Therefore the 81 x 41 grid system shown in Fig. 1 is used throughout these calculations.

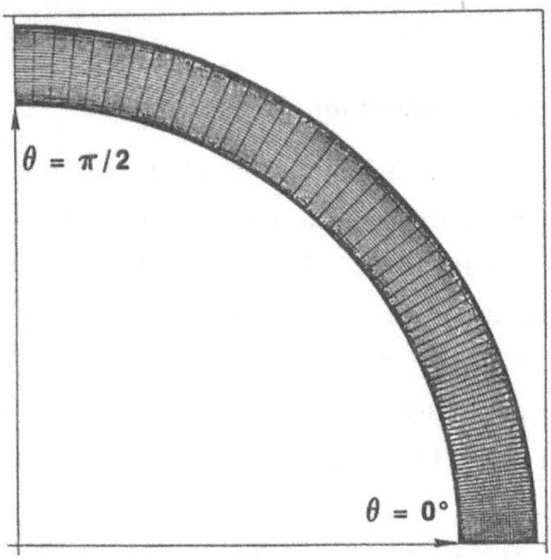

Figure 1. Computational grid system (81 x 41)

In the region near the pole, local flow is similar to that which occurs in the gap between a rotating disk and a stationary housing. In the region near the equator, local flow is similar to that of cylindrical Taylor-Couette flow. Because of the Eckman pumping effects, centrifugal forces drive fluid along the inner rotating sphere from the pole toward the equator. Near or at the equator plane (depending on flow mode), the fluid is deflected outward and moved back to the pole in the vicinity of the outer sphere. It should be noted that each flow mode has its own region of existence at steady state; see WIMMER [2] or Buhler [9]. For instance, at Re = 600, only 0-vortex flow exists; at Re = 800, 1- and 2-vortex flows exist; at Re = 1500, the three flow modes exist. The following discussions are based on the assumption that each flow mode exists at a given Reynolds number in each category. Routes leading to these flow modes are described below:

1. <u>Route to 0-vortex flow</u> - Using initial and boundary conditions depicted in Section 3, the curent numerical method converges to a

0-vortex flow. It has to be noted that the friction torque for both inner and outer spheres has to reach a constant at steady state. If the flow mode does not exist at a given Reynolds number, an equilibrium state (both inner and outer friction torques are equal) may be obtained during the process of the calculation; however, the friction torque is not a constant during a time interval. Eventually, the friction torque reaches a constant and then another steady flow mode is formed. An example will be given later.

2. Route to 1-vortex flow - Experimental observations indicated that vortex size near the equator of 1-vortex flow has the same order of the gap width's magnitude. The vortex near the equator is separated from the major portion of meridional flow through a dividing streamline. Local flow on either side of the streamline reveals symmetric character. For the numerical simulation of the 1-vortex flow, a radial line at one gap width distance away from the equator is chosen as a fictitious inner symmetric boundary; there, flow variables are treated as symmetry. Using conditions described in Section 3 and the inner symmetric treatment, a "rough" 1-vortex flow can be simulated. With the "rough solution" as the initial condition, and taking out the treatment at the specified radial line, the current numerical method converges to a 1-vortex flow. In the calculations, of [4] the whole domain ($\pi/2 \leq \theta \leq \pi/2$) is required since the transition process of the 1-vortex flow revealed symmetry breaking at the equator. With the current approach, only half domain is required and there is no need to control the rotating history of the inner sphere. The treatment at the chosen fictitious boundary during the initial calculation creates a necessary state for 1-vortex flow to exist.

3. Routes to 2-vortex flow - There are two routes that can lead to a steady 2-vortex flow. First a similar treatment as of 1-vortex flow is applied to the 2-vortex flow simulation. In this case, two radial lines are chosen as fictitious inner symmetric boundaries. The separation distance among the two radial lines and the equator is about one gap width each. Computation with conditions given in Section 3 and the inner symmetric treatment at the two radial lines can lead to a rough 2-vortex flow. Taking out the treatment at the two specified radial lines and using the rough solution as the initial condition, the current numerical method converges to a

steady 2-vortex flow. This treatment toward the final steady 2-vortex flow is called forced transition. Another route is called free transition. In this case, straightforward calculation using initial and boundary conditions given in Section 3 can converge to a steady 2-vortex flow. In the process of the free transition, it is observed that the friction torque for both inner and outer spheres is in an equilibrium state but is not a constant with respect to time during a period. Figure 2 shows an example of convergent histories for both forced and free transition for the case of Re = 800. In the case of free transition, the flow reaches an equilibrium state after 500 time steps (NT = 500). But friction torque varies with respect to time until NT = 1600. In the case of forced transition, the first 300 time steps were treated as having inner symmetric boundaries at two specified radial lines. After taking out the two fictitious inner boundaries and continuing the calculation, the flow converges to a steady 2- vortex flow. It is obvious that the case of the forced transition led to a much faster led to a much faster convergent rate toward a steady state.

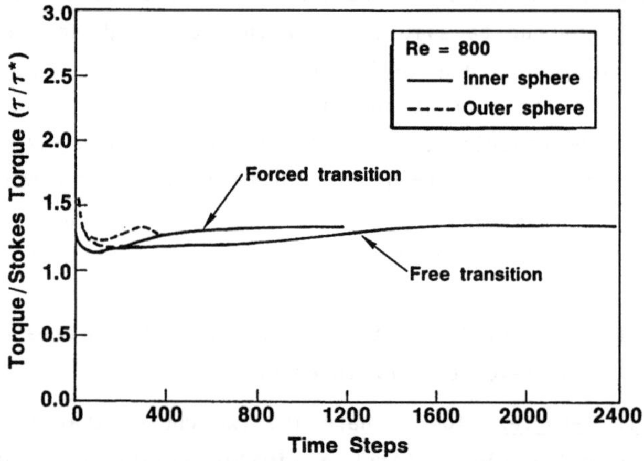

Figure 2. Convergent histories toward 2-vortex flow

The three steady flow modes behave quite differently near the equator. Figure 3 gives velocity vectors on meridional plane, and Fig. 4 gives circumferential velocity component at the equator for the case of Re = 1500. Taylor vortices for 1- and 2-vortex flows are clearly shown. The circumferential velocity profiles at the equator are quite different due to the Taylor vortices. Figures 5 and 6 give circumferential skin friction and azimuthal pressure gradient coefficients. Wild differences are

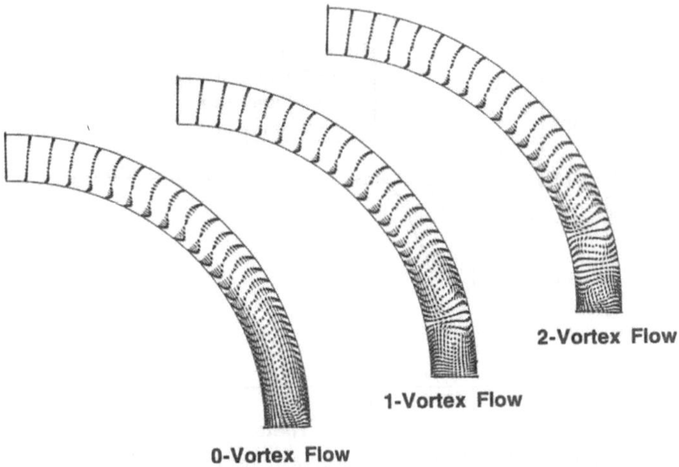

Figure 3. Velocity Vectors on Meridional Plane (Re = 1500, σ = 0.18)

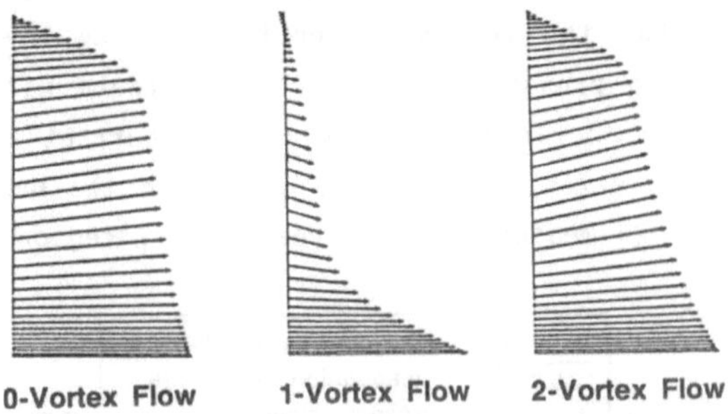

Figure 4. Circumferential Velocity Profiles at Equator (Re = 1500, σ = 0.18)

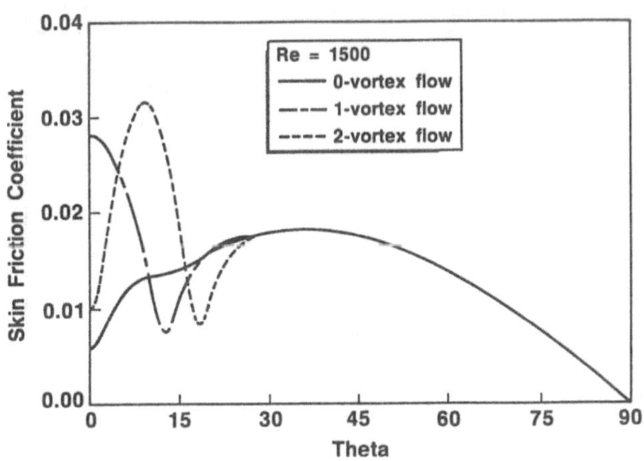

Figure 5. Circumferential Skin Friction Ceofficient on Inner Sphere Surface

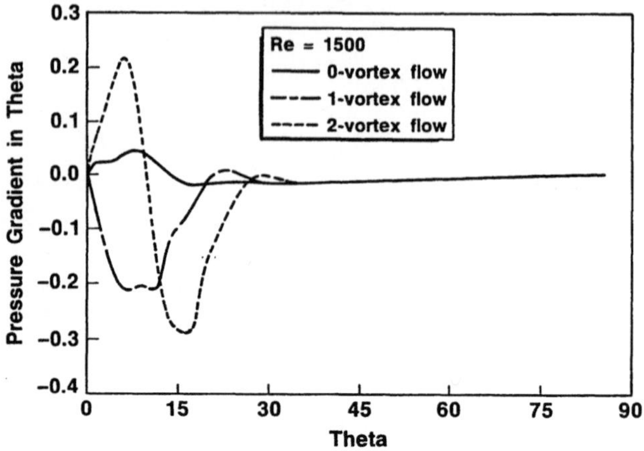

Figure 6. Azimuthal Pressure Gradient Coefficient on Inner Sphere Surface

clearly shown within $\theta \leq 40$ deg. From Fig. 3 and Fig. 6, it is interesting to note that the turning directions of the Taylor vortices are consistent with the pressure gradient in the theta direction along the inner sphere surface. In Fig. 7 and 8, the current results (friction torques and vortex sizes) agree quite well with the MARCUS [4] calculations and WIMMER [2] measurements. From Fig. 7, region of existence for

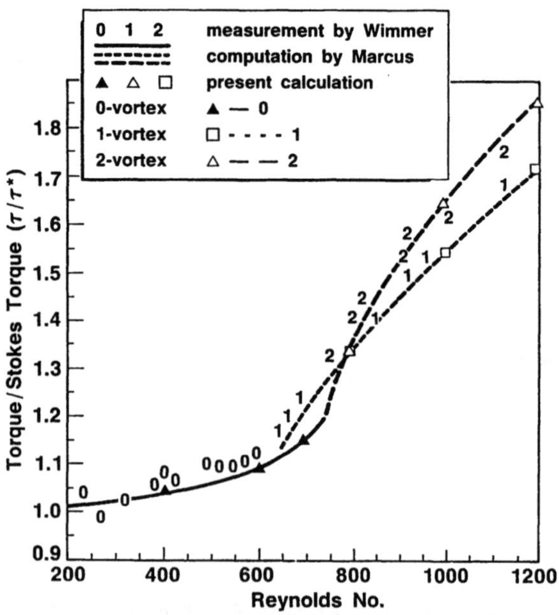

Figure 7. Friction torques

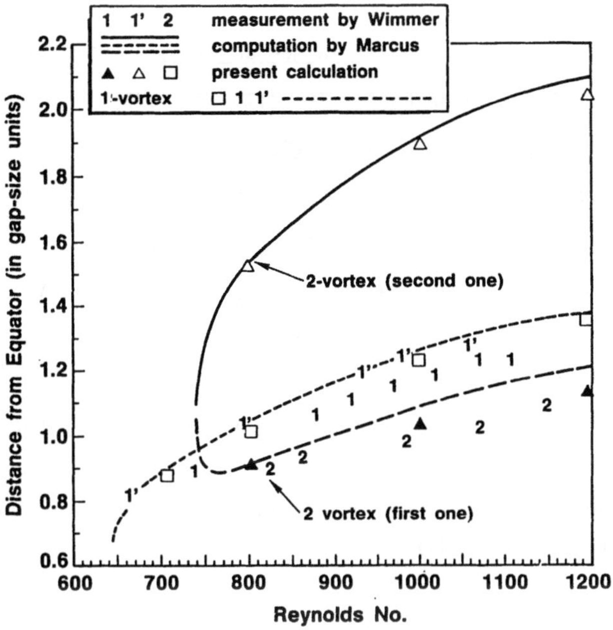

Figure 8. Vortex sizes

the three flow modes is clearly shown. The 0- and 2-vortex flows lie on the same solution branch, and the 1-vortex flow lies on a separate branch.

5. Conclusion

A new technique to simulate Taylor vortices in a spherical gap has been developed and tested. The nonunique steady flow modes associated with the Taylor vortices are simulated through special ways. Routes leading to 0-, 1- and 2-vortex flows are designed heuristically. Fictitious symmetric boundaries near the equator are imposed during a time of the calculation. The choice of the location of the fictitious boundaries is determined by either 1- or 2-vortex flow being simulated. The imposition of one or two fictitious symmetric boundaries during the initial calculation generates the state suitable for 1- or 2-vortex flow to exist. After taking out the fictitous boundaries, the flow settles down into its own attractor. By this method, the three steady flow modes can be simulated by using half domain, i.e., $0 \leq \theta \leq \pi/2$. The current numerical technique can converge to desired flows very fast.

References

1. O. Sawatzki and J. Zierep, Acta Mechanica 9, 13-25 (1970).
2. M. Wimmer, J. Fluid Mechanics 79, 317-335 (1976).
3. F. Bartels, J. Fluid Mechanics 119, 1-25 (1982).
4. P. S. Marcus and L. S. Tuckerman, J. Fluid Mechanics 185, 1-65 (1987).
5. G. Schrauf, J. Fluid Mechanics 166, 287-303 (1986).
6. E. Turkel, J. Computational Physics 72, 277-297 (1987).
7. D. Kwak, J. Chang, S. P. Shanks, and S. Chakaravarthy, AIAA J. 24 390-396 (1986).
8. W. R. Briley and H. McDonald, J. Computational Physics 24, 372-397 (1977).
9. K. Buhler and J. Zierep, Proc. 4th Inter. Sym. on Turbulent Shear Flows, 16.1-16.6 (1983).

Trajectories of Vortex Lines Beneath a Free Surface of Above a Plane

B. Yim

1.0 Introduction

When an airfoil moves in the air above the ground, trailing vortices are created as a wake. In the wake, we often see two long distinguishable tip vortex lines. If a hydrofoil moves beneath a water surface a similar situation may arise. In these cases, what shapes would the tip vortex lines have? To find out, the effects of the ground as well as the induced velocities of the vortex lines have to be taken into account.

As an approximation, an inviscid flow and the ideal vortex filament may be assumed for the real situation. Then the well known Biot-Savart Law may be applied for the vortex induced velocity [1]. However since the shape of the vortex line is not known a priori, and the locations of vortex singularities are not known except at the vortex generator such as the tip of the wing, the problem is not simple.

For a general wing problem in an infinite medium, the locations of trailing vortices are generally assumed to be known with the linearization concept.

In the present problem there may be three approaches to the solution. First, consider the problem as a steady problem with an established wake and find the shape of the wake. Second, consider the problem as an unsteady problem and at each time step from the beginning, find the wake shape. Third, consider the solution not near to the end points of the vortex line and then the two-dimensional approach may be acceptable. In fact, there exists a classical solution for a simple case [2,3] and the test results substantiate the acceptability of the third approach. In the first approach, the location of a vortex line which is semi-infinite is not known except at the tip. If the deviation of the location from a known straight line is assumed to be small an iteration method may be applicable.

In the second approach the problem is direct because each time the locations of vortex singularities will be known. When the time steps increase, the length of vortex lines increases, or the number of discretized vortex lines increases. Because each discretized vortex line changes the location at each time step, the coordinates of each discretized vortex line should be unknown variables whose number increases with time. Thus this approach is suitable only for a present day high speed computer with a large memory.

In the present study, the second and third approaches mentioned above are taken. Then the velocity of the vortex line is obtained by the Biot-Savart law in the three-dimensional space and by the complex potential in two-dimensional space. The trajectories of vortex lines are obtained by solving the simultaneous ordinary differential equations with the Runge-Kutta method. The computational results obtained by both the two- and three-dimensional analyses are compared.

2.0 Formulation of Problem

We consider that a vortex line is formed beneath a free surface from a point P(x,y,z) that moves with a constant velocity $\bar{u}$. The Cartesian coordinates 0-x, y, and z are defined as the x-axis in the direction of velocity $\bar{u}$, the z-axis perpendicular to the undisturbed free surface, positive upward. When u is small, the free surface may be considered as a solid surface and the mirror image with respect to the free surface can be considered. The problem becomes exactly the same as the vortex lines near the ground. When the vorticity at a point Q is $\bar{\zeta}_Q$ in a vortex line A, the velocity $\bar{q}$ at a point P can be found by the Biot-Savart law,

$$\bar{q} = \frac{1}{4\pi} \int_A \frac{\bar{\zeta}_Q \wedge \bar{r}}{r^3} \, d\tau$$

where $\bar{r} = \overline{PQ}$.

If A is a line vortex, $\bar{\zeta}_Q d\tau = k\bar{\ell} d\varrho$ with unit direction $\bar{\ell}$ along the vortex line and ϱ the length of the line, $\bar{q}$ becomes

$$\bar{q} = \frac{k}{4\pi} \int_A d\varrho \, \frac{\bar{\ell} \wedge \bar{r}}{r^3} \tag{1}$$

If we nondimensionalize Equation (1) by the velocity u, then k is of the dimension of length. Thus we nondimensionalize r and ϱ by $k/(4\pi u) = k_1$ and we can write with such nondimensional quantities

$$\bar{q} = \overline{ui} + \overline{vj} + \overline{wk} = \int \frac{d\varrho}{r^3} \bar{\ell} \wedge \bar{r}$$

With this nondimensionalization care is necessary that starting points at t = 0 should be nondimensionalized likewise with k_1.

Since $$\frac{Dx}{Dt} = u$$

$$\frac{Dy}{Dt} = v$$

$$\frac{Dz}{Dt} = w \qquad (2)$$

x,y,z can be obtained by solving these simultaneous ordinary differential equations by the Runge-Kutta technique with the initial condition. However, since the vortex point is moving with a constant velocity $\bar{u}$ leaving a vortex line behind, and images need to be taken care of, the trajectory of the vortex line which consists of all the discretized vortex elements $\Delta\bar{\zeta}_i$ (i = 1,2,...) is time dependent. Consider one typical $\Delta\bar{\zeta}$ with a constant vorticity along the length $\Delta\varrho$. Then each discretized vortex line contribution to the velocity field $\bar{q}_i$ can be obtained by a quadrature as a closed integration.

Consider three points as in Fig. 1 $Q_1(\xi_1, \eta_1, \zeta_1)$, $Q_2(\xi_2, \eta_2, \zeta_2)$ and P(x,y,z), where $\bar{\ell}\Delta\varrho = (\xi_2 - \xi_1)\bar{i} + (\eta_2 - \eta_1)\bar{j} + (\xi_2 - \xi_1)\bar{k} = \overline{Q_1Q_2}$. Then

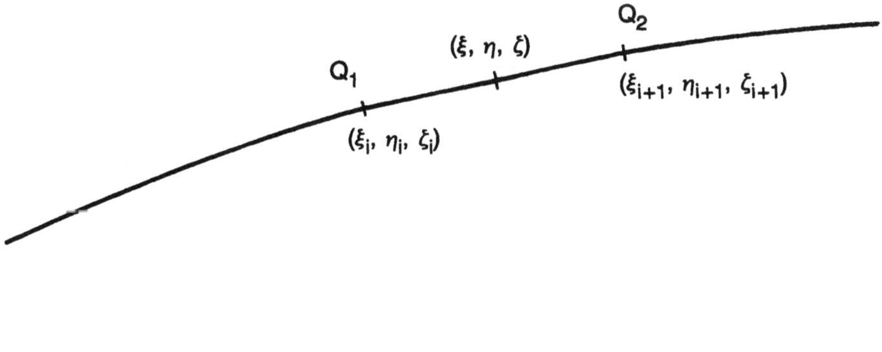

Fig. 1. Vortex line and field point P.

$$\bar{r} = (x-\xi)\bar{i} + (y-\eta)\bar{j} + (z-\zeta)\bar{k}$$

where $\varrho\ (\xi,\eta,\zeta) < \Delta\varrho$ and point (ξ,η,ζ) is on the segment Q_1Q_2 or

$$\varrho = \sqrt{\{(\xi-\xi_1)^2 + (\eta-\eta_1)^2 + (\zeta-\zeta_1)^2\}}.$$

If the direction cosines of $\bar{\ell}$ are ℓ_1, ℓ_2, and ℓ_3, then we can write

$$\xi - \xi_1 = \varrho\ell_1$$

$$\eta - \eta_1 = \varrho\ell_2$$

$$\zeta - \zeta_1 = \varrho\ell_3$$

where $(\ell_1, \ell_2, \ell_3) = \{(\xi_2-\xi_1), (\eta_2-\eta_1), (\zeta_2-\zeta_1)\}/\Delta\varrho$

and we have

$$\bar{q}_i = \int_{Q_1}^{Q_2} \frac{\bar{\ell}\wedge\bar{r}}{r^3}\, d\varrho = \int \begin{vmatrix} \bar{i} & \bar{j} & \bar{k} \\ \ell_1 & \ell_2 & \ell_3 \\ x-\xi_1-\varrho\ell_1 & y-\eta_1-\varrho\ell_2 & z-\zeta_1-\varrho\ell_3 \end{vmatrix} (\varrho^2 - 2\varrho b + c)^{-3/2}\, d\varrho$$

where $\quad b = \ell_1\,(x-\xi_1) + \ell_2\,(y-\eta_1) + \ell_3\,(z-\zeta_1)$

$$c = (x-\xi_1)^2 + (y-\eta_1)^2 + (z-\zeta_1)^2$$

or $$\bar{q}_i = \frac{1}{c-b^2}\left(\frac{\varrho-b}{\sqrt{(\varrho^2-2\varrho b+c)}}\right)\Bigg|_{\varrho=0}^{\varrho=\Delta\varrho} [\bar{i}\{\ell_2(z-\zeta_1)-\ell_3(y-\eta_1)\}$$

$$+ \bar{j}\{\ell_3(x-\xi_1)-\ell_1(z-\zeta_1)\} + \bar{k}\{\ell_1(y-\eta_1)-\ell_2(x-\xi_1)\}] \tag{3}$$

and $$\bar{q}(x,y,z) = \Sigma_i \int_{(\xi_i,\ \eta_i,\ \zeta_i)}^{(\xi_{i+1},\ \eta_{i+1},\ \zeta_{i+1})} (\bar{\varrho}\wedge\bar{r})\, r^{-3}\, d\varrho$$

Here we have to take care of a seemingly singular point on the vortex line. We can consider this as LOUKAKIS [4] did for a point on a circular curve. The induced velocity on a straight segment of vortex line can be evaluated by considering a point on an infinite straight line. We evaluate the induced velocity of the infinite vortex line except the part of the concerned short vortex line segment, and subtract this from the well known induced velocity of the infinite vortex line that is zero. From (3) it is easy to see all the velocity components due to the infinite line except the short segment where the co-location point is located. The induced velocity at a point on a vortex line due to its own vorticity is zero. In fact, this can be obtained from the result that the induced velocity due to a circular curve is inversely proportional to the radius of the circle.

When the vortex source advances with a constant velocity a new vortex line element is created. The first vortex element at $t = 0$ is denoted $\Delta\zeta_1$; then $\Delta\zeta_i$ is added with $i = 2, 3, \ldots$ as time passes. When the vortex line becomes long, each vortex element and the images influence each other, and the location of each vortex line changes. Thus the location of each vortex element is solved by the simultaneous ordinary differential equations (2). The numbers of unknowns and equations increase as time increases. Thus only a high speed computer can solve this problem. The Runge-Kutta sixth order method is used here. There is no limit in the number of initial vortex points as long as ample computer memory and computing time are available. The value of k_1 in general is small. Therefore the nondimensional value of the initial coordinates of the vortex point, y_0/k_1 and z_0/k_1, are large values unless y_0 and z_0 are very small. In this case the computation of trajectories is very stable and converges rapidly. When y_0/k_1 and z_0/k_1 have small values the computation becomes unstable.

3.0 Two-Dimensional Approach

Trajectories of trailing vortex lines behind a wing near the ground or beneath a free surface have been investigated both theoretically [2, 3] and experimentally [5, 6]. However, the theory has used the two-dimensional analysis.

When two-dimensional rectilinear vortex filaments are located at four points: $\zeta_j = \pm y \pm zi$ in the complex plane with the four combinations of $\pm$, it is tantamount to the situation of a wing with two trailing vortices beneath a free surface or above a ground. Since the complex potential is

$$w = \frac{k}{2\pi} \sum_{j=1}^{4} i \log (\zeta - \zeta_j)$$

we can obtain the complex velocity dw/dz. It is well known in the above problem that the induced velocity component at the vortex line is [2, 3]

$$\frac{dz}{dt} = \frac{k}{4\pi}\frac{z^2}{yr^2}$$

$$\frac{dy}{dt} = \frac{k}{4\pi}\frac{y^2}{zr^2}$$

where $r^2 = y^2 + z^2$.
Eliminating k by division, the differential equation for the trajectory becomes

$$\frac{dz}{z^3} + \frac{dy}{y^3} = 0$$

and the solution is

$$a^2\,(z^2 + y^2) = 4z^2y^2$$

where

$$a = \frac{-2\,z_0y_0}{\sqrt{(z_0{}^2 + y_0{}^2)}}$$

with the starting point (z_0, y_0) at $t = 0$. Therefore we can obtain the relation between the time and the vortex location by inserting the path relation back into the original differential equation

$$\frac{dz}{dt} = \frac{k}{4\pi}\frac{(4z^2 - a^2)^{3/2}}{4az^3}$$

Integrating this equation

$$\frac{k}{4\pi}t = \int_{z_0}^{z}\frac{4az^3\,dz}{(4z^2 - a^2)^{3/2}} = \frac{a(z^2 - a^2)}{(4z^2 - a^2)^{1/2}} - \frac{a(z_0{}^2 - a^2/2)}{(4z_0{}^2 - a^2)^{1/2}}$$

This result has been used to check experiments which are reasonably close [5,6]. Since the two-dimensional solution for this problem has been studied extensively, the three-dimensional results for this problem may better be compared with the two-dimensional ones. In Fig. 2 the two-dimensional results and the three-dimensional results are shown and compared. However, the physical configurations of two- and three-dimensional problems are a little different. That is, the two-dimensional result is the trajectory of a two-dimensional vortex point while the three-dimensional result is the shape of the

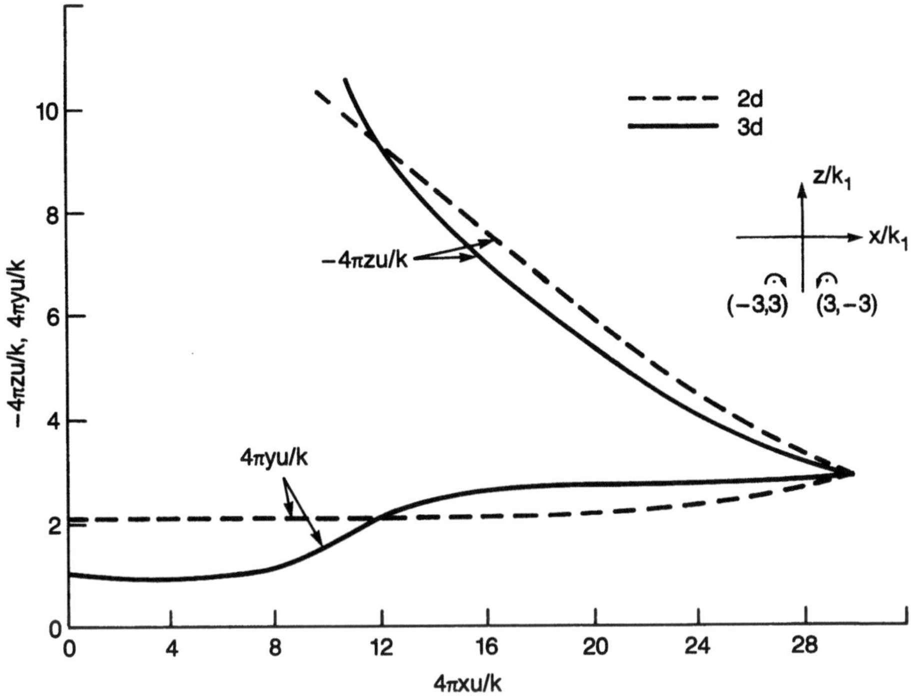

Fig. 2. Comparison of vortex trajectories at $4\pi tu^2/k = 30$.

vortex line at each time instant. It has been understood that the trailing vortex trajectory near the ground can be well predicted by the two-dimensional method. However, it seems to depend upon the parameters involved. The three-dimensional effect is very large in a large area surrounding the end points of vortex lines when k_1 is small and the distance from the boundary is large. In such circumstances, it takes a long time to establish the near two-dimensional phenomenon.

4.0 Two-Dimensional Analysis for 2N Vortices Beneath a Free Surface

Since predictions by the two-dimensional analysis for two vortices near the ground correlate well with experimental results, it is worth considering the flow due to 2N vortices with conditions similar to those for two vortices, by the two-dimensional analysis. Making use of the analysis of two vortices, we can write the velocity component at the vortex points (x_i,y_i) $i = 0, \ldots N-1$ by defining

$$-k_0 \frac{x_0^2}{yr_0^2} + k_1 2 \left[\frac{y_0 - y_1}{(x_0 - x_1)^2 + (y_0 - y_1)^2} - \frac{y_0 + y_1}{(x_0 - x_1)^2 + (y_0 + y_1)^2} \right.$$

$$\left. + \frac{y_0 + y_1}{(x_0 + x_1)^2 + (y_0 + y_1)^2} - \frac{y_0 - y_1}{(x_0 + x_1)^2 + (y_0 - y_1)^2} \right]$$

$$\equiv U(x_0, y_0, k_0; x_1, y_1, k_1)$$

$$\frac{dx_j}{dt} = \sum_{i-0}^{N-1}{}' \; U(x_j, y_j, k_j; x_i, y_i, k_i)$$

$$\frac{dy_j}{dt} = -\sum_{i-0}^{N-1}{}' \; U(y_j, x_j, k_j; y_i, x_i, k_i)$$

where $j = 0, i, 2, \ldots. N-1$

$$r_i^2 = x_i^2 + y_i^2$$

$\sum_{i=0}^{N-1}{}'$ indicates summation from $i = 0$ to $i = N-1$ except $i = j$. These simultaneous equations can be solved by the Runge-Kutta method, given the boundary condition at $t = 0$.

5.0 Comparisons of Two- and Three-Dimensional Results

It is understandable that the initial trajectory may be very close to a straight line parallel to the velocity of the vortex generator. In the beginning, the three-dimensional result may be considerably different from the two-dimensional one, because the two-dimensional problem is tantamount to that for the infinitely long vortex lines in the three-dimensional space, while the three-dimensional vortex line is just beginning. On the other hand, because the induced velocity components of a 3-D vortex decays as $1/(\text{distance})^2$, the trailing vortex lines not near the end points may be approximated by the two-dimensional lines.

Since vortex strength is incorporated in the boundary condition in the nondimensionalization process in the present results, the starting points given in the present examples as, z_0/k_1 and y_0/k_1, can be interpreted by a set of actual quantities which lead to the same values of z_0/k_1 and y_0/k_1 with the combination of values of z_0, y_0 and vortex strength or induced velocity which is proportional to the vortex strength. Although the representation is simple in the present approach, care is needed to compare two- and three-dimensional results because the same value of vortex strength does not necessarily give the same induced velocity in two- and three-dimensional spaces. For this reason, SARPKAYA [7, 8] nondimensionalized by the actual initial induced velocity component in the vertical direction, when comparing the two-dimensional theoretical results with his experimental results. If we do the same nondimensionalization, it is obvious that the present results in the two- and three-dimensional analyses will be very close.

A simple problem is chosen to test the accuracy of the program in Fig. 2, where two vortices with opposite vortex strength as shown in Fig. 2 advance with a nondimensional constant velocity 1 parallel to the x-axis at $z/k_1 = -3$, $y/k_1 = \pm 3$. The figure shows the vortex shape at time $t = 30\ k_1/u$. In Fig. 2, the slope of the trajectory for z/k_1 is equal to the z component of velocity and the slopes of the two- and three-dimensional curve are quite different. If we change the time scale to $t\ w_0/k_1$ in the two- and three-dimensional curves, the slopes will become 1 near $t = 0$ as in SARPKAYA'S results [7] and only the difference will be at the larger values of $t\ w_0/k_1$.

6.0 References

1. L.M. Milne-Thomson: Theoretical Hydrodynamics, Third Ed., The MacMillan Co., p. 521 (1953).
2. H. Lamb: Hydrodynamics, Dover Publication, pp. 224 (1945).
3. Greenhill: Quarterly Journal of Mathematics, Vol. XV, 10 (1878).
4. T.A. Loukakis: MIT, Dept. of Naval Architecture and Marine Engineering Report No. 71-7 (May 1971).
5. J.K. Harvey, and F.J. Perry: AIAA Journal, Vol. 9, No. 8, 1659-1660 (Aug 1971).
6. F.S. Dee, and O.P. Nicholas: CP 1065, British Aeronautical Research Council, London, England (Jan 1968).
7. T. Sarpkaya: Journal of Fluid Mechanics, Vol. 136, 85-109 (1983).
8. T. Sarpkaya, and D.O. Henderson, Jr.: AIAA-85-0445, AIAA 23rd Aerospace Science Meeting, Reno, Nevada (Jan 1985).

A Mathematical Model for Response of the Coronary Circulation to High Sustained Gravitational Force Fields

Sophia Huai Zhou, Emilia Mateeva, and Richard Collins

1. Introduction

The present generation of fighter aircraft is highly maneuverable, capable of pulling angular accelerations which can load the pilots in $+G_z$ caudalward levels rising to 15 G in less than one second. These levels of footward accelerative loading may be maintained for minutes, during which the systemic circulation drains rapidly toward the feet. Under these new non physiological conditions, the sympathetic nervous system attempts to compensate as mechanically-induced disturbances of the electrical excitation wave normally responsible for synchronous pumping contractions of the myocardium disorganize these contractions. Gross redistributions of the coronary blood volume may lead to myocardial ischemia and infarct. It is our purpose in this paper to formulate a new model for the **complete** coronary circulation which will provide improved insight into the response of the heart to such regimes of high sustained gravity.

The approach adopted here will be to model the unsteady quasi one-dimensional coronary blood flow in a branching network of distensible vessels subject to periodic collapse under the time- and space-dependent variations of the intramyocardial pressure (IMP) which loads the coronary vessels externally.

An excellent attempt in this direction was made by RUMBERGER & NEREM [10, 11], whose attention however was restricted to that portion of the coronary system associated mainly with the larger calibre coronary arteries, to the total exclusion of the drainage veins. Furthermore, the flow into the smaller collateral arterial side branches was assimilated into a "leakage" term appearing in the mass continuity equation for the main arteries from which the branches derive. These functions, including the all-important intramyocardial pressures external to the coronary vessels, were taken as exponential functions, the parameters of which were adjusted to produce computed solutions in approximate agreement with their experimental measurements on horses.

Further development is now required of a theoretical model which connects the crucial intramyocardial pressure field within the myocardium to the mechanics of the left ventricular contraction, particularly inasmuch as the latter is disorganized during the application of large accelerative force fields.

To this end, the imaginative approach of BEYER & SIDEMAN [3, 4] would appear to possess considerable potential in the modelling of both the passive and active control of the human coronary circulation.

In the following sections, a schematic representation is given of the anatomy of the **complete** coronary circulation and the best reported estimates to date of vessel wall elasticity. Next, a hybrid method-of-characteristics and Lax-Wendroff two-step finite-difference scheme is described for the computational solution of the coronary flow field. Only very preliminary results are available at this early stage of the program testing. Finally, the problems posed by the dearth of morphological and rheological data available lead the authors to propose the initiation of a series of careful experimental programs, for which it is suggested that the present computational model may provide a guiding framework for their planning and interpretation.

2. Formulation of the Model

The complete human coronary circulation is considered as a distensible branching network of piecewise straight vessels distributed within the wall (myocardium) of the heart. The network (fig. 1) extends from the main coronary artery, through the progressively narrowing and bifurcating medium and small arteries and arterioles, and into the capillary bed. Most models of the coronary circulation stop here. Our model goes on to include the venous vasculature, extending from the venules, small and medium veins into the large cardiac veins and coronary sinus, all of which empty mainly into the right atrium.

The coronary blood vessel are modelled as elastic segments, with modulus of elasticity varying markedly with vessel generation or distance along the branching network. Due to the relatively high intramyocardial pressures (IMP) developed within the heart wall during ventricular contraction, the coronary vessels are much stiffer than say the intracranial vessels (HAYASHI et al. [8]), thus rendering them more resistant to collapse. Notwithstanding this higher relative stiffness; however, the coronary vessels have been observed to collapse, or to spasm, under certain physiological conditions. Accordingly, the pressure-area relations which are used here to model the elastic properties of the

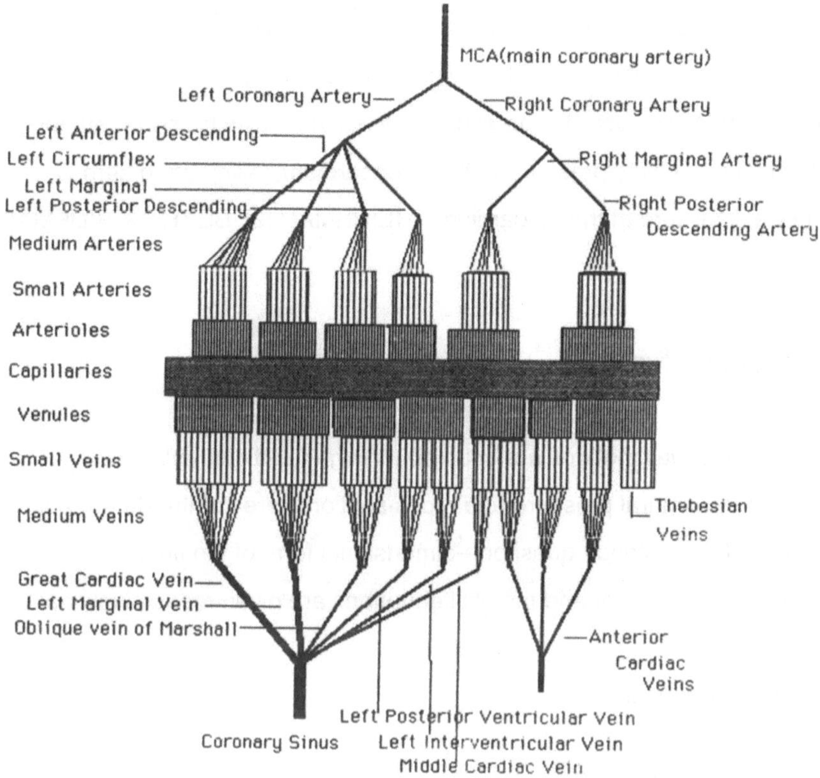

Fig. 1: Schematic representation of the coronary system.

vasculature also account for such states of partial or total collapse.

The pressure boundary conditions used here reflect the close proximity of: a) the main coronary artery to the aortic root at the inlet to the coronary circulation, and b) the fact that the coronary sinus and Thebesian and anterior cardiac veins empty into the right atrial cavity situated distally at the outlet of the coronary circulation. Similarly, the time variation of intramyocardial pressure is taken to be synchronous with the left ventricular cavity pressure at the endocardium, falling off to zero (relative) pressure at the epicardium (HEINEMAN [9], BEYER & SIDEMAN [3, 4]).

3. Governing Equations

The fluid and wall motions are coupled through functional relations between transmural pressure and cross-sectional area of the deformable vessels. The unsteady quasi one-dimensional form of the fluid continuity, momentum and the wall equations are given respectively as:

Governing Equations

$$\frac{\partial A}{\partial t} + \frac{\partial}{\partial x}(AV) = 0$$

$$\frac{\partial V}{\partial t} + \frac{\partial}{\partial x}\left(\frac{V^2}{2} + \frac{P}{\rho}\right) = - F + G_z$$

$$P = P(A) \quad \dots \text{ transmural pressure}$$

with

$$F = \frac{8\pi}{A}\frac{\mu}{\rho}V, \quad G_z = \frac{d\phi}{dx} .$$

Wall Relations

For $p \geq 0$, $A/A_0 \geq 1$

$$p = \frac{1}{\alpha}\left(A - A_0 \right) \quad \dots \ \alpha = \text{compliance.}$$

For $p \leq 0$, $0.8 < A/A_0 < 1$

$$p = K_p\left[\alpha_0 + \alpha_1\left(\frac{A}{A_0}\right) + a_2\left(\frac{A}{A_0}\right)^2\right]$$

with $\alpha_0 = 20.9625$, $\alpha_1 = -55.9$,

$\alpha_2 = 34.9375$.

For $p \leq 0$, $A/A_0 < 0.8$

$$p = -K_p\left(\frac{A}{A_0}\right)^{-\frac{3}{2}} \quad \ldots \quad \text{similarity law}$$

Their relationships are represented graphically in Fig. 2 below.

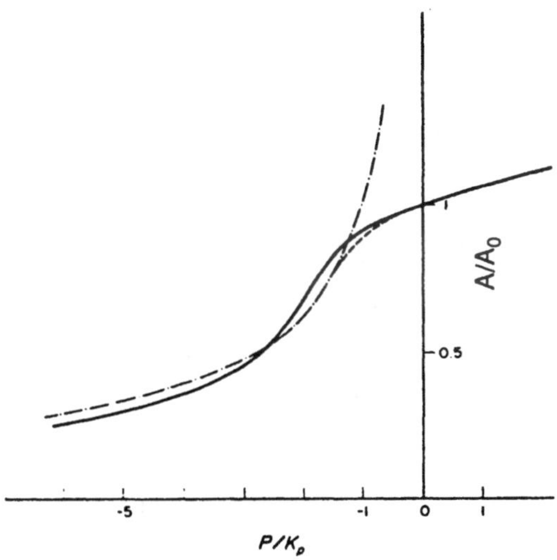

Fig. 2: Pressure-area relation for coronary vessels (with permission).
—— Experiments (Shapiro 1977).
— · — · Similarity law $(P/K_p)^{-2/3}$.
---- Third-order polynomial bridge.

Here F is a laminar friction factor based for simplicity on fully developed Poiseuille flow, while G_z denotes a conservative force field, expressible as the gradient of a scalar potential ϕ . For negative values of the transmural pressure, corresponding to time intervals during which the intramyocardial pressures (IMP) exceed the intraluminal (blood) pressures in certain vessels, provision has been made for partial collapse ($p \leq 0$, $A/A_0 < 1$). These expressions are based on experimental results of SHAPIRO [12] which are faired smoothly into a $P^{-2/3}$ self-similarity law for P highly negative through the use of a third-order bridging polynomial (Fig. 2).

4. Boundary Conditions

As mentioned above, boundary conditions on pressure are imposed at the upstream (proximal) and downstream (distal) extremities of the complete coronary circulation:

a) *Proximal coronary pressure:* the right and left coronary arteries (Fig. 1) arise in the depths of the right and left pouches of the aortic valve from a bulge (aortic sinus) in the aortic wall. Accordingly, it is reasonable to take the proximal coronary artery pressure to be equal the aortic root pressure (Fig. 3).

b) *Distal coronary pressure:* the coronary sinus (Fig. 1) drains into the right atrium as do the smaller veins. Accordingly, the common distal boundary condition is taken to be the right atrial pressure (Fig. 3).

c) *Intramyocardial pressure (IMP):* influences blood flow through the coronary vessels by external myocardial compression during systole (contraction) and by pulsatile variations. IMP has been shown by BAIRD et al. [2] and HEINEMAN et al. [9] to be synchronous with left ventricular pressure. Its variation through the depth of the heart wall has been computed by BEYER et al. [3] using the Laplace law (for membranes) applied pragmatically to a thick-walled prolate spheroid. We have combined these two findings into an approximate characterization of the IMP as a continuous function of time and depth within the myocardial wall (Fig. 4).

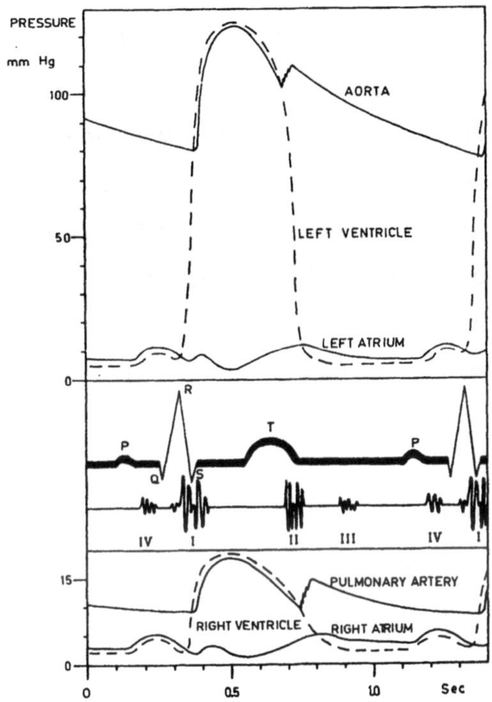

Fig. 3: Timing of cardiac events with respect to aortic and cardiac chamber and the corresponding electrocardiogram (from BURTON [5]) with permission).

5. Morphology and Mechanical Properties of the Coronary Vasculature

These data are in large part lacking in the published literature, and most particularly for the veins. The most detailed measurements of vessel lengths and diameters have been reported by ZAMIR et al. [17]. They were obtained from arterial casts of many autopsied human hearts . Unfortunately, these measurements were not carried out for the coronary veins. Nor are the arterial measurements related to vessel depths within the myocardium; nor for that matter, do they refer to the anatomical names of the vessel measured. Many measurements are reported by other authors for the coronary vasculature of various animals such as dogs and horses by GREGG et al. [7], SPAAN [13, 14], RUMBERGER et al. [10, 11], ARTS et al. [1], TOMOIKE et al. [15], VAN DER MEER et al. [16] and HEINEMAN et al. [9]. Clearly, this paucity of anatomical measurements of the *human* coronary vasculature places a heavy burden on our efforts to model that circulation mathematically. Indeed, the situation concerning material properties of the vessel walls, such as the modulus of elasticity or compliance, is even dimmer! These serious shortcomings will have to be made up with additional experimental programs in the future. For the time being, our best gathered data are set out in Tables I and II.

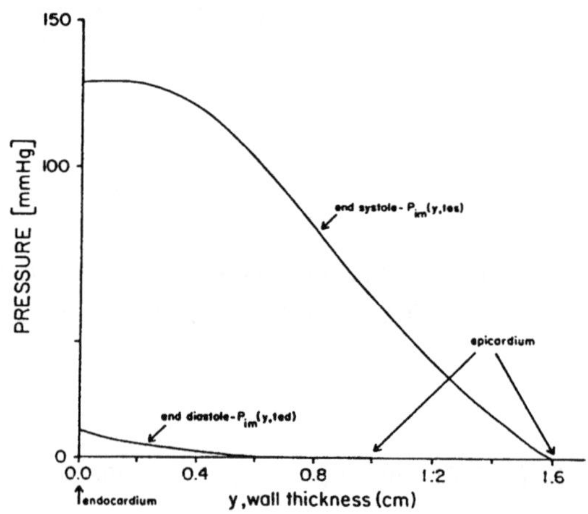

Fig. 4 a): Characterization of the intramyocardial pressure based on results of BEYER & SIDEMAN [4] (with permission).

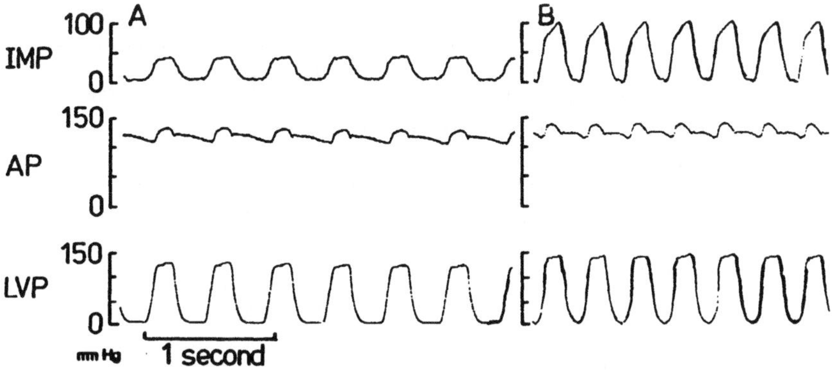

Fig. 4 b): Characterization of the intramyocardial pressure based on results of HEINEMAN [9] .

Table I: Anatomical measurements of the coronary vessels.

Generation	No. of Vessels	D (cm)	R (cm)	Area (cm^2)	Total Area (cm^2)	Length (cm)	Source
1 (MCA)	1	.44	.22	.152	.152	2.2	D,L --- Marcus
2 (L.+R.C.A.)	2	.4(L) .37(R)	.2	.126	.252	2.2	D,L --- Marcus
3 Large Arteries)	6	.3	.15	.071	.424	2.7	D ---- Marcus L -- Zamir
4 (Medium Arteries)	180	.11	.055	.009	1.70	2.5	D,L -- Zamir
5 (Small Arteries)	3×10^3	.044	.022	1.52×10^{-3}	4.56	2.4	D -- Zamir
6 (Arterioles)	4.1×10^6	.002	.001	$.31\times10^{-5}$	12.87	.6	
7 (Capillary)	8.9×10^7	.0006	.0003	$.28\times10^{-6}$	25.11	.18	
8 (Venules)	3.5×10^6	.022	.011	$.7\times10^{-5}$	13.30	.9	
9 (Small Veins)	4×10^3	.042	.021	.501	5.54	2.0	
10 (Medium Veins)	180	.15	.075	.018	3.19	3.0	A -- Hood
11 (Large Veins)	9	.4	.2	.126	1.131	4.0	A -- Hood
12 (CS+ACV)	2	.7 .25	.35 .125	.385 .049	.434	3.32	D,L -- Esperanee (dog)

Table II: Mechanical properties of the coronary vessels.

Generation	No. of Vessels	D (cm)	Total Volume (cm ^{3})	Elastic Modulus $\times 10^6$ dye/cm ^{2}	Compliance $\times 10^{-6}$ cm ^{4}/dye	Source
1 (MCA)	1	.44	.33	5.8	.524	
2 (L.+R.C.A.)	2	.4(L) .37(R)	.55	6.2	.406	α -- Gow
3 Large Arteries)	6	.3	1.15	6.6	.172	α-- Gow
4 (Medium Arteries)	180	.11	4.24	7.0	.013	
5 (Small Arteries)	3×10^3	.044	6.39	7.2	$.17\times10^{-4}$	
6 (Arterioles)	4.1×10^6	.002	7.73	7.4	$.17\times10^{-5}$	
7 (Capillary)	8.9×10^7	.0006	4.53	8	$.2\times10^{-6}$	
8 (Venules)	3.5×10^6	.022	11.97	6	$.7\times10^5$	
9 (Small Veins)	4×10^3	.042	11.08	4	$.348\times10^{-3}$	
10 (Medium Veins)	180	.15	9.54	3	.0826	
11 (Large Veins)	9	.4	4.52	2	1.512	
12 (CS+ACV)	2	.7 .25	1.44	1.2	12.833	

6. Numerical Solution

A hybrid method of characteristics combined with a two-step Lax-Wendroff finite difference scheme has been used to integrate the system of governing fluid/wall equations with their boundary conditions in the x-t plane. The partial differential equations must be written in conservative form. Past experience shows that the yet unknown initial conditions can be guessed roughly; these will converge to reproducible values after just a few computed cardiac cycles of periodic flow.

Three types of points are calculated in the (x,t) plane:

a) *Internal points:* The governing equations are expressed in finite difference form on a staggered two-step grid, where the superscripts n denote time-step, and i distance increment.

Half-step in t; i = 1, 2, ... , M -1.

$$A_{i+\frac{1}{2}}^{n+\frac{1}{2}} = \frac{1}{2}\left(A_i^n + A_{i+1}^n\right) - \frac{\Delta t}{2\Delta x}\left(A_{i+1}^n V_{i+1}^n - A_i^n V_i^n\right)$$

$$V_{i+\frac{1}{2}}^{n+\frac{1}{2}} = \frac{1}{2}\left(V_i^n + V_{i+1}^n\right) - \frac{\Delta t}{2\Delta x}\left[\frac{\left(V_{i+1}^n\right)^2}{2} + \frac{P_{i+1}^n}{\rho} + \phi_{i+1}^n\right] + \frac{\Delta t}{2\Delta x}\left[\frac{\left(V_i^n\right)^2}{2} + \frac{P_i^n}{\rho} + \phi_i^n\right] - \frac{\Delta t}{2} F_{i+\frac{1}{2}}^n$$

$$P_{i+\frac{1}{2}}^{n+\frac{1}{2}} = P\left(A_{i+\frac{1}{2}}^{n+\frac{1}{2}}\right)$$

where $F_{i+\frac{1}{2}}^n = \frac{8\pi\mu}{\rho}\frac{1}{2}\left(\frac{V_{i+1}^n}{A_{i+1}^n} + \frac{V_i^n}{A_i^n}\right)$

Full step in t; i = 1, 2, ..., M-1

$$A_i^{n+1} = A_i^n - \frac{\Delta t}{\Delta x}\left(A_{i+\frac{1}{2}}^{n+\frac{1}{2}} V_{i+\frac{1}{2}}^{n+\frac{1}{2}} - A_{i-\frac{1}{2}}^{n+\frac{1}{2}} V_{i-\frac{1}{2}}^{n+\frac{1}{2}}\right)$$

$$V_i^{n+1} = V_i^n - \frac{\Delta t}{\Delta x}\left[\frac{\left(V_{i+\frac{1}{2}}^{n+\frac{1}{2}}\right)^2}{2} + \frac{P_{i+\frac{1}{2}}^{n+\frac{1}{2}}}{\rho} + \phi_{i+\frac{1}{2}}^{n+\frac{1}{2}} - \frac{\left(V_{i-\frac{1}{2}}^{n+\frac{1}{2}}\right)^2}{2} - \frac{P_{i-\frac{1}{2}}^{n+\frac{1}{2}}}{\rho} - \phi_{i-\frac{1}{2}}^{n+\frac{1}{2}}\right] - \Delta t\, F_i^{n+\frac{1}{2}}$$

$$P_i^{n+1} = P(A_i^{n+1})$$

$$\text{where } F_i^{n+1/2} = \frac{8\pi\mu}{\rho}\frac{1}{2}\left(\frac{V_{i+1/2}^{n+1/2}}{A_{i+1/2}^{n+1/2}} + \frac{V_{i-1/2}^{n+1/2}}{A_{i-1/2}^{n+1/2}}\right)$$

b) *Branch points:* The characteristic relations for second-order hyperbolic systems of partial differential equations are employed to relate flow values on either side of a branch point, ensuring continuity of flow and pressure at the bifurcation. The latter may be easily generalized to allow for pressure drops across bifurcations, once the angles of bifurcation are known from more detailed anatomical data. The bifurcations considered here are not necessarily symmetric, either in branching angle or in the diameters and lengths of the two daughter vessels emanating from each parent vessel. The finite-difference algorithms are outlined schematically below:

Branch Points

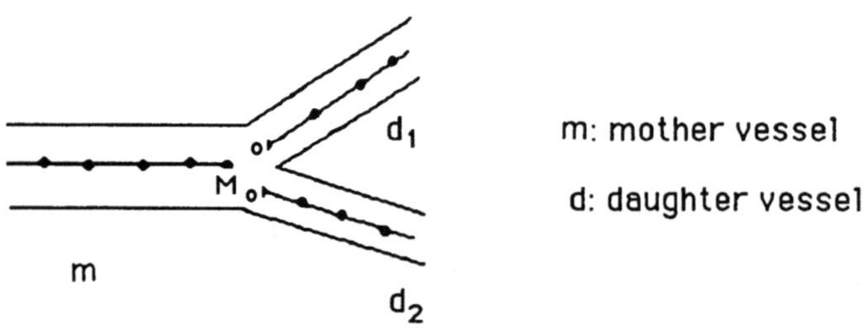

Mass Continuity:

$$(AV)_m = (AV)_{d_1} + (AV)_{d_2}$$

Pressure Continuity:

$$P_m = P_{d_1} = P_{d_2}$$

Characteristic equations: $A\,dV \pm c\,dA = +Af\,dt$

along the characteristic directions

$$\frac{dx}{dt} = V \pm c$$

where $c^2 = \frac{A}{\rho}\frac{\partial P}{\partial A}$

and $f = -\frac{8\pi\mu}{\rho}\frac{V}{A} + \frac{d\phi}{dx}$.

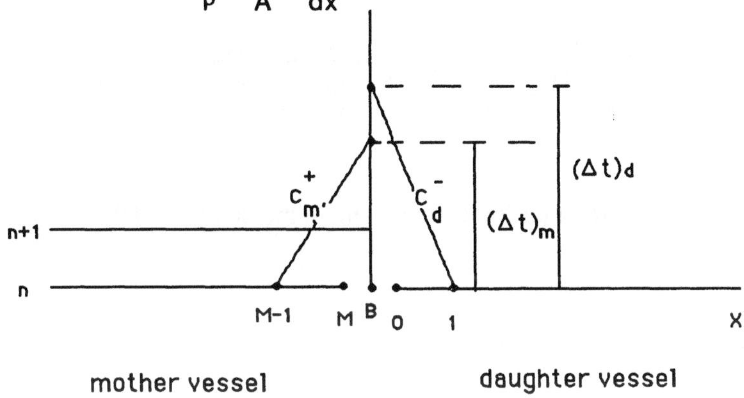

(1) Extending the characteristics:

c_m^+ from point M - 1 (mother)

$c_{d_1}^-$ from point 1 (daughter 1)

$c_{d_2}^-$ from point 1 (daughter 2)

We establish the following relations:

Along the c_m^+ characteristic: $\frac{dx}{dt} = V + c$

$$x_B - x_{M-1} = \left[\left(V_{M-1}^n \right)_m + \left(c_{M-1}^n \right)_m \right] (\Delta t)_m$$

Along the c_d^- characteristic: $\frac{dx}{dt} = V - c$

$$x_0 - (x_1)_{d_1} = \left[\left(V_1^n\right)_{d_1} - \left(c_1^n\right)_{d_1} \right] (\Delta t)_{d_1}$$

$$x_0 - (x_1)_{d_2} = \left[\left(V_1^n\right)_{d_2} - \left(c_1^n\right)_{d_2} \right] (\Delta t)_{d_2}$$

$$(\Delta t)_m = \frac{x_B - x_{M-1}}{\left(V_{M-1}^n\right)_m + \left(c_{M-1}^n\right)_m} \; ; \; (\Delta t)_{d_1} = \frac{x_0 - (x_1)_{d_1}}{\left(V_1^n\right)_{d_1} - \left(c_1^n\right)_{d_1}} \; ; \; (\Delta t)_{d_2} = \frac{x_0 - (x_1)_{d_2}}{\left(V_1^n\right)_{d_2} - \left(c_1^n\right)_{d_2}}$$

Accordingly, we adopt for the time increment the smallest of the above Δt as:

$$\Delta T = \min \left\{ (\Delta t)_m , (\Delta t)_{d_1} , (\Delta t)_{d_2} \right\}.$$

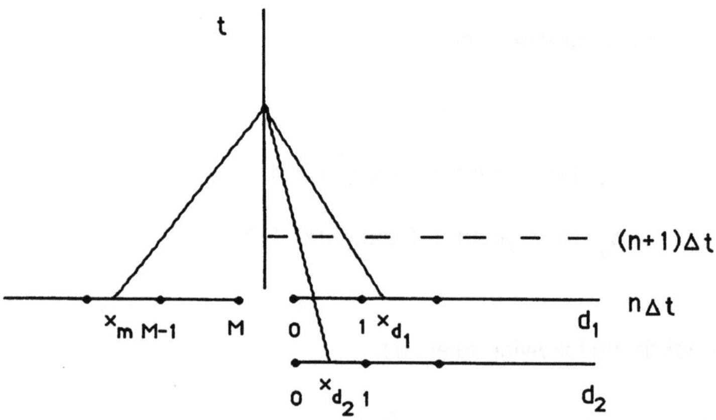

(2) Extending the characteristic directions for ΔT, we determine x_m, x_{d_1} and x_{d_2} as :

$$x_m = x_M - \left[(V^n_{M-1})_m + (c^n_{M-1})_m \right] \Delta T$$

$$x_{d_1} = (x_0)_{d_1} - \left[(V^n_1)_{d_1} - (c^n_1)_{d_1} \right] \Delta T$$

$$x_{d_2} = (x_0)_{d_2} - \left[(V^n_1)_{d_2} - (c^n_1)_{d_2} \right] \Delta T .$$

(3) After determining the intervals (Δx) at points x_m , x_{d_1} and x_{d_2} , we calculate the values of the functions $A(x, n\,\Delta t)V$ and $P(x, n\,\Delta t)$ at x_m , x_{d_1} , x_{d_2} by linear interpolation.

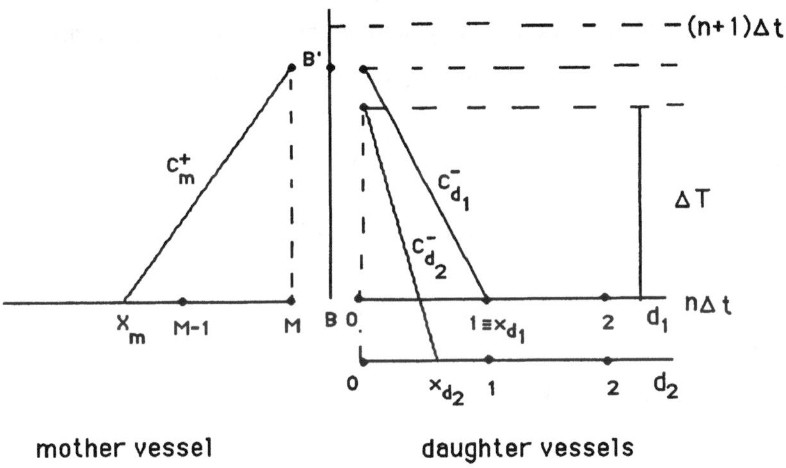

(4) We wish to compute the values of A, V and P at the point B' from the characteristic equations.

$$A\,dV \pm c\,dA = A\,f\,dt$$

$$c^2 = \frac{A}{\rho}\frac{\partial P}{\partial A}$$

$$f = -\frac{8\pi\mu}{\rho}\frac{V}{A} + \frac{d\phi}{dx}\,; \quad \frac{d\phi}{dx} = G_z \cos\theta\,,$$

or

$$A\,dv \pm c\,dA = \Gamma_1 V + \Gamma_2 A$$

where $\Gamma_1 = -\dfrac{8\,\pi\,\mu}{\rho}\,dt$, $\Gamma_2 = G_z \cos\theta\,dt.$

Using the abbreviated notation:

$$A(x_m, n\,\Delta t) = A_m \qquad A(x_m, n\,\Delta t + \Delta T) = A_m(B') = A_m'.$$

$$A(x_{d_1}, n\,\Delta t) = A_{d_1} \qquad A\left((x_0)_{d_1}, n\,\Delta t + \Delta T\right) = A_{d_1}(B') = A_{d_1}'$$

$$A(x_{d_2}, n\,\Delta t) = A_{d_2} \qquad A\left((x_0)_{d_2}, n\,\Delta t + \Delta T\right) = A_{d_2}(B') = A_{d_2}'.$$

we obtain the solution at each (non-symmetric) branch point in the form of a system of six nonlinear simultaneous algebraic equations:

$$A_m' (V_m' - V_m) + c_m (A_m' - A_m) = \Gamma_1 V_m' + \Gamma_2 A_m'$$

$$A_{d_1}' (V_{d_1}' - V_{d_1}) - c_{d_1} (A_{d_1}' - A_{d_1}) = \Gamma_1 V_{d_1}' + \Gamma_2 A_{d_1}'$$

$$A_{d_2}' (V_{d_2}' - V_{d_2}) - c_{d_2} (A_{d_2}' - A_{d_1}) = \Gamma_1 V_{d_2}' + \Gamma_2 A_{d_2}'$$

$$A_m' V_m' = A_{d_1}' V_{d_1}' + A_{d_1}' V_{d_2}'$$

$$P(A_m') = P(A_{d_1}')$$

$$P(A_m') = P(A_{d_2}').$$

for the cross-section A', velocity V' and transmural pressures P' across the bifurcation. The system is solved numerically by iteration.

7. Results and Discussion

Preliminary results of the numerical computations confirm that the flows developed in the human coronary circulation are highly sensitive to:

a) *vascular architecture*; that is, vessel diameters, lengths and branching angles, the latter reflected in corresponding pressure drops across the bifurcations, and

b) *vessel compliance* (or alternatively, *modulus of elasticity*) which can itself vary several-fold from one end to the other of the coronary vasculature. In addition, vessel wall compliance varies markedly with transmural pressure, most particularly for negative transmural pressures which produce partial collapse of the vessel walls. The coronary vessels are especially susceptible to such states of collapse during ventricular systole. Systolic contraction of the myocardial muscle fibres leads to rapid and substantial increases in intramyocardial pressure (IMP). The accompanying external compression of the coronary blood vessels can reduce the flow to zero or nearly zero in some vessels, inducing retrograde flows (CHILIAN et al. [6]) in the segments proximal to the periodic close-down. The IMP also varies with depth through the myocardial wall.

In general, such large changes in effective modulus of elasticity alter the speed of propagation of pressure waves in the blood-filled branching network, with concomitant alterations in the instantaneous calibres of the vessels. Accordingly, both the blood flow distributions and the perfusion pressures can change significantly.

It is our present dilemma that accurate data for the compliance or elasticity of the human coronary vessels, particularly the veins, are virtually non existent! Furthermore, *in vivo* measurements of IMP in humans as a function of myocardial wall depth and time are largely lacking.

Nevertheless, the present mathematical model has been formulated with sufficient generality to permit incorporating improved morphological and material property data if and when they become available. In the interim, the model serves as a framework for planning of the requisite experimental protocols for supplying these lacking data and for establishing, through computational parameter sensitivity studies, the acceptable limits of accuracy and reproducibilty which the experimental measurements must meet.

8. Conclusions

This work represents the first attempt known to the authors to formulate a nonlinear mathematical model of the *complete* human coronary circulation which takes into account the vascular morphology and material properties of the blood vessels. The full input data required to implement this model are presently lacking in the international literature. It is intended that the present model provide a general theoretical framework for organizing new experiments and analyzing the results there from, with a view to shedding new insight into the passive and active control of the human coronary circulation.

The present study was motivated by a particular need to evaluate cardiovascular response under conditions of high sustained gravitational force fields. It is well known that, in such environments, the sympathetic central nervous system modifies significantly the response of the heart in a well-intentioned effort to compensate for such unnatural alterations of a human's normal gravitational environment. However, the results can be disastrous, leading to cardiac arrhythmia, coronary insufficiency, endocardial hemorrhage and indeed, to myocardial fibrillation. The present work represents the first step in a phased medium-term research program directed toward the development of an improved understanding of *aerospace and space physiology* of the human cardiovascular system.

Acknowledgement

This work has been supported in part by an NSERC Operating Grant, and in part by DSS Contract W7711-7-7029 resulting from acceptance of a proposal submitted under the UPP. The technical assistance of Dr. Deming ZHUANG in the technical preparation of the mathematical equations of the final typescript is gratefully acknowledged.

References

1. T. Arts, R.S. Reneman: Interaction between Intramyocardial Pressure (IMP) and Myocardial Circulation. J. Biomech. Engrg. 107, 51-54 (1985).
2. R. J. Baird, R. T. Manktelow, P. A. Shah, F. M. Ameli: Intramyocardial pressure - A study of its regional variations and its relationship to intraventricular pressure. J. Thorac. & Cardiovasc. Surg. 59, 810-823 (1970).
3. R. Beyar, S. Sideman: A computer study of the left ventricular performance based on fiber structure, sarcomere dynamics, and transmural electrical propagation velocity. Circ. Res. 55 , 358-375 (1984).
4. R. Beyar, S. Sideman: Time-dependent coronary blood flow distribution in left ventricular wall. Am. J. Physiol. 252, (Heart Circ. Physiol. 21), H417-H433 (1987).
5. A. C. Burton: Physiology and Biophysics of the Circulation, p.137, fig.69 (Year Book Medical Publishers, Inc., Chicago 1965).
6. W. M. Chilian, M. L. Marcus: Phasic Coronary Blood Flow Velocity in Intramural and Epicardial Coronary Arteries. Circ. Res. 50, 775-781 (1982).
7. D. E. Gregg, R. E. Shipley: Studies of the venous drainage of the heart.

Am. J. Physiol. 151, 13-25 (1947).

8. K. Hayashi, Y. Igarashi, K. Takamizawa : Mechanical Properties and Hemodynamics in Coronary Arteries. In New Approaches in Cardiac Mechanics, ed. K. Kitamura et al. , pp. 285-294 (Japan Scientific Press and Breach Scientific Publishers, Tokyo 1986).

9. F. W. Heineman, J. Grayson : Transmural distribution of intramyocardial pressure measured by micropipette technique. Am. J. Physiol. 249 (Heart Circ. Physiol. 18) : H1216-1223 (1985)

10. J. A. Rumberger, R. M. Nerem: A method-of-characteristics calculation of coronary blood flow. J. Fluid Mech. 82, 429-448 (1977).

11. J. A. Rumberger, R. M. Nerem, W. W. Muir: Coronary artery pressure development and wave transmission characteristics in the horse. Cardiovasc. Res. 13, 413-419 (1979).

12. A. H. Shapiro: Steady flow in collapsible tubes. J. Biomech. Engrg. 99, 126-147 (1977). (See Fig. 3 in Collins et al. : Blood flow in the lung. J. Biomech. 12, 373-395 (1979).

13. J. A. E. Spaan: Coronary Diastolic Pressure-Flow Relation and Zero Flow Pressure Explained on the Basis of Intramyocardial Compliance. Circ. Res. 56, 293-309 (1985).

14. J. A. E. Spaan , N. P. W. Breuls, J.D. Laird: Diastolic-Systolic Coronary Flow Differences are Caused by Intramyocardial Pump Action in the Anesthetized Dog. Circ. Res. 49, 584-593 (1981).

15. H. Tomoike, H. Ootsubo, K. Sakai, Y. Kikuchi, M. Nakamura: Continuous measurement of coronary artery diameter in situ. Am. J. Physiol. 24, (Heart Circ. Physiol. 9): H73-H79 (1981).

16. J. J. van der Meer, R. S. Renemann, H. Schneider, J. Wieberdink : A technique for estimation of intramyocardial pressure in acute and chronic experiments . Circ. Res. 4, 132-140 (1970).

17. M. Zamir, H. Chee: Segment Analysis of Human Coronary Arteries. Blood Vessels 24, 76-84 (1987).

A Class of Discontinuities Caused by 'Delta Functions'

You-lan Zhu

1. Introduction

In mechanics, there are a number of discontinuties, such as shocks, contact discontinuities, detonations, deflagrations, interfaces of phase-change etc. Some of them appear only due to the nonlinearity of problems and others, for example, interfaces of phase-change, are caused mainly by 'δ-functions.' In fact, the two-phase Stefan problem can be described as follows:

$$\frac{\partial}{\partial t}b(T) = \frac{\partial}{\partial x}(a(T)\frac{\partial T}{\partial x}),$$

where $a(T)$ and $b(T)$ are heat-conduction coefficient and internal energy. Clearly, $a(T)$ and $b(T)$ are discontinuous at the melting temperature T^*. The discontinuity of $a(T)$ and $b(T)$ causes appearance of a type of 'special' weak-discontinuity—the interface of phase-change. Since $\partial\, b(T)/\partial\, t$ and $\partial\, a\,(T)/\partial\, x$ are δ-functions at $T = T^*$, we say that the interfaces of phase-change are caused by δ-functions. One of the models of combustion problem can be expressed by the following system:

$$\frac{\partial}{\partial t}G(U) + \frac{\partial}{\partial x}F(U) = 0,$$

where $U, G(U), F(U)$ are n-dimensional vectors. One component of $G(U)$, which is related to internal energy, is discontinuous when the unburned gas changes into the burned gas. This means that $G(U)$ is discontinuous on a line in the 'phase-space.' Clearly, the corresponding component of $\partial\, G(U)/\partial\, t$ is a δ-function on the line. Therefore, detonation and deflagration can also be understood as discontinuities caused by δ-functions. In this paper, we would like to discuss a new class of discontinuities caused by δ-functions.

2. Results

We consider the problem of one dimensional flow in a tube with variable cross-section areas. This problem can be described by

$$(\rho\, A)_t + (\rho\, u\, A)_x = 0, \tag{1}$$

$$(\rho\, u\, A)_t + (\rho\, u^2 A + pA)_x = pA_x, \tag{2}$$

$$(EA)_t + ((E + p)uA)_x = 0, \tag{3}$$

where

$$E \equiv \frac{1}{2}\rho u^2 + \frac{p}{\gamma - 1}$$

stands for the total energy, ρ for the density, u for the gas velocity, p for the pressure, γ for the adiabatic constant and $A(x)$ for the area of cross-section at point x. If the area of cross-section changes suddenly at some points (see Fig. 1), $dA(x)/dx$ has the characters of δ-functions at those points. The problem is what kind of singularity is caused by δ-functions in this case. If $A(x)$ is discontinuous at $x = x^*$ and if there is

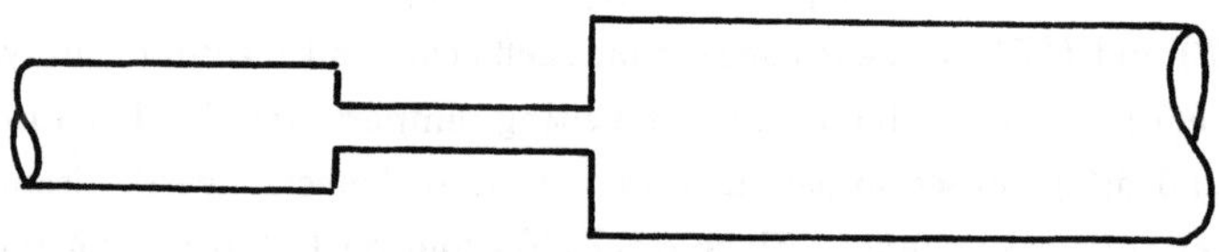

Figure 1

a strong discontinuity, then the physical quantities on the two sides should satisfy the following jump condition at $x = x^*$:

$$\rho_1 u_1 A_1 = \rho_r u_r A_r, \tag{4}$$

$$\rho_1 u_1^2 A_1 + (A_1 + A_r)p_1/2 = \rho_r u_r^2 A_r + (A_1 + A_r)p_r/2, \tag{5}$$

$$(E_1 + p_1)u_1 A_1 = (E_r + p_r)u_r A_r, \tag{6}$$

where the quantities with the subscripts 1 and r represent the physical quantities on the left and the right sides of discontinuity respectively. Equations (4)-(6) can be rewritten as

$$\frac{\rho_r}{\rho_1}\frac{u_r}{u_1} = \frac{A_1}{A_r} \tag{7}$$

$$\frac{u_r}{u_1} + \frac{p_r}{p_1}\frac{(1+\frac{A_r}{A_1})}{2\gamma M_1^2} = 1 + \frac{\frac{A_r}{A_1}}{2\gamma M_1^2}, \tag{8}$$

$$\frac{1}{2}\left(\frac{u_r}{u_1}\right)^2 + \frac{p_r}{p_1}\frac{u_r}{u_1}\frac{\frac{A_r}{A_1}}{(\gamma-1)M_1^2} = \frac{1}{2} + \frac{1}{(\gamma-1)M_1^2}, \tag{9}$$

where $M_1^2 = u_1^2\rho_1/\gamma p_1$. Therefore $\rho_r/\rho_1, u_r/u_1$ and p_r/p_1 depend only on A_r/A_1 and M_1. Let $U = u_r/u_1$, $P = p_r/p_1$, $RA = A_r/A_1$, $H = 1 + RA/(2\gamma M_1^2)$, $D = 1/[(\gamma-1)M_1^2]$, and $F = RA/[(\gamma-1)M_1^2]$, (8) and (9) can further be changed into the form

$$U + HP = 1 + H, \tag{10}$$

$$\frac{1}{2}U^2 + PUF = \frac{1}{2} + D. \tag{11}$$

Therefore P satisfies

$$\frac{1}{2}(1+H-HP)^2 + P(1+H-HP)F = \frac{1}{2} + D$$

or

$$AP^2 + BP + C = 0, \tag{12}$$

where

$$\begin{cases} A = \frac{1}{2}H^2 - HF, \\ B = (1+H)(F-H), \\ C = \frac{1}{2}(1+H)^2 - \frac{1}{2} - D. \end{cases} \tag{13}$$

From (12), (10), (7), we easily obtain the following expressions:

$$P = \frac{-B \pm \sqrt{B^2 - 4AC}}{2A}, \tag{14}$$

$$U = 1 + H - HP, \tag{15}$$

$$R \equiv \frac{\rho_r}{\rho_1} = \frac{1}{RAU}, \tag{16}$$

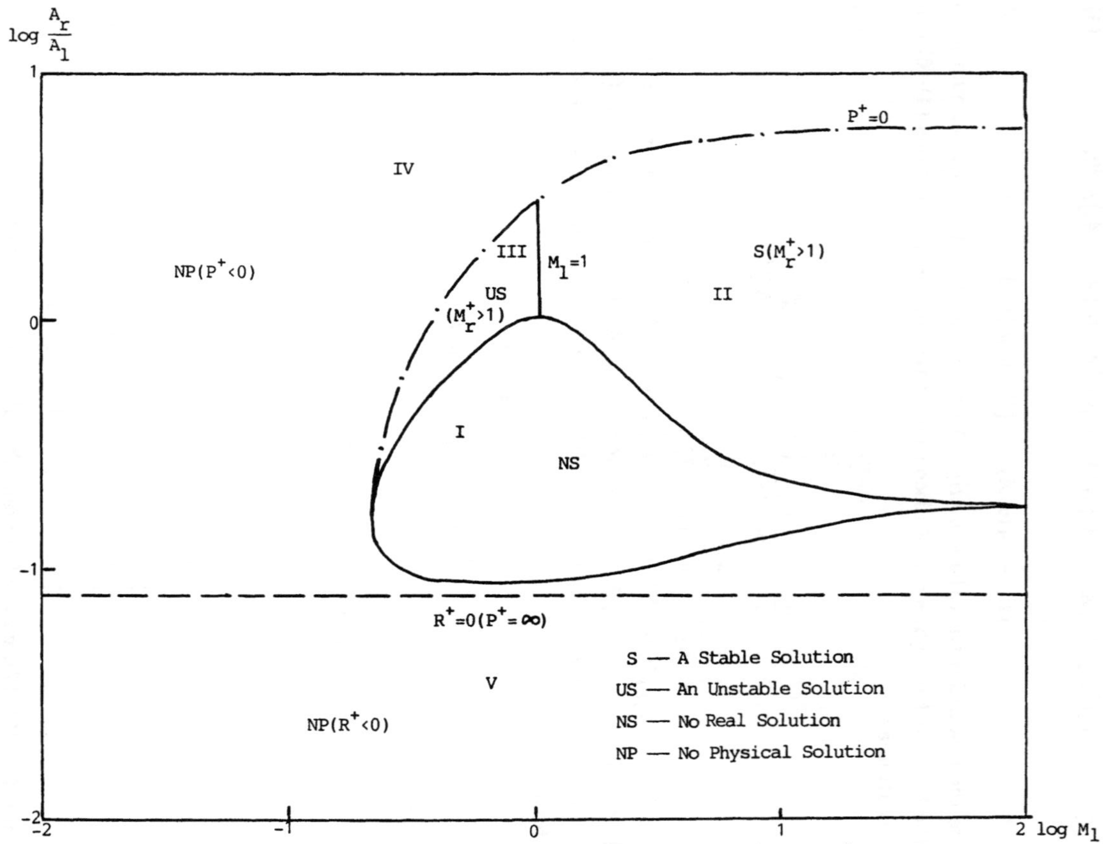

Figure 2 Structure of the Class of 'Plus' Solutions ($\gamma = 1.4$)

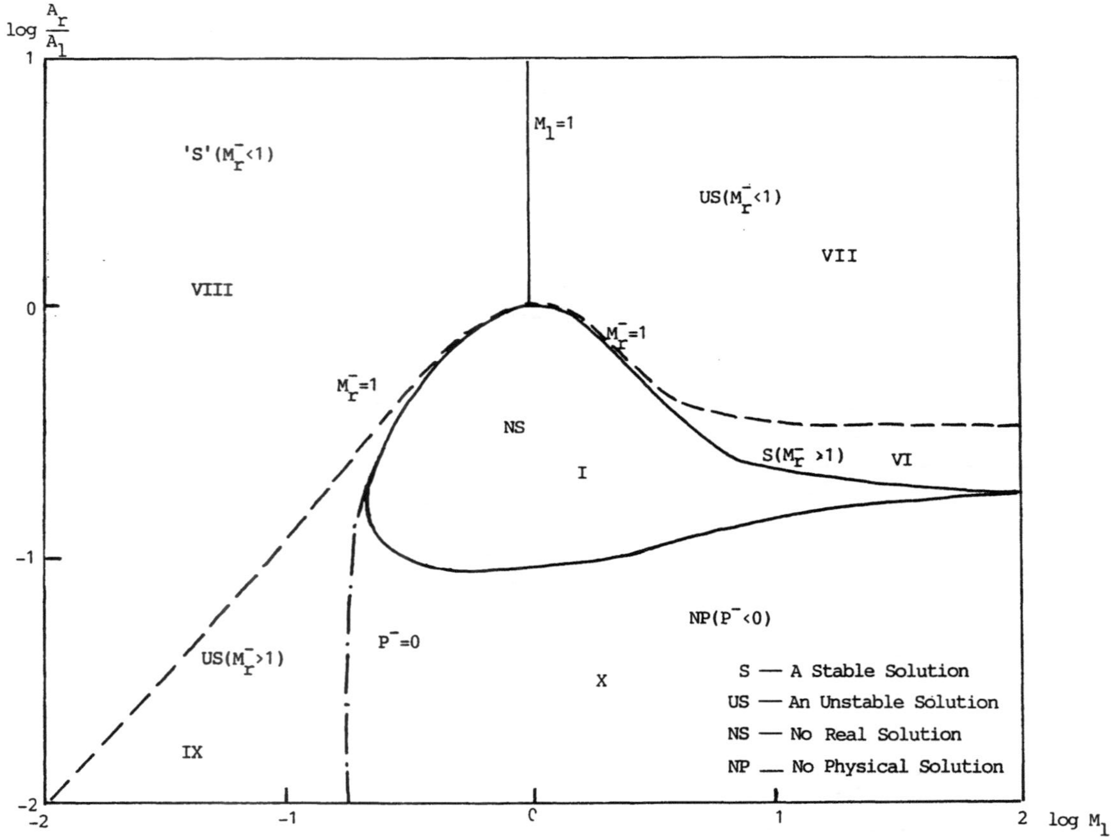

Figure 3 Structure of the Class of 'Minus Solutions' ($\gamma = 1.4$)

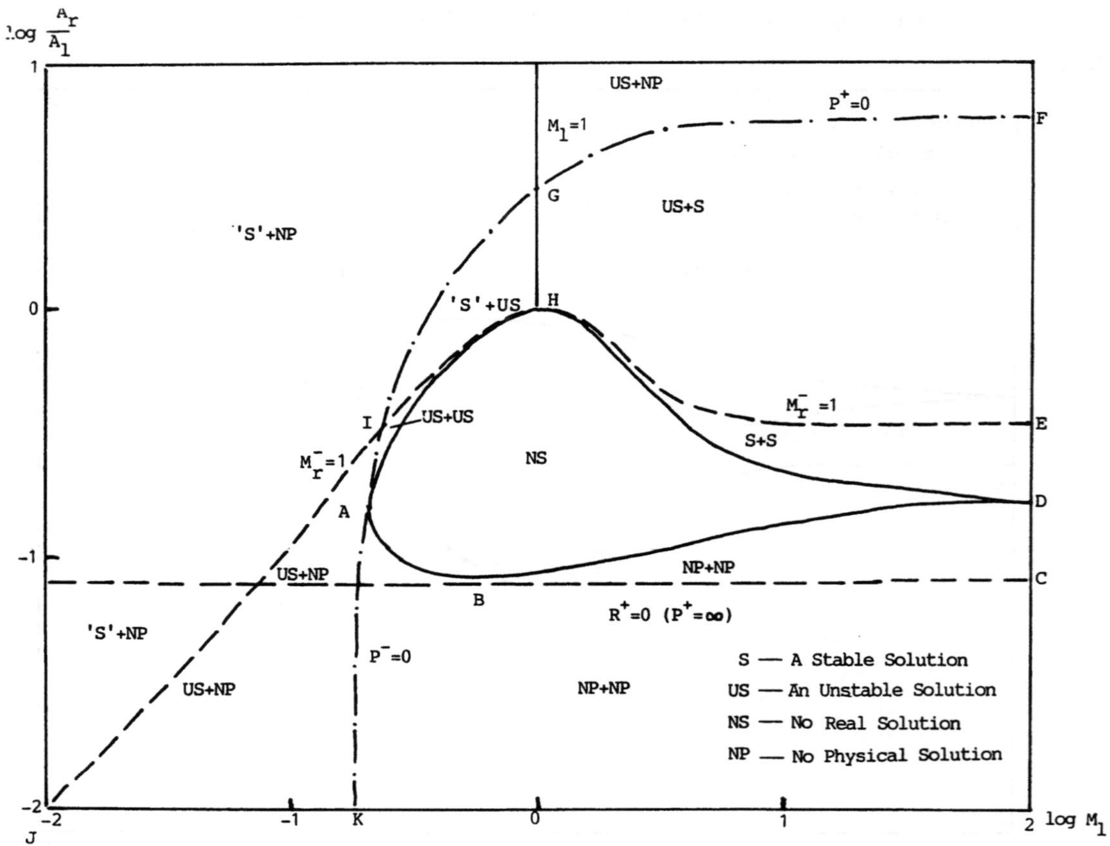

Figure 4 Structure of the Class of Solutions (γ = 1.4)

Coordinates of the Special Points in Figure 4

$$A:\quad \frac{A_r}{A_1} = \frac{\gamma-1}{\gamma+1},\quad M_1 = \frac{1}{2}\sqrt{\frac{\gamma-1}{\gamma+1}}$$

$$B:\quad \frac{A_r}{A_1} = \frac{\gamma+1-2\sqrt{\gamma}}{\gamma-1},\quad M_1 = \sqrt{\frac{-4\gamma(\gamma+1)+(\gamma^2+6\gamma+1)\sqrt{\gamma}}{(\gamma-1)(\gamma(\gamma+3)-(3\gamma+1)\sqrt{\gamma})}}$$

$$C:\quad \frac{A_r}{A_1} = \frac{\gamma-1}{3\gamma+1},\quad M_1 = \infty$$

$$D:\quad \frac{A_r}{A_1} = \frac{\gamma-1}{\gamma+1},\quad M_1 = \infty$$

$$E:\quad \frac{A_r}{A_1} = \frac{1}{-(2\gamma+1)+2\gamma\sqrt{\frac{\gamma+1}{\gamma-1}}},\quad M_1 = \infty$$

$$F:\quad \frac{A_r}{A_1} = \frac{\gamma+1}{\gamma-1},\quad M_1 = \infty$$

$$G:\quad \frac{A_r}{A_1} = -(2\gamma+1)+2\gamma\sqrt{\frac{\gamma+1}{\gamma-1}},\quad M_1 = 1$$

$$H:\quad \frac{A_r}{A_1} = 1,\quad M_1 = 1$$

$$I:\quad \frac{A_r}{A_1} = \frac{1}{-(2\gamma+1)+2\gamma\sqrt{\frac{\gamma+1}{\gamma-1}}},\quad M_1 = \sqrt{\frac{\gamma-1}{2\gamma^2+5\gamma+1-2(\gamma^2-1)\sqrt{\frac{\gamma+1}{\gamma-1}}}}$$

$$J:\quad \frac{A_r}{A_1} = 0,\quad M_1 = 0$$

$$K:\quad \frac{A_r}{A_1} = 0,\quad M_1 = \frac{1}{2}\sqrt{\frac{\gamma-1}{\gamma(\gamma+1)}}$$

Figure 4 (continued)

solution is also not real at that point. Therefore in region I the 'minus' solution is also complex. The remainder is divided into two parts by the line $P^- = 0$. In region $XP^- < 0$, so the solution has no physical meaning. The other part is divided into four regions VI-IX by the lines $M_1 = 1$ and $M_r^- = 1$. Regions VI and VIII satisfy conditions (1) and (2), but Regions VII and IX do not satisfy condition (2). Therefore only in regions VI and VIII there probably is a reasonable solution. In region VIII, $M_1 < 1$ and $M_r^- > 1$ and it seems that discontinuites should not appear in subsonic regions, consequently, we think that only in region VI there is a reasonable solution.

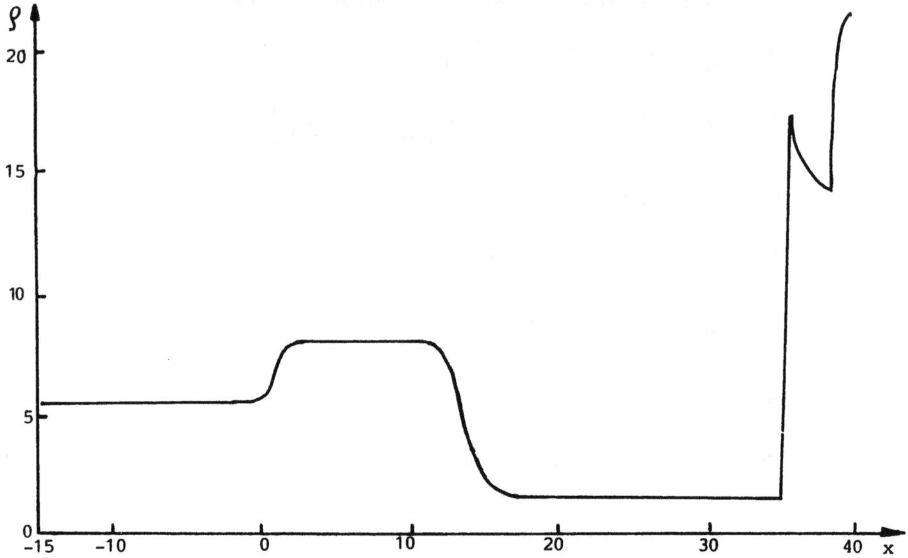

The shape of the corresponding tube

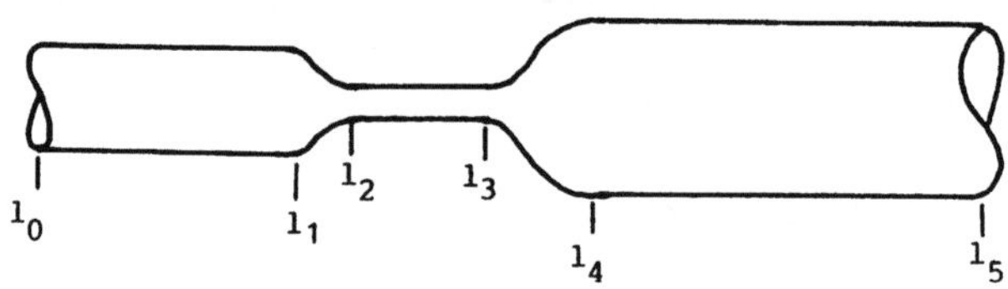

Figure 5 Density Distribution
($l_1 = 0, l_2 = 2, l_3 = 12, l_4 = 16$)

Putting Figs. 2 and 3 together, we can obtain Fig. 4 in which all important lines are drawn and the coordinates of all important points are given and which clearly shows how many stable solutions exist at each point. From Fig. 4 we have the following conclusion.

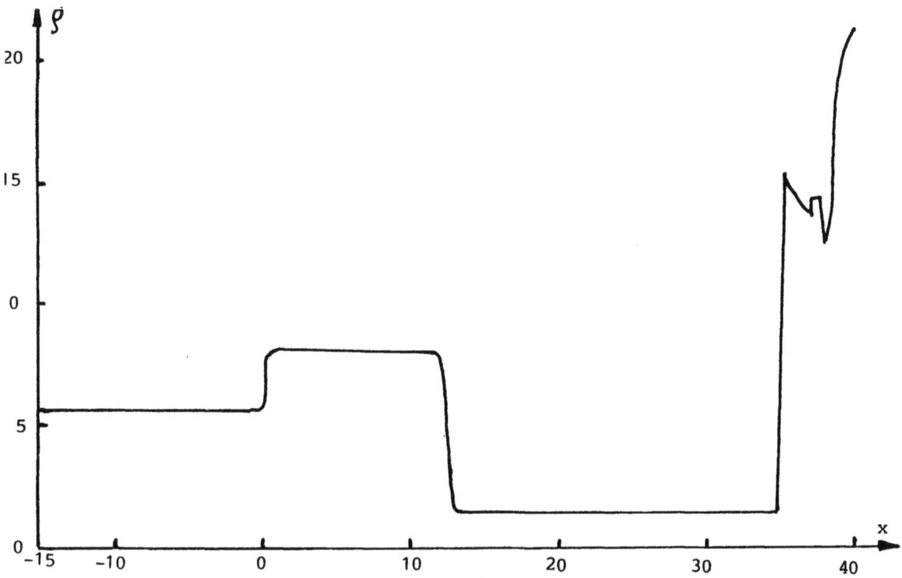

Figure 6 Density Distribution

The shape of tube similar to that in Figure 5 but ($l_1 = 0$, $l_2 = 0.5, l_3 = 12$, $l_4 = 13$)

A discontinuity on which the physical quantities satisfy (4)-(6) and conditions (1) and (2) hold may exist but does not always exist when M_1 and RA are given. For a certain point on the $M_1 - RA$ plane there may exist one or two or no solution. If $M_1 \geq 1$ and there exists a solution which satisfies conditions (1) and (2), such a discontinuity most likely exists. In order to clear up whether our conclusion is true, we have computed a series of solutions[1] for a series of tubes whose limit is a tube with discontinuous cross-section-areas at some points. We call those solutions U_i and in Figs. 5-7 three density distributions are given. Meanwhile we have calculated the solution in the tube with discontinuous areas of cross-section at some points using the jump conditions (4)-(6) at those points. We call the solution U^* and the density distribution is shown in Fig. 8. From Figs. 5-8 we known that U_i tends to U^* as the shape of tube corresponding to U_i gradually approaches to that of the tube with discontinuous areas. Therefore such a type of discontinuity does exist.

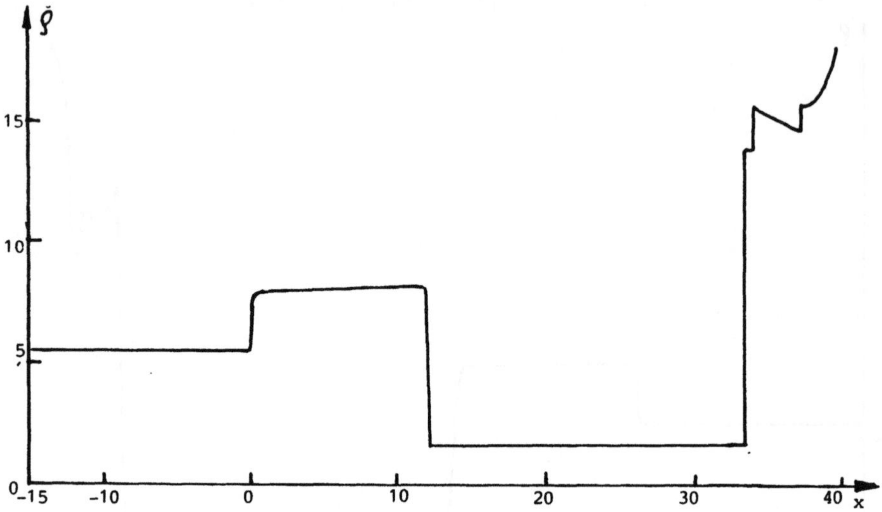

Figure 7 Density Distribution
(The shape of tube similar to that in Figure 5 but ($l_1 = 0, l_2 = 0.25, l_3 = 12, l_4 = 12.5$)

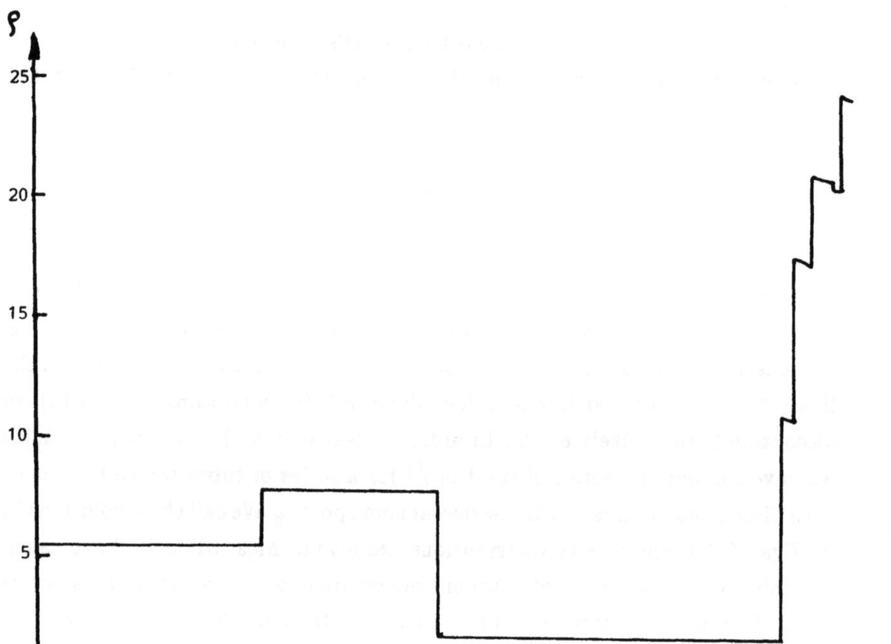

Figure 8 Density Distribution
The shape of tube similar to that in Figure 5 but ($l_1 = l_2 = 0, l_3 = l_4 = 12$)

References

1. X.-h. Wu and Y.-1. Zhu, Computation of Flow Field In Tubes With Rapidly Changed Cross-section Areas (to appear).

MIX
Papier aus verantwortungsvollen Quellen
Paper from responsible sources
FSC® C105338
FSC
www.fsc.org

If you have any concerns about our products,
you can contact us on
ProductSafety@springernature.com

In case Publisher is established outside the EU,
the EU authorized representative is:
Springer Nature Customer Service Center GmbH
Europaplatz 3, 69115 Heidelberg, Germany

Printed by Libri Plureos GmbH
in Hamburg, Germany